The Release of Thermonuclear Energy by Inertial Confinement

Ways Towards Ignition

The Release of Thermonuclear Energy by Inertial Confinement

Ways Towards Ignition

Friedwardt Winterberg

University of Nevada, USA

World Scientific

NEW JERSEY • LONDON • SINGAPORE • BEIJING • SHANGHAI • HONG KONG • TAIPEI • CHENNAI

Published by

World Scientific Publishing Co. Pte. Ltd.
5 Toh Tuck Link, Singapore 596224
USA office: 27 Warren Street, Suite 401-402, Hackensack, NJ 07601
UK office: 57 Shelton Street, Covent Garden, London WC2H 9HE

British Library Cataloguing-in-Publication Data
A catalogue record for this book is available from the British Library.

THE RELEASE OF THERMONUCLEAR ENERGY BY INERTIAL CONFINEMENT
Ways Towards Ignition

ISBN-13 978-981-4295-90-1
ISBN-10 981-4295-90-6

Printed in Singapore by B & Jo Enterprise Pte Ltd

Dedicated to

Otto Hahn and Fritz Strassmann
Discoverers of Nuclear Fission

Edward Teller
Father of Man-made Thermonuclear Fusion

Willard Bennett
Founder of Magnetic Plasma Confinement

and

Wernher von Braun
who first thought about nuclear
rocket propulsion

Preface

My interest in space flight can be traced back to the time I was about 10 years old, when as a birthday gift I got a popular book about the feasibility of space flight. There I heard for the first time about Oberth and Goddard, and of the possibility to reach the moon with a multistage rocket. It was the same time when Hahn and Strassmann had announced the discovery of nuclear fission with the possibility of an atomic bomb by a fission chain reaction.

Having been born in Germany in 1929, I received my PhD in physics under Heisenberg in 1955. Inspired by the 15 Megaton hydrogen bomb test conducted in 1952 by the United States, I have been since 1954 deeply interested in the non-fission ignition of thermonuclear reactions by inertial confinement. At this time all fusion research in US was still classified, but I had quite independently discovered the basic principles of inertial confinement, the Guderley convergent shock wave, and the Rayleigh imploding shell solutions. In 1956 I presented my findings in Goettingen, at a meeting at the Max Planck Institute, which was organized by von Weizsäcker. The abstracts of this meeting still exist and are kept in the library of University of Stuttgart.

Due to the fact that in 1958, I had delivered at the 2nd United Nations Conference on the Peaceful Use of Atomic Energy, a paper which turned out to be the importance for Nerva-type nuclear rocket reactors, I was invited by the US government under "Operation Paperclip" to come to the United States. In San Diego I met Ted Taylor and Freeman Dyson, who were working on the famous "Orion" nuclear bomb propulsion concept. This concept is generally credited to Ulam, but as I know from conversations I had with Heisenberg, a similar idea was presented to Heisenberg by Wernher von Braun, who had visited Heisenberg in Berlin in or around 1942. Because of my idea to use the Guderly convergent shock wave solution for thermonuclear ignition, Ted Taylor and Freeman Dyson were interested in my joining their group. But because at that time this work was classified and I was not yet a US citizen, this was not possible.

About 10 years later, in 1967, I saw a new possibility for the non-fission ignition of thermonuclear micro-explosions by intense relativistic electron and ion beams, driven by a high voltage Marx generator. This ignition concept could be used not only for the controlled release of energy by nuclear fusion, but also for the propulsion of a space craft, replacing the pusher plate of the

Orion concept with a magnetic mirror, reflecting the plasma-fireball of the thermonuclear micro-explosion.

This idea was adopted by the British Interplanetary Society in their 1978 Project Daedalus starship study, replacing the neutron-rich deuterium-tritium (DT) thermonuclear explosive with a neutron-poor DHe^3 explosive. Unlike the DT reaction where 80% of the released energy goes into neutrons, most of the energy in D-He^3 reaction goes into alpha particles which can be deflected by a magnetic mirror. But since He^3 is not abundantly available everywhere, it was proposed to "mine" it from the atmosphere of Jupiter.

Studies to propel a spacecraft with the matter-antimatter annihilation reaction have also been made, but because of the enormous technical problem to produce antimatter in appreciable quantities, this can be delegated into the realm of science fiction, as are space-warp drives, space flight through wormholes and other fantasies not supported by one shred of experimental evidence. Only the idea to use nano-gram amounts of antimatter for the ignition of fission-fusion micro-explosions appears to have some credible potential, but even there the production and storage of nano-gram quantities of antimatter poses serious technical problems.

We have no reason to expect that new fundamental laws in physics, which could lead to a breakthrough in propulsion, are still awaiting us. Very much as America was discovered only once, it is quite well possible that all the fundamental laws of physics relevant for propulsion have been discovered, challenging our imagination to find out if they are sufficient to invent propulsion systems which might ultimately bring us to earthlike planets of nearby solar systems.

I will end this preface with an imaginary talk by Ted Taylor to Freeman Dyson as it has been recorded by George Dyson, the son of Freeman Dyson, in his book "Project Orion — The True Story of the Atomic Spaceship", followed by a dream of Ted Taylor. "Freeman's hope for the Orion had rested on the fact that there seems to be no law of nature forbidding the construction of fission-free bombs", and on the belief that "improvements in the design of the nuclear devices (by reducing the fraction of total yield due to fission) might achieve reduction factors of 10^2 to 10^3. This belief in small, fission-free bombs has largely evaporated". "One exception is Ted (Taylor). He remains convinced that small, clean bombs could propel Orion, but he still fears, more than ever that such devices would be irresistible as weapons, until we outgrow the habit of war. There are lots of different routes to that final result of a very clean bomb", he says. "Could you make a one-kiloton explosion in which the fission yield was zero, which is bad news on the proliferation front, but could turn Orion into something quite clean?" "Freeman thinks Ted is wrong — and Ted hopes Freeman is right". I for my part think Freeman is wrong.

Many years later shortly before his death, Ted reports: "I had a dream last night, about a new form of nuclear weapon, and I am really scared of it". He tells us that when he woke up, he wrote down his dream, and it appeared scientifically sound and feasible. What was it? We never will know with certainty but I have a guess. It is the possibility of chemical super-explosives (explained in Chapter 11.6), powerful enough to ignite a thermonuclear bomb.

Overview

With the dramatic advance of laser and electric pulse power techniques, the fissionless ignition of a small thermonuclear explosive device becomes a real possibility, with many of the ideas developed for large thermonuclear explosive devices very likely to be useful for thermonuclear microexplosions. A fission-triggered thermonuclear explosive device is governed by "the tyranny of the critical mass" (F. Dyson), which means that the magnitude of the total explosion (fission plus fusion), is at least as large as the explosion of the fission trigger. With fissionless trigger devices this tyranny can be broken, opening up the prospect for small thermonuclear explosive devices, suitable for the commercial extraction of energy from the fusion of light nuclei where about a million times less energy is set free than in fission-triggered large thermonuclear explosive devices. For thermonuclear microexplosions the so called "first wall problem" is much less serious than for magnetic plasma confinement devices. But even large thermonuclear explosive devices have more than just military applications. There are the "plowshare" applications, for the building of canals, the mining of planetary bodies, and ultimately for the deflection of cometary or asteroidal impactors.

I try to give an overview and introduction into the entire problem, including elements in nuclear physics, plasma physics, laser and electric pulse power techniques. At the forefront today is the search for fissionless small thermonuclear explosive devices. The fact that this goal has not been reached almost 50 years after the first successful fission-triggered large thermonuclear explosion (Mike test), demonstrates how much more difficult it is. The material is presented in an easy way to make it accessible even to the non-physicist with an engineering background. Considering the very large ground I had to cover, I could not give every individual the proper credit in the list of references, which I restricted to additional reading material, and I must apologize for presenting many of my own ideas, some of them never previously published.

Unlike other texts which are mostly focused on the plasma physics aspect, this volume is focused on the various ways to achieve ignition. Ignition becomes easier with increasing energy, and it is rather easy with the large pulse power of a fission explosion. In the absence of large pulse power, the plasma physics of the thermonuclear ignition and burn becomes very important, requiring a very detailed analysis to determine the minimum energy and power required for ignition. The situation resembles the physics of the internal combustion engine. With its complex turbulent mixing of air and fuel not yet fully understood, it nevertheless works, because it is sufficiently large. The same should be true for fission triggered thermonuclear explosions. It can therefore be expected that sufficiently large pulse power, electric, photonic or otherwise, would greatly ease the ignition of thermonuclear microexplosions. It is this mostly "engineering" hardware aspect of large pulse power, which is not treated with the required respect in other texts on inertial confinement fusion and it is the purpose of this volume to help close this gap.

I would like to express my sincere thanks to Edward Teller, the father of man-made thermonuclear fusion, for his permission to include his photograph. It was taken at the time he had made his breakthrough discovery. It was to Hans Bethe as great a surprise as the discovery of nuclear fission. I would also like to express my thanks to my teacher Werner Heisenberg, who had told me about Wernher von Braun's first idea to use nuclear energy for rocket propulsion.

Contents

List of Figures

List of Tables

Chapter 1

Introduction

With the discovery of nuclear fission by Hahn and Strassmann in 1938, and the subsequent development of a nuclear fission explosive, the idea to use this explosive as a "match" for a much larger thermonuclear explosive must have occurred to many physicists. However, simply surrounding the fission explosive with a thermonuclear fusion explosive would blow apart the fusion explosive long before it can "catch fire". Following up on initial work by Ulam, this problem was solved by Teller in 1951. In the so-called Teller-Ulam configuration the intense ~50 million degree blackbody radiation released by an exploding fission bomb is used to implode and ignite a thermonuclear explosive. This is possible because even though the radiation has a large energy flux density, its pressure is still small enough for its confinement inside a cavity (Hohlraum) made up of solid walls. But, whatever the configuration to achieve thermonuclear ignition might be, the inertial forces must hold together the thermonuclear explosive long enough to lead to a large thermonuclear gain (ratio of thermonuclear energy output to ignition energy input). One therefore speaks of inertial confinement, as distinct from magnetic confinement (or gravitational confinement in stars). Magnetic confinement is, of course, the principal competitor for the controlled release of thermonuclear energy under terrestrial conditions.

One fundamental problem of thermonuclear burn, in comparison to chemical burn, is that the reaction cross sections are about 100 million times smaller and the ignition temperatures about 100 thousand times larger. With an energy release per reaction about a million times larger, the ignition of thermonuclear burn is about 10 million times more difficult.

The hydrogen bomb hohlraum ignition concept also evolved as the preferred fissionless (laser or particle beam) fusion concept through the ignition of thermonuclear microexplosions. However, a minority of researchers hold the view, that the radiation implosion hohlraum concept, as good as it may be for large thermonuclear explosive devices, does not make use of the lessons learned from magnetic confinement research, and that some of the difficulties encountered with hohlraum concepts could be alleviated by strong magnetic fields.

The basic principle underlying inertial confinement fusion concepts can be summarized as follows: The thermonuclear fuel with the lowest ignition temperature is a stoichiometric mixture of deuterium and tritium undergoing the reaction

$$\mathrm{D} + \mathrm{T} \longrightarrow \mathrm{n} + \alpha + \varepsilon_0 \tag{1.1}$$

where $\varepsilon_0 = 17.6$ MeV is the energy released, with 80% going into neutrons and the remaining 20% into α-particles. Defining an energy multiplication factor

$$F = E_{out}/E_{in} \tag{1.2}$$

where E_{in} is the input energy required for ignition, E_{out} the fusion energy released, and the conversion efficiency

$$\varepsilon = E_T/E_{in} \tag{1.3}$$

for the conversion of the input energy into thermal energy E_T of the DT with T the ignition temperature, one thus has

$$E_{out} = \frac{F}{\varepsilon} E_T \; . \tag{1.4}$$

The thermal energy to be deposited into a DT sphere of radius r (k Boltzmann constant) is

$$E_T = \frac{4\pi}{3} r^3 \, 3nkT \tag{1.5}$$

where n is the atomic number density of DT, for liquid (or solid) DT equal to $n = n_0 = 5 \times 10^{22}$ cm^{-3}. The fusion energy released in the DT sphere is

$$E_{out} = \frac{4\pi}{3} r^3 \left(\frac{n^2}{4}\right) \langle \sigma \mathrm{v} \rangle \varepsilon_0 \tau \tag{1.6}$$

where $\langle\sigma \mathrm{v}\rangle$ is the average nuclear reaction cross section ion velocity product, and τ the inertial confinement time

$$\tau = r/a \tag{1.7}$$

and

$$a = \sqrt{\frac{3kT}{M}} \tag{1.8}$$

the thermal expansion velocity. Inserting (1.5) and (1.6) into (1.4) one has

$$n\tau = \frac{F}{\varepsilon}\frac{12kT}{\varepsilon_0\langle\sigma \mathrm{v}\rangle}\,. \tag{1.9}$$

For $T \cong 10^8$ °K one has $\langle\sigma \mathrm{v}\rangle \cong 10^{-15}$ cm^3/s, and one obtains the Lawson criterion

$$n\tau \cong 6\frac{F}{\varepsilon} \times 10^{13}\ \mathrm{cm}^{-3}\ \mathrm{s}\,. \tag{1.10}$$

Expressing τ by (1.7) and (1.8), and using (1.9), the input energy then is

$$E_{in} = \frac{E_T}{\varepsilon} = \frac{F^3}{\varepsilon^4}\frac{4\pi kT}{n^2}\left(\frac{3kT}{M}\right)^{3/2}\left[\frac{12kT}{\varepsilon_0\langle\sigma \mathrm{v}\rangle}\right]^3 . \tag{1.11}$$

Inserting $\langle\sigma \mathrm{v}\rangle \cong 10^{-15}$ cm^3/s, and $n_0 = 5 \times 10^{22}$ cm^{-3} (density of liquid DT) one has

$$E_{in} \cong 2\frac{F^3}{\varepsilon^4}\left(\frac{n_0}{n}\right)^2 [\mathrm{MJ}]\,. \tag{1.12}$$

For breakeven $F = 1$. If $\varepsilon = 1$ one finds that $E_{in} \cong 2$ MJ, a rather modest amount of energy. The radius of the DT sphere is obtained from (1.5), putting $E_T = E_{in}$, and one finds that $r \cong 0.1$ cm. With $a = \sqrt{3kT/M} \cong 10^8$ cm/s one has $\tau \approx 10^{-9}$ s.

Therefore, even though the minimum input energy is rather small, it must be supplied in $\sim 10^{-9}$ s, onto an area of $\sim 10^{-2}$ cm^2, with a power $P \approx E_{in}/\tau$ $= 10^{15}$ Watt, and a power flux density $\Phi = P/r^2 = 10^{17}$ Watt/cm^2.

Inserting (1.7) into (1.10) with $a \approx 10^8$ cm/s, one has

$$nr \cong 6 \left(\frac{F}{\varepsilon}\right) \times 10^{21} \ [\mathrm{cm}^{-2}] \tag{1.13}$$

or by expressing n in terms of the density ρ ($n = L\rho/A$, L Avogadro number, $A = 2.5$ for the DT mixture)

$$\rho r \cong 4 \left(\frac{F}{\varepsilon}\right) \times 10^{-2} \ [\mathrm{g/cm}^2] \ . \tag{1.14}$$

It is customary to assume $F/\varepsilon \approx 3$ as a lower limit, hence

$$\rho r \geq 0.1 \ [\mathrm{g/cm}^2] \ . \tag{1.15}$$

In these simple estimates we have ignored self-heating by the charged α particles from the fusion reaction. Unlike the uncharged neutrons the α particles can be stopped even in a small DT sphere. The stopping range of a 3.5 MeV α particle is

$$\lambda \cong \frac{n_0}{n} T^{3/2} \times 10^{-12} \ [\mathrm{cm}] \tag{1.16}$$

for $T \cong 10^8$ °K

$$\lambda \cong n_0/n \ [\mathrm{cm}] \ . \tag{1.17}$$

The density for liquid DT is ~ 0.1 g/cm^3, hence one has

$$\lambda\rho \cong 0.1 \ [\mathrm{g/cm}^2] \ . \tag{1.18}$$

For a substantial stopping of the fusion α particles, we may assume that $r \cong 10\lambda$, replacing (1.15) by

$$\rho r \geq 1 \ [\mathrm{g/cm}^2] \ . \tag{1.19}$$

Whereas (1.15) is the minimal condition for the thermonuclear burn without self-heating, condition (1.19) is required for a propagating burn, i.e. the ignition of a thermonuclear detonation wave.

A different situation can arise in the presence of strong magnetic fields, if the ion Larmor radius is smaller than the stopping length. In this case a propagating thermonuclear burn is possible even for $\rho r < 1$ [g/cm^2]. For large thermonuclear explosive devices (1.19) is always satisfied, but the presence of a magnetic field can become important for fissionless triggered thermonuclear microexplosions.

Also important for thermonuclear microexplosions is to go to higher than solid densities. From (1.12) it follows that

$$E_{in} \propto \frac{1}{n^2} \,. \tag{1.20}$$

Combined with $nr = const.$ one has

$$P = E_{in}/\tau \propto 1/n \tag{1.21}$$

$$\Phi = P/\pi r^2 \propto n \tag{1.22}$$

$$\tau \propto 1/n \,. \tag{1.23}$$

The reduction in energy needed for ignition is offset in part by the difficulty of reaching higher densities, requiring precompressing the thermonuclear fuel, which also needs energy.

Finally, we have to address the problem of ignition. An exploding fission bomb has a temperature of the order $\sim 5 \times 10^7$ °K. According to the Stefan-Boltzmann law this implies a radiation flux $\Phi = \sigma T^4 \approx 4 \times 10^{19}$ Watt/cm^2, larger than the minimum above estimated value $\Phi \cong 4 \times 10^{17}$ Watt/cm^2. This explains why fission explosives can be used for thermonuclear ignition. By comparison, the radiation flux of chemical high explosives is negligible, but the fluid dynamic energy flux $\Phi = \varepsilon \mathrm{v}$, where ε is the energy density of the chemical and v the explosion velocity, is not so small. Typical values are $\varepsilon \cong 3 \times 10^4$ J/cm^3 and v $\cong 3 \times 10^5$ cm/s, resulting in $\Phi \cong 10^{10}$ Watt/cm^2. This is quite large, but for thermonuclear ignition still falls short by several orders of magnitude. However, by "focusing" this energy in a convergent shock wave where the temperature rises in proportion to $r^{-0.8}$, with r the distance from the center of convergence, conditions for thermonuclear ignition are approached.

There is a huge gap between chemical and nuclear energy densities. For chemical energy one has 10^{11}-10^{12} erg/cm^3, but for nuclear energy the energy densities are of the order 10^{18} erg/cm^3, (about 6 orders of magnitude larger).

This gap can be bridged with the energy density in particle beams, from laser beams all the way up to fast moving solid projectiles, with the goal to produce these beams with the required intensity needed for ignition. A breakthrough here can lead to a breakthrough in thermonuclear microexplosion ignition.

Finally, we would like to mention that the energy for ignition must be supplied in a time shorter than the radiation loss time. For blackbody radiation and a sphere of radius r this time τ_B is given by ($\sigma = 5.75 \times 10^{-5}$ erg/cm^{2}s °K^4):

$$\left(\frac{4\pi}{3}r^3\right)\frac{3nkT}{\tau_B} = 4\pi r^2 \sigma T^4 \tag{1.24a}$$

or by

$$\tau_B = \frac{nkr}{\sigma T^3} = 2.4 \times 10^{-12}\frac{nr}{T^3}\ [\mathrm{s}]\ . \tag{1.24b}$$

For the example $T = 5 \times 10^7$ °K, one finds that

$$\tau_B \sim 2 \times 10^{-13} r\ [\mathrm{s}]\ . \tag{1.24c}$$

τ_B should not be shorter than $\sim 10^{-9}$ s, which makes $r \gtrsim 50$ m, much too large to be of practical interest. For a hydrogen plasma and small plasma dimensions the situation is fortunately much better, because for densities and dimensions of interest for inertial confinement fusion, the plasma is far from being in thermodynamic equilibrium, radiating essentially by free-free transitions bremsstrahlung, with the plasma transparent for this radiation. There the radiation loss time is

$$\tau_R = \frac{3nkT}{\varepsilon_r}\ , \tag{1.25}$$

with

$$\varepsilon_r = bn^2\sqrt{T}\ ,\ b = 1.42 \times 10^{-27}\ [\mathrm{cgs}]$$

or

$$\tau_R = 2.9 \times 10^{11}\frac{\sqrt{T}}{n}\ [\mathrm{s}] \tag{1.26}$$

independent of r. For the example $n = 5 \times 10^{22}$ cm^{-3}, $T = 10^8$ °K one finds $\tau_R \simeq 4 \times 10^{-8}$ s, longer than the expansion time $\tau = r/a$ given by (1.7), with $a \simeq 10^8$ cm/s, which is $\tau \sim 10^{-9}$ s.

It is also instructive to compare these times with the heat conduction loss time of a "tamped", i.e. in a metallic shell encapsulated, DT sphere. There one has

$$\tau_c = r^2/\chi \simeq 2.1 \times 10^{-10} \frac{nr^2}{T^{5/2}} \text{ [s]} \tag{1.27}$$

In the presence of a strong magnetic field H, generated by a large current passing through the sphere and directed perpendicular to the flow of heat, the time is

$$\tau_{c\perp} = \frac{r^2}{\chi_\perp} \simeq 1.76\,(Hr)^2 \frac{\sqrt{T}}{n} \text{ [s]} \tag{1.28}$$

where χ is the heat conduction coefficient in a plasma without a magnetic field, and $\chi_\perp$ with a magnetic field. Without a magnetic field and for the same numbers one has $\tau_c \simeq 5 \times 10^{-9}$ s. But with a magnetic field equal to $H \simeq 5 \times 10^6$ G, one has $\tau_{c\perp} \simeq 6 \times 10^{-8}$ s. For the about 10 times larger temperatures where the fusion reaction rate reaches a maximum, one has $\tau_c \sim 2 \times 10^{-11}$ s, and $\tau_{c\perp} \simeq 2 \times 10^{-7}$ s, which demonstrates the advantage of a tamp for a magnetized DT plasma sphere.

1.1 Bibliography for Chapter 1

Physics of High Energy Density, edited by P. Caldirola and H. Knoepfel, Academic Press, New York, 1971.

N. G. Basov and O. N. Kroklin, 3rd Intern. Conf. Quantum Electronics, Paris 1963, edited by P. Grivet and N. Bloembergen; (Dunot, Paris 1964).

F. Winterberg, Phys. Rev. **174**, 212 (1968).

J. H. Nuckolls, L. Wood, A. Thiessen and G. B. Zimmerman, Nature **239**, 139 (1972).

K. A. Brueckner and S. Jorna, Rev. Mod. Phys. **46**, 325 (1974).

J. G. Linhart, Plasma Physics, EURATOM, Brussels, 1969.

J. D. Lindl, R. L. McCory and E. M. Campbell, Physics Today **45** (9), 32 (1992).

J. D. Lindl, in "International School of Plasma Physics Piero Caldirola: Inertial Confinement Fusion" (1988), edited by A. Caruso and E. Sindoni (Editrice Compositoxi, Bologna, Italy 1988), pp. 617-631.

J. D. Lindl, Inertial Confinement Fusion, AIP Press, Springer, New York 1998.

R. P. Drake, High-Energy-Density Physics, Springer, Berlin-Heidelberg, 2006.

Chapter 2

Nuclear Fission and Fusion Reactions

2.1 Nuclear Binding Energies

In first approximation an atomic nucleus can be viewed as a spherical drop made up of nuclear matter. The radius of a nucleus is

$$R = R_0 A^{1/3}, \; R_0 = 1.4 \times 10^{-13} \text{ cm} \tag{2.1}$$

where A is the atomic number. (2.1) expresses the important result, that the nuclear volume is proportional to the number of neutrons and protons.

For small values of A the binding energy E turns out to be proportional to A, and one has $E/A \cong$ 6-8 MeV. As in a liquid drop, each nucleon, be it a proton or neutron, interacts with a limited number of adjacent nucleons. The forces between them saturate, with an additional nucleon, contributing about 6-8 MeV to the binding energy, independent of the number of the nucleons already present. And as in a liquid drop, there must be a negative surface energy contribution proportional to $R^2 \propto A^{2/3}$, because particles near the surface have unsaturated valences.

For Z protons and N neutrons, with $Z + N = A$, the volume dependent part of the binding energy should only be a function of the concentrations Z/A resp. N/A:

$$\frac{E^{volume}}{A} = f\left(\frac{N}{A}\right) . \tag{2.2}$$

Furthermore, if there is symmetry of the forces between the protons and neutrons, the function in (2.2) can only depend on the difference $N - Z$, and must be an even function of this difference:

$$\begin{aligned} \frac{E^{volume}}{A} &= f\left[\frac{N-Z}{N+Z}\right]^2 \\ &= -a + b\left[\frac{N-Z}{N+Z}\right]^2 + \ldots \end{aligned} \tag{2.3}$$

Finally, the electrostatic repulsion between the protons must be taken into account. This effect, small for light nuclei, becomes important for large nuclei. For a uniformly charged sphere of radius R this leads to a positive energy by the Coulomb repulsion:

$$E_{el.} = \frac{3e^2}{5R_0}\frac{Z^2}{A^{1/3}}\,. \tag{2.4}$$

Because of the electrostatic repulsion between protons, the energy minimum at $N = Z$ for light nuclei is shifted to $N = 1.4Z$ for heavy nuclei. The surplus of neutrons over protons in heavy nuclei has important consequences for nuclear chain reactions with neutrons.

Summarizing the different contributions to the binding energy one has

$$\frac{E}{A} = -a + b\left(\frac{N-Z}{N+Z}\right)^2 + cA^{-1/3} + \frac{3}{5}\frac{e^2}{R_0}\frac{Z^2}{A^{4/3}} \tag{2.5}$$

where $a = 15.74$, $b = 22$, $c = 16.5$. In Fig. 2.1 the binding energy per nucleon E/A as a function of $N + Z$ is plotted for the stable (resp. almost stable) nuclei of the periodic system of the elements found in nature. The binding curve has a minimum at $A \approx 50$, in the vicinity of iron. The curve shows that energy can be set free (a) by the fusion of light nuclei with $A \leq 50$, and (b) by the fission of heavy nuclei with $A > 50$. The curve also shows that more energy per nucleon can be released by the fusion of light nuclei than by the fission of heavy nuclei.

The forces between the protons and neutrons are short range and can be approximately described by the Yukawa potential

$$V = \frac{g}{r}e^{-\kappa r} \tag{2.6}$$

where $\kappa = 1/R_0$ and g is the strong coupling constant. The latter is analogous to the electric charge e of a proton (or electron), where the potential is the

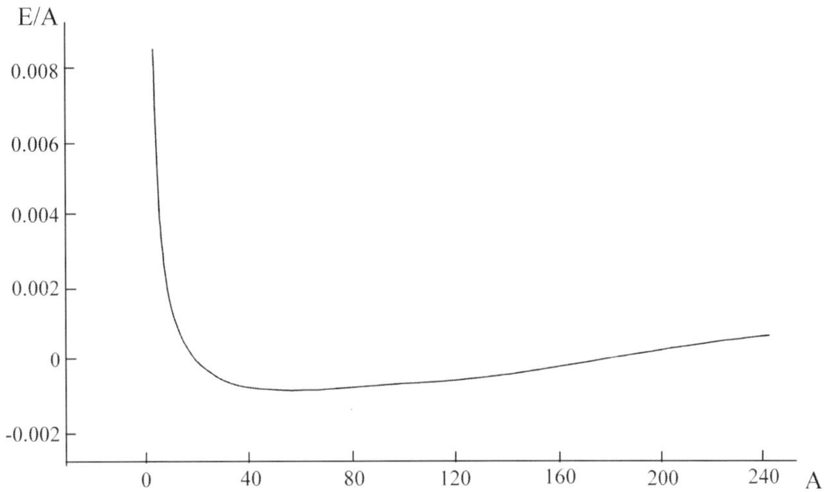

Figure 2.1: Nuclear binding energies per nucleon E/A in MeV as a function of A.

Coulomb potential $V = e/r$, with the fine structure constant $e^2/\hbar c = 1/137$ a measure of the strength of the electromagnetic coupling constant. By comparison, for the strong coupling constant one has $g^2/\hbar c \approx 10$.

The nuclear forces of the form (2.6), are a mixture of ordinary (Wigner-) and exchange (Majorana-, Heisenberg-, and Bartlett-) forces, with the mixtures so chosen as to make the forces saturate. The Yukawa potential (2.6) is really only true for a point charge (as is the Coulomb potential), but since the nucleons behave more like extended objects with a radius equal to $R_0 = 1.4 \times 10^{-13}$ cm, the interaction is better described by a potential well, and most simply by a square well of depth $V_0 \cong 20$ MeV and radius R_0. By the superposition of the potential wells of all nuclei, one obtains a potential with a depth of 20-30 MeV and a radius $R = R_0 A^{1/3}$.

From the uncertainty principle for a nucleon of mass M inside this well

$$MR\text{v} \cong \hbar \tag{2.7}$$

one can estimate its velocity

$$\text{v} \cong \frac{\hbar}{MR} \approx \frac{c}{10} \tag{2.8}$$

where $c = 3 \times 10^{10}$ cm/s is the velocity of light. By comparison, an electron in the lowest orbit of a hydrogen atom has the velocity $v/c = 1/137$. Consequently, the kinetic energy of a nucleon in a nucleus is $\cong 1.836 \times 10^3 (137/10)^2 \cong 3.5 \times 10^5$ times larger than the kinetic energy of the electron in an atom. With the kinetic energy of an electron in the hydrogen atom of the order 10 eV, the kinetic energy of a nucleon in a nucleus is of the order MeV. This roughly explains why there is much more energy stored in atomic nuclei than in electronic shells.

2.2 Nuclear Reactions

A nuclear reaction between two colliding nuclei has the following sequence of events:

1. After colliding inelastically, the two nuclei merge into one larger nucleus, called the compound nucleus.

2. A short time later, typically of the order $R/\mathrm{v} \approx 10^{-21}$ s ($R \approx 10^{-12}$ cm, $\mathrm{v} \approx c/10$), the compound nucleus either decays into several other nuclei, or it decays into its ground state under the emission of gamma radiation.

The reaction can be either exo- or endothermic. In many reactions, which have been studied, a small nucleus collides with a much larger one. In most of these cases the atomic number of the large nucleus changes only very little. An exception to this rule is nuclear fission, where the compound nucleus splits up into two large fragments.

For high collision energies the nuclear reaction cross section is equal to the geometric cross section

$$\sigma = \pi R^2 \tag{2.9}$$

with $R \approx 10^{-12}$ cm, one has $\sigma \approx 10^{-24}$ cm^2. A cross section of 10^{-24} cm^2 is called a barn.

For lower energies the cross section can become much larger. This happens near a nuclear resonance energy, the energy of an excited nuclear state of the compound nucleus. In the vicinity of the resonance energy, the lifetime of the compound nucleus can become much larger than $R/\mathrm{v} \approx 10^{-21}$ s, and is instead of the order $\hbar/\Delta E$, where ΔE is the width of the resonance. (This is a consequence of the time energy uncertainty relation $\Delta E \Delta t \cong \hbar$). As a result, the cross section can become many times larger than the geometric cross section.

Near a resonance with the energy E_0, the cross section $\sigma(P, Q)$ for a compound nucleus P decaying into the reaction products Q is given by the Breit-Wigner formula

$$\sigma(P, Q) = \pi \lambda_P^2 (2\ell + 1) \frac{\Gamma^P \Gamma^Q}{(E - E_0)^2 + \Gamma^2/4} . \tag{2.10}$$

In this equation $\lambda_P = \hbar / M A \mathrm{v} = \hbar / \sqrt{2MAE}$, where v is the relative collision velocity, E the collision energy, M the proton mass, MA the reduced atomic weight of the colliding nuclei with atomic number A_1, A_2, with $A = A_1 A_2 / (A_1 + A_2)$. Γ^P is the reaction width (in energy) of the colliding nuclei forming the compound nucleus P, and Γ^Q the width of the compound nucleus to decay into the reaction products Q. Furthermore, $\Gamma = \Gamma^P + \Gamma^Q$ is the total width of the reaction. The factors Γ^P/Γ and Γ^Q/Γ then represent the probabilities (1) for the formation of the compound nucleus and (2) for its subsequent decay. Finally ℓ is the angular momentum quantum number of the colliding nuclei. According to quantum mechanics a state of angular momentum number ℓ has $2\ell + 1$ possible orientations under which the compound nucleus can be formed. For thermonuclear reactions only the state with zero angular momentum is important. The reason is that for non-zero angular momentum the reacting nuclei have to overcome a repulsive centrifugal potential in addition to the repulsive Coulomb potential, decreasing the probability Γ^P/Γ for the formation of the compound nucleus by a much larger factor than the increase of the cross section by the factor $2\ell + 1$.

Because the probability to overcome the Coulomb barrier is small, one has $\Gamma^P \ll \Gamma^Q$ and $\Gamma \cong \Gamma^Q$. We then define a transmission coefficient for the probability to penetrate the Coulomb barrier

$$T(E) \equiv \frac{\Gamma^P}{\Gamma} . \tag{2.11}$$

With it one can write for (2.10)

$$\sigma(P, Q) = \pi \lambda_P^2 T(E) \frac{\Gamma \Gamma^Q}{(E - E_0)^2 + \Gamma^2/4} . \tag{2.12}$$

For energies $E \ll E_0$ this is approximately

$$\sigma(P,Q) \cong \pi \lambda\!\!\!^{-}{}_P^2 T(E) \frac{\Gamma \Gamma^Q}{E_0^2} \,. \tag{2.13}$$

According to the time-energy uncertainty relation one has

$$\Delta E \approx E_0 \approx M A \mathrm{v}^2 \approx \frac{\hbar}{\Delta t} = \Gamma \tag{2.14}$$

and according to the position-momentum uncertainty relation, with $R = R_0 (A_1 + A_2)^{1/3}$, one has

$$M A R \mathrm{v} \cong \hbar \,. \tag{2.15}$$

From (2.14) and (2.15) one obtains

$$\Gamma \cong \frac{\hbar^2}{M A R^2} \,. \tag{2.16}$$

One therefore has

$$\frac{\Gamma \Gamma^Q}{E_0^2} \cong \frac{\Gamma^Q}{\Gamma} \cong \frac{\Gamma^Q M A R^2}{\hbar^2} \,. \tag{2.17}$$

For the sake of simplicity we drop the index Q by putting $\Gamma \equiv \Gamma^Q$ as the width for the nuclear reaction towards Q, and arrive at

$$\sigma = \pi \lambda\!\!\!^{-}{}_P^2 T(E) \frac{\Gamma M A R^2}{\hbar^2} \,. \tag{2.18}$$

Next we have to compute the transmission coefficient $T(E)$. The situation is illustrated in Fig. 2.2. Classically, for $E > Z_1 Z_2 e^2/R$ the energy is larger than the Coulomb barrier and $T(E) = 1$, but for $E < Z_1 Z_2 e^2/R$ the energy is not large enough to overcome the Coulomb barrier and there $T(E) = 0$, where Z_1,Z_2 are the charge numbers of the two colliding nuclei. But because of the quantum mechanical tunnel effect there is a finite penetration probability even for $E < Z_1 Z_2 e^2/R_0$.

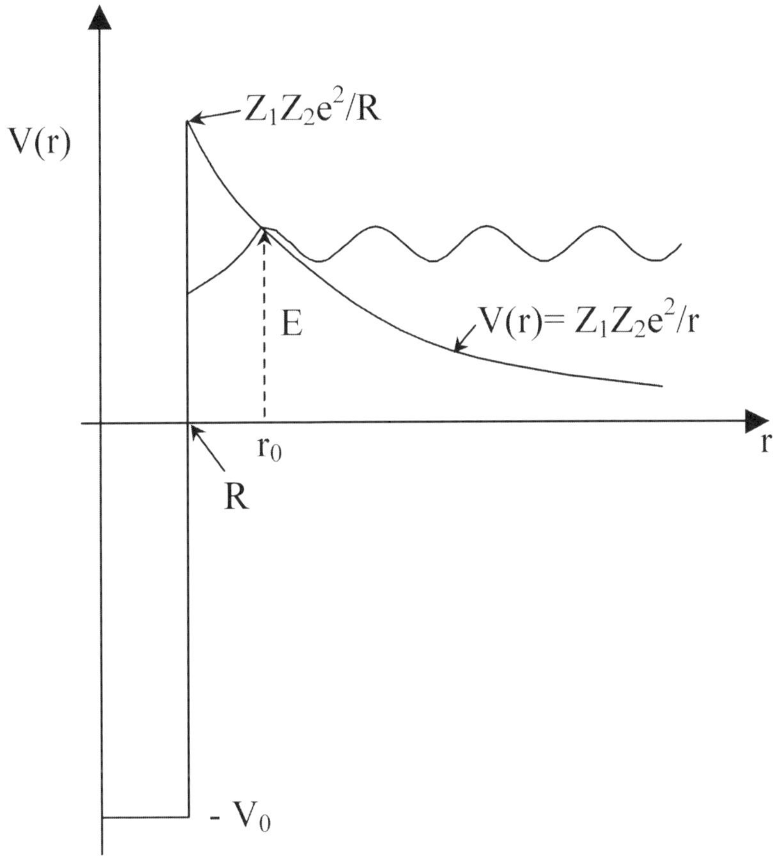

Figure 2.2: The nuclear and Coulomb potential for two colliding nuclei.

The transmission coefficient can be computed by the single particle Schrödinger equation for the reduced mass MA of the colliding nuclei:

$$\nabla^2\Psi + \frac{2MA}{\hbar^2}(E-V)\Psi = 0 \tag{2.19}$$

where

$$\begin{aligned} V(r) &= -V_0\,, & r &< R \\ &= Z_1Z_2e^2/r\,, & r &> R\,. \end{aligned}$$

For zero angular-momentum s-waves one has $\Psi = \Psi(r)$. Putting $\Psi = u/r$, the equation (2.19) becomes

$$\frac{d^2u}{dr^2} + \frac{2MA}{\hbar^2}(E-V)u = 0 \tag{2.20}$$

To compute the transmission coefficient one needs the solution of (2.20) for $u(r)$ from the classical turning point $r = r_0 = Z_1Z_2e^2/E$ to $r = R$. If $V(r)$ is a slowly varying function of r, the decrease in the amplitude of the wave function can be calculated by the WKB method. In this approximation one has in the region $R < r < r_0$ where $E < V$

$$u(r) \sim \exp\left[\pm\int\sqrt{\frac{2MA}{\hbar^2}(E-V)}\,dr\right]. \tag{2.21}$$

The transmission coefficient is the decrease in the square of the wave function amplitude from $r = r_0$ to $r = R$. It is given by

$$T(E) = \exp\left[-2\int_R^{r_0}\sqrt{\frac{2MA}{\hbar^2}(V-E)}\,dr\right]. \tag{2.22}$$

With $V = Z_1Z_2e^2/r$ and $E = Z_1Z_2e^2/r_0$ this is

$$T(E) = \exp\left[-\frac{2\sqrt{MAZ_1Z_2}\,e}{\hbar}\int_R^{r_0}\left(\frac{1}{r}-\frac{1}{r_0}\right)^{1/2}dr\right]. \tag{2.23}$$

The integral in (2.23) can be evaluated with the substitution $r = r_0\cos^2\phi$:

$$\int_R^{r_0}\left(\frac{1}{r}-\frac{1}{r_0}\right)^{1/2}dr = \sqrt{r_0}\left[\phi_0 - \frac{1}{2}\sin 2\phi_0\right] \tag{2.24}$$

where $\cos^2\phi_0 = R/r_0$. In nuclear reactions of importance for thermonuclear processes one has $R/r_0 \ll 1$ whereby

$$2\phi_0 - \sin 2\phi_0 \cong \pi - 4\sqrt{R/r_0} \tag{2.25}$$

Making this approximation one finally obtains what is called the Gamow factor:

$$T(E) \cong \exp\left[-\frac{\pi e^2\sqrt{2MA}\, Z_1 Z_2}{\hbar\sqrt{E}} + \frac{4e\sqrt{2MAZ_1Z_2R}}{\hbar}\right] \tag{2.26}$$

From (2.26) one sees that $T(E) = 1$ for $E = (\pi/4)^2\, Z_1 Z_2 e^2/R$, rather than $E = Z_1 Z_2 e^2/R$ as in the classical limit. The reason for the discrepancy is that the WKB method is not very accurate near the classical turning point.

With the Gamow factor one obtains for (2.18)

$$\sigma\left(E\right) \cong \left(\frac{\pi}{2}\right)\left(\frac{\Gamma}{E}\right) R^2 \exp\left[\frac{4e\sqrt{2MAZ_1Z_2R}}{\hbar} - \frac{\pi e^2\sqrt{2MA}\, Z_1 Z_2}{\hbar\sqrt{E}}\right] \tag{2.27}$$

where $R = R_0\left(A_1 + A_2\right)^{1/3}$. From (2.15) and (2.16) one has

$$\Gamma = \hbar \mathrm{v}/R\,. \tag{2.28}$$

The cross section (2.27) is therefore of the form (a, b, constants)

$$\sigma\left(\mathrm{v}\right) = \left(\frac{a}{\mathrm{v}}\right) e^{-b/\mathrm{v}}\,. \tag{2.29}$$

For nuclear reactions with uncharged neutrons $T\left(E\right) = 1$. There one has for the cross section

$$\sigma(\mathrm{v}) = \frac{a}{\mathrm{v}}\,. \tag{2.30}$$

For large velocities $e^{-b/\mathrm{v}} \to 1$ and there one has $\sigma \propto 1/\mathrm{v}$, both for neutrons and charged particles. The dependence of the cross sections for charged nuclei-nuclei and uncharged neutron-nucleus collisions is displayed in arbitrary units in Fig. 2.3. In the limit of large collision velocities the cross sections become equal the geometric cross section

$$\sigma = \pi R^2 = \pi R_0^2\left(A_1 + A_2\right)^{2/3} \tag{2.31}$$

Nuclear reactions of importance for thermonuclear processes are:

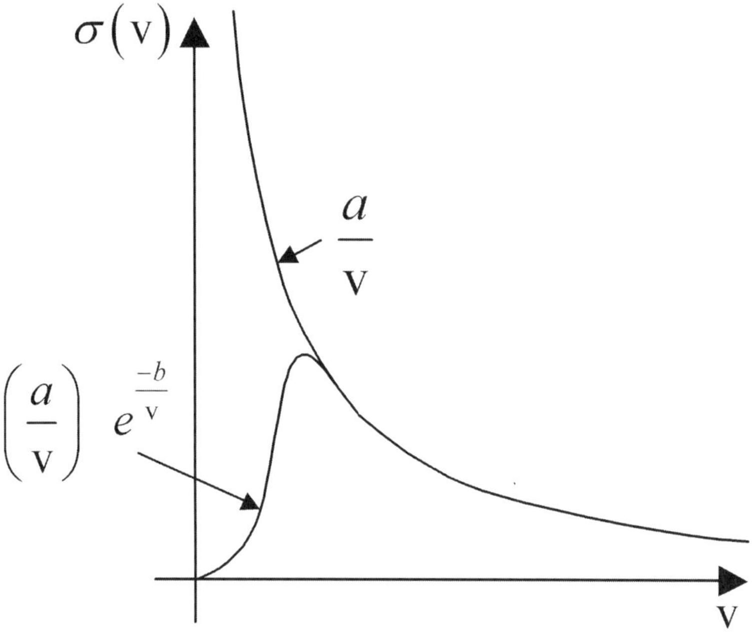

Figure 2.3: Nuclear reaction cross sections for nucleus-nucleus reaction $\sigma = (a/\mathrm{v})e^{-b/\mathrm{v}}$, and for neutron-nucleus reaction $\sigma = a/\mathrm{v}$, in arbitrary units as a function of the collision velocity v.

$$\mathrm{D + T} \longrightarrow \mathrm{He^4 + n + 17.6\ MeV} \tag{1}$$

$$\mathrm{D + D} \begin{cases} \nearrow \mathrm{T + H + 4\ MeV} & 50\% \\ \searrow \mathrm{He^3 + n + 3.25\ MeV} & 50\% \end{cases} \tag{2}$$

$$\mathrm{D + He^3} \begin{cases} \nearrow \mathrm{He^4 + H + 18.4\ MeV} & \\ \searrow \mathrm{Be^7 + n + 3.4\ MeV} & \text{small } \% \end{cases} \tag{3}$$

$$\mathrm{D + Li^6} \begin{cases} \nearrow \mathrm{Li^7 + n + 3.4\ MeV} & \\ \searrow \mathrm{2He^4 + 22.4\ MeV} & \text{small } \% \end{cases} \tag{4}$$

$$\mathrm{T + T} \longrightarrow \mathrm{He^4 + 2n + 11.3\ MeV} \tag{5}$$

$$\mathrm{H + B^{11}} \longrightarrow \mathrm{3He^4 + 8.7\ MeV} \tag{6}$$

$$\mathrm{H + N^{15}} \longrightarrow \mathrm{He^4 + C^{12} + 4.84\ MeV} \tag{7}$$

In reaction (2) the two branches occur with about equal probability, whereas the probability for the second branch of reaction (3) and (4) is small.

For the first three most important reactions one has

$$\mathrm{D + T} \longrightarrow \mathrm{(He^4 + 3.6\ MeV) + (n + 14.1\ MeV)} \tag{1a}$$

$$\mathrm{D + D} \begin{cases} \nearrow \mathrm{(T + 1.0\ MeV) + (H + 3.0\ MeV)} \\ \searrow \mathrm{(He^3 + 0.8\ MeV) + (n + 2.45\ MeV)} \end{cases} \tag{2a}$$

$$\mathrm{D + He^3} \longrightarrow \mathrm{(He^4 + 3.7\ MeV) + (H + 14.7\ MeV)} \tag{3a}$$

For the fraction f of the energy, which goes into charged fusion products, one has

DT: $f = 0.2$
DD: $f = 0.66$
DHe^3: $f = 1.0$

In Fig. 2.4 the dependence of the cross sections as a function of the collision energy is displayed for some important reactions. For the HB^{11} reaction the cross section is quite different from the one given by (2.27). The reason is that the approximation $E \ll E_0$ where E_0 is the energy of a nuclear resonance is there invalid, and in fact the rather large value of $\sigma(E)$ for the HB^{11} reaction at $\cong$ 600 keV is due to a resonance at this energy.

Of all the reactions the DT reaction has the largest cross section, with its maximum also at the lowest energy. Next in line is the DHe^3 reaction. In between 10 and 100 keV it is the DD reaction, which has the largest cross section of all naturally occurring substances. The DHe^3, HB^{11} and HN^{15} reactions are distinguished by their property not to produce neutrons.

In thermonuclear processes the neutron induced reaction

$$\mathrm{Li}^6 + \mathrm{n} \longrightarrow \mathrm{He}^4 + \mathrm{T} + 4.8\ \mathrm{MeV} \tag{8}$$

plays an important role for the "dry" hydrogen bomb closed chain process

$$\begin{array}{ccccccccc} & & \mathrm{T} & + & \mathrm{D} & \longrightarrow & \mathrm{He}^4 & + & \mathrm{n} + 17.6\ \mathrm{MeV} \\ & & \uparrow & & & & & & \downarrow \\ 4.8\ \mathrm{MeV} & + & \mathrm{T} & + & \mathrm{He}^4 & \longleftarrow & \mathrm{Li}^6 & + & \mathrm{n} \end{array} \tag{9}$$

In order for it to work no neutrons can be lost. In reality there will always be some neutron losses due to the finiteness of the geometry. These losses must be compensated for by some neutron multiplying substance, for example U238 (fast fission) or Be^9 (n, 2n reaction).

2.3 Fission Chain Reactions

The fission chain reaction in a nuclear fission explosion can be described with a one-velocity neutron diffusion model. If n is the number of neutrons per unit volume and v_0 the neutron velocity, the undirectional neutron flux is given by

$$\phi = n\mathrm{v}_0 \tag{2.32}$$

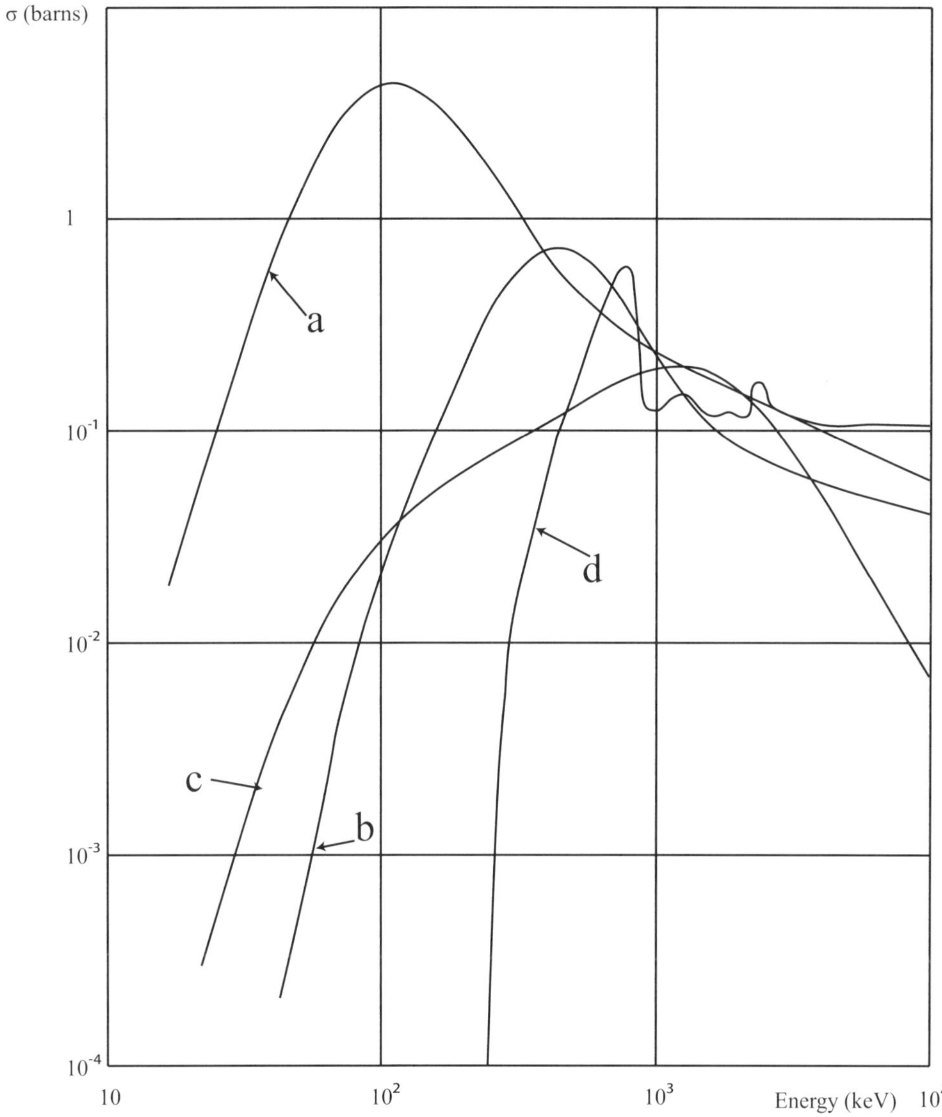

Figure 2.4: Some cross sections of important thermonuclear reactions (σ is in barns, 1 barn = 10^{-24} cm^2: **(a)** $D+T \longrightarrow$ He^4+n; **(b)** D + He^3 $\longrightarrow$ He^4 + p; **(c)** D + D $\longrightarrow$ T + p or He^3 + n; **(d)** p + B^{11} $\longrightarrow$ 3He^4.

For fission neutrons $v_0 \approx 10^9$ cm/s. In the diffusion approximation the neutron current vector is

$$\mathbf{j} = -D \,\text{grad}\, \phi \tag{2.33}$$

where $D = 1/3N\sigma_s$ is the neutron diffusion coefficient, N is the atomic number density of the fission explosive and σ_s the nuclear scattering cross section.

The neutron balance is governed by the equation

$$\frac{\partial n}{\partial t} + \text{div}\,\mathbf{j} = S \tag{2.34}$$

where S is a neutron source term, which in a fission chain reaction is

$$S = (\nu - 1)\, N\sigma_f\, \phi \tag{2.35}$$

In (2.35), $\nu > 1$ is the number of neutrons set free in a nuclear fission reaction and σ_f the fission cross section. With the definition

$$B_0^2 = 3\sigma_s\sigma_f N^2\,(\nu - 1) \tag{2.36}$$

one obtains from (2.32-2.36)

$$\nabla^2\phi + B_0^2\phi = \frac{1}{D v_0}\frac{\partial \phi}{\partial t} \tag{2.37}$$

For an infinitely large fissile assembly $\nabla^2\phi = 0$ with the solution of (2.37)

$$\phi = \phi_0 e^{\lambda_0 t} \tag{2.38}$$

where

$$\lambda_0 = N\sigma_f v_0\,(\nu - 1) \tag{2.39}$$

For metallic uranium one has $N = 4.5 \times 10^{22}$ cm^{-3}. The cross section is with (2.9) given by $\sigma_f = \pi R^2 = 2.3 \times 10^{-24}$ cm^2. One furthermore has $v_0 = 2 \times 10^9$ cm/s for ~2 MeV fission neutrons, and $\nu \simeq 2.5$, hence $\lambda_0 \simeq 3 \times 10^8$ s^{-1}. The exponential growth time $\tau = 1/\lambda_0$ is equal to $\simeq 3 \times 10^{-9}$ s.

To compute the critical size above which the neutron chain is growing, we set in (2.37) $\partial\phi/\partial t = 0$ and have

$$\nabla^2\phi + B_0^2\phi = 0\,. \tag{2.40}$$

For a spherical assembly (2.40) has the solution

$$\phi = A\frac{\sin B_0 r}{r} \tag{2.41}$$

According to transport theory (going beyond the diffusion approximation), one has to set $\phi = 0$ at the distance $d = 0.71/N\sigma_s$, measured from the surface of the sphere with radius R_0. Therefore, putting $\phi = 0$ at $R_0 + d$ one obtains from (2.41) the critical radius

$$R_0 = \frac{\pi}{B_0} - d = \frac{1}{N}\left[\frac{\pi}{\left[3\sigma_s\sigma_f(\nu - 1)\right]^{1/2}} - \frac{0.71}{\sigma_s}\right] \tag{2.42}$$

To obtain the time dependent solution we put

$$\phi = A\frac{\sin Br}{r}e^{\lambda t} \tag{2.43}$$

and insert (2.43) into (2.37). One obtains

$$B^2 = \left[\frac{\pi}{R + d}\right]^2 = B_0^2 - \frac{\lambda}{D\mathrm{v}_0}\ . \tag{2.44}$$

Here R is larger than the critical radius R_0. From (2.44) one obtains (by using the expression for B_0 and neglecting d):

$$\lambda = \lambda_0\left[1 - \left(\frac{R_0}{R}\right)^2\right] \tag{2.45}$$

As expected, if follows from (2.45) that a chain reaction in a finite assembly rises less rapidly. For a slightly above critical radius setting $R = R_0 + \Delta R$, with $\Delta R/R_0 \ll 1$, one obtains for the e-fold growth time

$$t_e = \frac{1}{\lambda} \simeq \frac{1}{\lambda_0}\frac{R_0}{2\Delta R}\ . \tag{2.46}$$

If, for example, $\Delta R/R_0 = 0.05$, then $t_e = 10/\lambda_0$. This means that a 5% increase of the radius above the critical radius would increase the e-fold time for the fission chain by a factor of 5 compared to the growth time for an infinite assembly. For $\sigma_s \simeq \sigma_f = 2.7 \times 10^{-24}$ cm^2, one obtains from (2.42)

$R_0 \simeq 7.5$ cm for the critical radius of an uranium sphere. A smaller critical radius is possible, either by surrounding the uranium sphere with a neutron reflector, for example gold, or by compressing it above solid state density with high explosives.

To start the neutron avalanche, a neutron source must be provided. A polonium-beryllium (α, n) neutron source is ideal, because polonium is a pure α-particle emitter. The source can be activated in the state of highest criticality, by separating the polonium from the beryllium with an α-particle absorbing foil, broken in the last moment when the subcritical parts of the uranium sphere come together. Alternatively a small accelerator tube, accelerating deuterons onto a LiD target, setting off neutrons by the DD nuclear reaction, can also be used.

Still smaller critical masses are possible with transuranic elements like americium or curium, which release more neutrons per fission, and/or have larger cross sections, but these elements are expensive to produce.

2.4 Thermonuclear Reactions

If an assembly of nuclei having the potential to undergo nuclear reactions is heated up to high temperatures, the thermal motion of the nuclei can become large enough to make nuclear reactions by overcoming the mutual Coulomb repulsion between them. As the cross sections of some endothermic reactions displayed in Fig. 2.4 show, this requires energies above ~ 10 keV, corresponding to a temperature of $\sim 10^8$ °K.

To compute the thermonuclear reaction rate we assume that the first nuclear species is at rest, with the second species moving against the first one with the velocity $\mathrm{v} = (2E/MA)^{1/2}$, where E is the mutual collision energy and MA the reduced mass of the colliding nuclei. If the first species with number density n_1 is assumed to be at rest, it has with regard to the second species the macroscopic cross section $\Sigma = n_1\sigma(E)$. The second species, on the other hand, has with regard to the first species an undirected particle flux $\phi = n_2\mathrm{v}$, resulting in the reaction rate between both:

$$N = \Sigma\phi = n_1 n_2 \sigma \mathrm{v} \tag{2.47}$$

This however, is only correct if all the nuclei have the same velocity. In reality, the nuclei have a Maxwell velocity distribution. Therefore, because σ

is a function of E, the product $\sigma\mathrm{v}$ must be averaged over a Maxwell velocity distribution.

For a gas of particles with a mass m and temperature T, the Maxwell velocity distribution is given by (k Boltzmann constant):

$$dn = 4\pi n \left(\frac{m}{2\pi kT}\right)^{3/2} e^{-m\mathrm{v}^2/2kT}\mathrm{v}^2\, d\mathrm{v}\,. \tag{2.48}$$

In (2.48) n is the total number of particles per unit volume and dn the fraction within the velocity interval between v and v + dv. Putting $\mathrm{v} = (2E/MA)^{1/2}$ and $n = n_2$ one obtains from (2.48) the differential particle flux of species 2:

$$\mathrm{v}\, dn_2 = \frac{4n_2}{(2\pi MA)^{1/2}(kT)^{3/2}}\, e^{-E/kT} E\, dE\,. \tag{2.49}$$

The reaction rate then follows by multiplying (2.49) with $n_1\sigma(E)$ and by integrating over E:

$$N = \frac{4n_1 n_2}{(2\pi MA)^{1/2}(kT)^{3/2}} \int_0^\infty e^{-E/kT}\sigma(E)\, E\, dE\,. \tag{2.50}$$

Comparing (2.50) with (2.47) it follows that the average of $\sigma\mathrm{v}$ is given by

$$\langle\sigma\mathrm{v}\rangle = \frac{4}{(2\pi MA)^{1/2}(kT)^{3/2}} \int_0^\infty e^{-E/kT}\sigma(E)\, E\, dE \tag{2.51}$$

To compute integral (2.51) one has to insert the expression for $\sigma(E)$ given by (2.27).

The integrand is a function of E and can be written as follows

$$f(E) = a\exp\left[-\frac{E}{kT} - \frac{b}{E^{1/2}}\right] \tag{2.52}$$

where

$$a = (\pi/2)\,\Gamma R^2 \exp\left[\frac{4e\,(2MAZ_1Z_2R)^{1/2}}{\hbar}\right]$$

$$b = \frac{\pi e^2 (2MA)^{1/2} Z_1 Z_2}{\hbar}$$

The integrand is the product of two functions, (1) the Maxwell velocity distribution multiplied with the collision velocity, and (2) the collision velocity dependent cross section. Both functions go through a maximum at a certain velocity, with the product function having a sharp maximum in between (see Fig. 2.5). The integral over the product is largest if both functions optimally overlap. This explains why the reaction rate first rises with the temperature, but falls off above a certain temperature.

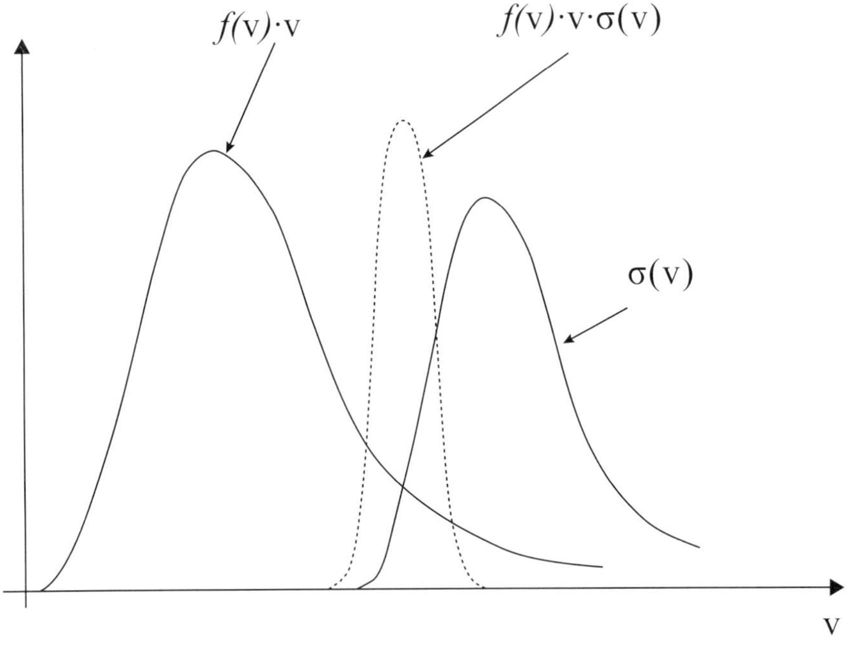

Figure 2.5: The Maxwell velocity distribution $f(\mathrm{v})$ multiplied with the collision velocity v, the nuclear cross section $\sigma(\mathrm{v})$ and the product of both.

Since the product function has a sharp maximum, the integral $\int_0^\infty f(E)\,dE$ can be evaluated with the saddle point approximation method. The maximum of $f(E)$ is positioned at

$$E_m = \frac{\left[\pi e^2 (MA)^{1/2} Z_1 Z_2 kT\right]^{2/3}}{\left(\sqrt{2}\hbar\right)^{2/3}} \tag{2.53}$$

Expanding $f(E)$ around $E = E_m$ one has

$$f(E) = f(E_m) + \frac{(E - E_m)^2}{2} f''(E_m) + \cdots \tag{2.54}$$

With the definitions $\alpha \equiv f(E_m)$, $\beta \equiv f''(E_m)$ one obtains

$$\int_0^\infty f(E)\, dE = \alpha \int_0^\infty \left[1 + \frac{(E - E_m)^2}{2} \frac{\beta}{\alpha} + \cdots\right] dE \tag{2.55}$$

which can be approximated by

$$\int_0^\infty f(E)\, dE \simeq \alpha \int_0^\infty \exp\left[\frac{(E - E_m)^2}{2} \frac{\beta}{\alpha}\right] dE\,. \tag{2.56}$$

Because of the sharpness of the maximum the integral can be extended from $-\infty$ to $+\infty$. In this case the integral can be evaluated using the formula $\int_{-\infty}^{+\infty} e^{-x^2}\, dx = \sqrt{\pi}$ and one finds

$$\int_0^\infty f(E)\, dE = \sqrt{\frac{2\pi\alpha^3}{-\beta}} \tag{2.57}$$

Inserting the values of α and β obtained from the function $f(E)$ at $E = E_m$, one finally has

$$\begin{aligned} \langle \sigma \mathrm{v} \rangle = \frac{(2\pi)^{4/3}}{3^{1/2}} \frac{e^{2/3} Z_1^{1/3} Z_2^{1/3} \Gamma R^2}{(MA\hbar)^{1/3} (kT)^{3/2}} \exp\Bigg[\frac{4e\,(2MAZ_1Z_2R)^{1/2}}{\hbar} - \\ - 3 \left(\frac{\pi^2 e^4 MAZ_1^2 Z_2^2}{2\hbar^2 kT}\right)^{1/3}\Bigg] \end{aligned} \tag{2.58}$$

This formula takes a much simpler form by putting

$$k_1 = \frac{4}{3^{5/2}} \frac{\hbar \Gamma R^2}{MAe^2 Z_1 Z_2} \exp\left[\frac{4e}{\hbar} (2MAZ_1Z_2R)^{1/2}\right]$$

$$k_2 = 3 \left(\frac{\pi^2 e^4 MAZ_1^2 Z_2^2}{2\hbar^2 k}\right)^{1/3}$$

whereby one has

$$\langle \sigma \mathrm{v} \rangle = k_1 \frac{\left(k_2 T^{-1/3}\right)^2}{\exp\left[k_2 T^{-1/3}\right]} \tag{2.59}$$

Introducing the variables

$$\left.\begin{aligned} x &= k_2 T^{-1/3} \\ y &= \langle \sigma \mathrm{v} \rangle / k_1 \end{aligned}\right\} \tag{2.60}$$

one can write for (2.59)

$$y = x^2 e^{-x} . \tag{2.61}$$

This function has a maximum at $x = 2$ where $y = 0.545$. It thus follows that

$$\langle \sigma \mathrm{v} \rangle_{max} = 0.545\, k_1 \tag{2.62}$$

reached at the temperature

$$T_0 = (k_2/2)^3 . \tag{2.63}$$

Replacing $\sigma \mathrm{v}$ by $\langle \sigma \mathrm{v} \rangle$ in the reaction rate formula (2.47) one has

$$N = n_1 n_2 \langle \sigma \mathrm{v} \rangle \tag{2.64}$$

and for the thermonuclear energy production rate

$$\varepsilon_f = N \varepsilon_0 = n_1 n_2 \langle \sigma \mathrm{v} \rangle \varepsilon_0 \tag{2.65}$$

where ε_0 is the energy set free in a nuclear reaction. In case of a stoichiometric two-component thermonuclear explosive one has $n_1 = n_2 = n/2$ and

$$\varepsilon_f = \left(n^2/4\right) \langle \sigma \mathrm{v} \rangle \varepsilon_0 \tag{2.66}$$

But if the explosive is composed of identical nuclei, for example deuterium, the reaction rate is twice as large. There one has

$$\varepsilon_f = \left(n^2/2\right) \langle \sigma \mathrm{v} \rangle \varepsilon_0 \tag{2.67}$$

In table 2.1 the thermonuclear reaction rate constants for a number of important reactions have been put together. Even though (2.27) is not correct for the HB^{11} reaction, the formula (2.59) with appropriately chosen constants can there still be used. In Fig. 2.6 the $\langle \sigma \mathrm{v} \rangle$ values for some thermonuclear reactions are displayed.

Reaction	ε_0 [MeV]	k_1 [cm^{3}sec]	k_2 [$^\circ$K$^{1/3}$]	T_0 [$^\circ K$]	$\langle\sigma v\rangle_{max}$ [cm^3/sec]
DT	17.5	1.8×10^{-15}	1.8×10^3	8.0×10^8	$\simeq 10^{-15}$
DD	3.2	9.2×10^{-17}	3.0×10^3	3.6×10^9	5.0×10^{-17}
DHe^3	18.3	4.5×10^{-16}	3.0×10^3	3.5×10^9	2.5×10^{-16}
HB^{11}	8.7	1.3×10^{-15}	2.8×10^3	3.0×10^9	7.4×10^{-16}

Table 2.1: Thermonuclear reaction rate constants.

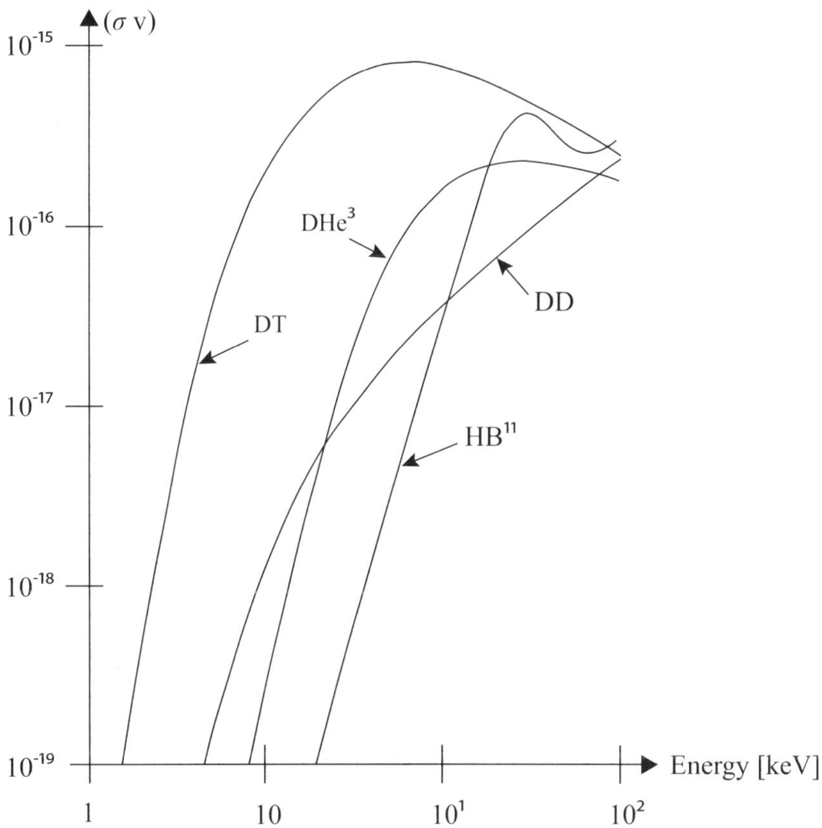

Figure 2.6: $\langle\sigma v\rangle$ values for some thermonuclear reactions.

2.5 Fusion Chain Reactions

In a fusion chain reaction the energetic charged fusion product nuclei give other nuclei a large kick, whereby their kinetic energy becomes larger than their thermal energy, enabling them to make nuclear reactions with other nuclei. The condition for a fusion chain reaction is roughly expressed by the inequality

$$n\sigma > \frac{1}{E_0}\left|\frac{dE}{dx}\right| \tag{2.68}$$

where n is the particle number density, σ the nuclear reaction cross section, dE/dx the stopping power and E_0 the initial kinetic energy of the charged fusion product. It is convenient to introduce a stopping power cross section $\sigma_s = |dE/dx|/nE_0$, whereby (2.68) becomes

$$\sigma > \sigma_s \,. \tag{2.69}$$

The stopping (resp. slowing down) of the energetic fusion product is primarily caused by collisions with electrons. In a low temperature nondegenerate electron gas the stopping cross section for ions of charge Ze, mass M and energy E_0 is

$$\sigma_s = \sigma_s^{(0)} = \frac{2\pi M Z^2 e^4}{m E_0^2} \ln \Lambda \tag{2.70}$$

where m is the electron mass and $\ln\Lambda \simeq 10$ the Coulomb logarithm. A typical value is $\sigma_s \sim 10^{-21}$ cm^2, about 10^3 times larger than a nuclear reaction cross section $\sigma \sim 10^{-24}$ cm^2. Under these conditions inequality (2.69) is not satisfied, but this is possible at high densities or temperatures.

For high densities electron degeneracy cannot be neglected. Roughly the following can there be said: If the Fermi energy E_f of a degenerate electron gas approaches E_0, then only the fraction $(1 - E_f/E_0)$ of the electrons contribute to the stopping cross section, because for $E < E_f$ the ions are not scattered by the electrons. With E_f rising with the 2/3 power of n it follows that for $n > 10^{28}$ cm^{-3}, corresponding to $\sim 10^6$ times solid density, a fusion chain reaction should become possible.

For high temperatures the stopping cross section can be obtained from the stopping power in this limit:

$$\sigma_s = \frac{4}{3\sqrt{\pi}}\left(\frac{mE_0}{MkT}\right)^{3/2} \sigma_s^{(0)} \,. \tag{2.71}$$

For the typical value $E_0 \sim 10^{-6}$ erg and $M \sim 10^{-24}$ g one has $\sigma/\sigma_s^{(0)} \sim 10^{-3}$. Therefore, to satisfy (2.69) one must have $kT \geq 500$ keV. A combination of high densities and temperatures may be optimal, but raising the temperature too high can destroy the electron degeneracy.

A fusion chain reaction requires an assembly with a dimension larger than

$$\ell = 1/n\sigma \,. \tag{2.72}$$

For $\sim 10^3$ times solid density or $n = 5 \times 10^{25}$ cm^{-3}, (proposed for microexplosion fusion assemblies), one would need $\ell \simeq 2 \times 10^{-2}$ cm. For a spherical assembly ℓ would have to be set equal the radius of the assembly. For a fusion chain reaction to spread from $r \leq 1/n\sigma$ to $r > 1/n\sigma$ would require a larger radius.

The situation is decisively better in a pinch discharge, where the magnetic field can radially entrap the charged fusion products. In this one-dimensional configuration, the pinch discharge channel would only have to be longer than $\ell = 1/n\sigma$.

To increase the energy output by a fusion chain reaction the assembly can even be subcritical (i.e. $\ell < 1/n\sigma$), because for a subcritical energy multiplication factor $k < 1$, the total energy output is increased by the factor $1 + k + k^2 + k^3 + \cdots = 1/(1-k)$. If for example $k = 1/2$, the energy output would be increased by the factor $1/(1 - 1/2) = 2$.

2.6 Fission-Fusion Chain Reactions

If fissile material is mixed (homogeneous or inhomogeneous) with neutron-producing thermonuclear material, and if the density and temperature are sufficiently high, thermonuclear fusion reactions releasing neutrons will make fission reactions, raising the temperature of the mixture. Since thermonuclear processes rise with a high power of the temperature in a range where $\langle \sigma \mathrm{v} \rangle$ has not yet reached its maximum, the higher temperature will increase the neutron production rate of the thermonuclear material, accelerating the fission reaction rate and so on. This coupling of the fission and fusion process through the release of heat and rise in temperature shall be called a fission-fusion chain reaction. It effectively increases the neutron multiplication factor thereby reducing the critical mass. This is especially true for high densities.

To analyze this process we are considering a mixture of fissile (U233, U235, PU239) and fusionable (DT) material. For a given pressure the atomic number densities in the fissile and fusionable material shall be N_U and N_h. Introducing a mixing parameter x, $0 < x < 1$, with $(1-x)N_U$ fissile and xN_h fusionable nuclei per unit volume, the neutron chain reaction in a mixture of infinite extension is determined by the equation

$$\frac{1}{\mathrm{v}_0}\frac{\partial \phi}{\partial t} = (\nu - 1)\,(1-x)\,N_U \sigma_f \phi + S \tag{2.73}$$

(v_0 velocity of fission neutrons, ν fission neutron multiplication factor, σ_f fission cross section, ϕ neutron flux), where (for DT)

$$S = \frac{1}{4}\,x^2 N_h^2 \langle \sigma \mathrm{v} \rangle \tag{2.74}$$

is the source term of the DT fusion reaction neutrons.

We are interested in the temperature range from 1 keV to 10 keV (10^7 °K to 10^8 °K). There $\langle \sigma \mathrm{v} \rangle$ rises rapidly with the temperature dependence (T in keV):

$$\langle \sigma \mathrm{v} \rangle \simeq 1.1 \times 10^{-20}\, T^{4.37}\,. \tag{2.75}$$

With (2.74) and (2.75), (2.73) becomes

$$\frac{1}{\mathrm{v}_0}\frac{\partial \phi}{\partial t} = (\nu - 1)\,(1-x)\,N_U \sigma_f \phi + 2.75 \times 10^{-21} x^2 N_h^2\, T^{4.37}\,. \tag{2.76}$$

Next we need a relation between T and ϕ. As long as $N_e kT > aT^4$ (N_e electron number density, $a = 7.76 \times 10^{-15}$ erg/cm^3 °K^4 Stefan-Boltzmann constant), the heat released by the fission and fusion reactions goes mostly into kinetic particle energy. If this inequality is not satisfied the heat goes mostly into blackbody radiation, and because of the T^4 dependence the temperature rises there only slowly. From the condition $N_e kT > aT^4$, resp. $N_e > (a/k)\,T^3$ it follows that for $T = 10^7$ °K (1 keV) $N_e > 5 \times 10^{22}$ cm^{-3} and for $T = 10^8$ °K (10 keV) that $N_e > 5 \times 10^{25}$ cm^{-3}. For the intermediate temperature $T = 5 \times 10^7$ °K only $N_e > 5 \times 10^{24}$ cm^{-3} is needed.

The energy released by the fission process per cm^3 and sec is

$$\varepsilon_f\,(1-x)\,N_U \sigma_f \phi$$

where ε_f =180 MeV = 2.9×10^{-4} erg is the fission energy. The energy released in the DT fusion reaction per cm^3 and sec is

$$\varepsilon_\alpha S = \frac{1}{4}\varepsilon_\alpha x^2 N_h^2 \langle \sigma \mathrm{v} \rangle = 2.75 \times 10^{-21} \varepsilon_\alpha x^2 N_h^2\, T^{4.37}$$

where ε_α = 17.6 MeV = 2.8×10^{-5} erg is the fusion reaction energy. With this heat source the temperature increase in the mixture is

$$\begin{aligned} &3k\left[g\,(1-x)\,N_U + xN_h\right]\frac{\partial T}{\partial t} \\ &= \varepsilon_f\,(1-x)\,N_U\sigma_f\phi + 2.75 \times 10^{-21}\varepsilon_\alpha x^2 N_h^2\, T^{4.37} \end{aligned} \tag{2.77}$$

where g is the degree of ionization of the fissile material at the temperature T, with $g \approx 10$ a likely value.

Expanding $f(T) = T^{4.37}$ around $T = T_0$ (> 1 keV) into a Taylor series one has

$$T^{4.37} = const. + 4.37\, T_0^{3.37} T \tag{2.78}$$

Inserting (2.78) into (2.76) and (2.77) one obtains

$$\frac{\partial \phi}{\partial t} = \alpha_1 \phi + \beta_1 T + \gamma_1 \tag{2.79}$$

$$\frac{\partial T}{\partial t} = \alpha_2 \phi + \beta_2 T + \gamma_2 \tag{2.80}$$

where

$$\begin{aligned} \alpha_1 &= (\nu - 1)\,(1-x)\,N_U\sigma_f \mathrm{v}_0 \\ \beta_1 &= 1.2 \times 10^{-20}\, \mathrm{v}_0 x^2 N_h^2\, T_0^{3.37} \\ \alpha_2 &= \frac{\varepsilon_f\,(1-x)\,N_U\sigma_f}{3k\left[g\,(1-x)\,N_U + xN_h\right]} \\ \beta_2 &= \frac{1.2 \times 10^{-20}\varepsilon_\alpha x^2 N_h^2\, T_0^{3.37}}{3k\left[g\,(1-x)\,N_U + xN_h\right]} \end{aligned}$$

Furthermore, γ_1, γ_2 are constants the values of which are of no interest.

Eliminating ϕ from (2.79) and (2.80) one obtains

$$\ddot{T} - (\alpha_1 + \beta_2)\,\dot{T} + (\alpha_1\beta_2 - \alpha_2\beta_1)\,T + \alpha_1\gamma_2 - \alpha_2\gamma_1 = 0 \tag{2.81}$$

The general solution of (2.81) is the sum of a particular solution of the inhomogeneous equation and the general solution of the homogeneous equation. A particular solution of the inhomogeneous equation is

$$T = \frac{\alpha_1 \gamma_2 - \alpha_2 \gamma_1}{\alpha_2 \beta_1 - \alpha_1 \beta_2} = const. \tag{2.82}$$

into which the constants γ_1, γ_2 enter, though these constants do not enter into the solution of the homogeneous equation

$$T = const.\, e^{\lambda t} \tag{2.83}$$

where

$$\lambda = \frac{\alpha_1 + \beta_2}{2} + \left[\left(\frac{\alpha_1 + \beta_2}{2} \right)^2 + \alpha_2 \beta_1 - \alpha_1 \beta_2 \right]^{1/2} . \tag{2.84}$$

For $x = 0$, that is for a pure fission assembly, one has

$$\lambda = \lambda_0 = (\nu - 1)\, N_U \sigma_f \mathrm{v}_0 \tag{2.85}$$

the same as in (2.39).

If there would be no coupling with the fusion process one would have

$$\lambda_1 = \alpha_1 = \lambda_0 (1 - x) \ . \tag{2.86}$$

If there is a coupling with the fusion process, one can define the ratio

$$f = \frac{\lambda}{\lambda_0} \tag{2.87}$$

from which an effective neutron multiplication factor ν^* can be obtained by putting

$$\nu^* - 1 = (\nu - 1)\, f \ . \tag{2.88}$$

Introducing the auxiliary function

$$F(x) = \frac{x^2}{g + \left[x/(1-x) \right] N_h / N_U} \frac{N_h^2}{N_U^2} \tag{2.89}$$

one can write

$$\alpha_2 = 2.1 \times 10^8 \varepsilon_f \sigma_f \left(N_U/N_h\right)^2 \; F(x)/x^2 \, , \tag{2.90}$$

$$\beta_2 = 2.5 \times 10^{-12} \varepsilon_\alpha T_0^{3.37} N_U \; F(x)/(1-x) \, . \tag{2.91}$$

One sees that for $x \sim 0.5$ and 1 keV $< T <$ 10 keV one has

$$\alpha_1 \beta_2 \ll \alpha_2 \beta_1$$

$$\left(\frac{\alpha_1 + \beta_2}{2}\right)^2 \ll \alpha_2 \beta_1$$

it therefore follows that approximately

$$\lambda \gtrsim \left(\alpha_2 \beta_1\right)^{1/2} \tag{2.92}$$

with the definition (2.87) and with $N_h/N_U \simeq 43$ (see chapter 3.10), one has for $f = f(x)$:

$$f(x) \gtrsim 11.2 \left[\frac{x^2(1-x)}{1+3.3x}\right]^{1/2} T_0^{1.68} \, . \tag{2.93}$$

This function has a maximum at $x = 0.57$ where

$$f(x) \gtrsim 2.48 \, T_0^{1.68} \, . \tag{2.94}$$

According to (2.36) and (2.88) B_0^2 is increased by multiplying it with $f(x)$.

If for $f = 1$ the critical radius is $R_0 = \pi/B_0$, it becomes for $f > 1$ equal to $R = R_0/\sqrt{f}$. And if for $x = 1$ the critical mass is M_0, it becomes for $f > 1$ and $0 < x < 1$ equal to M:

$$M/M_0 = (1-x) \, f^{-3/2} = 0.43 f^{-3/2} \, . \tag{2.95}$$

If in addition the fission-fusion assembly is at the same time compressed above solid density $\rho > \rho_0$, the critical mass is further reduced by $(\rho_0/\rho)^2$. This is in particular true if the heating is done by compression, through hypervelocity impact for example. There it may be advantageous to separate the DT from the fissile material by surrounding it with a shell of DT. As an example we take $T_0 = 1$ keV where $f = 2.48$ and obtain $M/M_0 \simeq 0.1$, which is a $\sim$ tenfold reduction of the critical mass. But for $T_0 = 2$ keV where $f = 8.55$

one already has $M/M_0 \simeq 2 \times 10^{-2}$. Because of (2.88) with $\nu = 2.5$ the first example implies that $\nu^* = 3.72$ and the second that $\nu^* = 12.8$.

If the critical mass without the described fission-fusion process is ~ 10 kg, it would in the second example be reduced to $\simeq 200$ g. At an impact pressure of $\sim 2 \times 10^{13}$ dyn/cm^2, the fissile material is compressed to about 5 times solid density, further reducing the critical mass ~ 25 fold down to $\sim$ 10 g.

2.7 Bibliography for Chapter 2

G. Gamow and E. Teller, Physical Review, **53**, 608 (1938).

G. Gamow and C. L. Critchfield, Theory of Atomic Nucleus and Nuclear Energy-Sources, Oxford at the Clarendon Press 1949.

E. Sänger, Astronautica Acta **1**, 61 (1955).

S. Glasstone and M. C. Edlund, The Elements of Nuclear Reactor Theory, D. Van Nostrand Company, Inc. New York 1952.

M. Gryzinskii, Physical Review **111**, 900 (1958).

F. Winterberg, Nuclear Science and Engineering **59** , 68 (1976).

G. H. Miley, Fusion Energy Conversion, American Nuclear Society, 1976.

Chapter 3

The Thermonuclear Plasma

3.1 Ionization Temperature

At the high temperatures in the core of thermonuclear explosions, all matter is in the state of a plasma, made up of positive ions and negative electrons. The ionization though is not necessarily complete, which is the state where the nuclei are stripped of all their electrons. This is especially true in the presence of heavy elements, as it is the case in a fission-fusion plasma.

A lower limit for the total ionization temperature can be obtained from the binding energy of an electron in the lowest Bohr orbit of a Z-times charged nucleus:

$$E_0 = \frac{me^4}{2\hbar^2} Z^2 = 2.2 \times 10^{-11} Z^2 \text{ [erg]} \tag{3.1}$$

(m electron mass, e electron charge, $\hbar = h/2\pi$, h Planck's constant). The energy required for complete ionization is larger, because not only has the innermost electron to be removed but also the less strongly bound outer electrons. Taking this into account the total ionization energy is

$$E_i \simeq 2.2 \times 10^{-11} Z^{2.42} \text{ [erg]} \,. \tag{3.2}$$

The average kinetic particle energy of the plasma at temperature T is $\frac{3}{2}kT$. Since in a collision momentum is conserved, the kinetic particle energy to cause ionization by collision has to be twice as large as E_i. Hence one has to set $\frac{3}{2}kT_i = 2E_i$, and one obtains for the ionization temperature

$$T_i = 2.1 \times 10^5 Z^{2.42} \text{ [°K]} \tag{3.3}$$

For hydrogen ($Z = 1$), the temperature for complete ionization would be $T_i = 2.1 \times 10^5$ °K and for uranium ($Z = 92$), $T_i \simeq 10^{10}$ °K. In a thermonuclear plasma made up of light elements one always has $T > T_i$, but in the plasma of a fission explosion one has $T_i \lesssim 10^8$ °K and there one is far away from complete ionization.

For partially ionized plasma the degree of ionization can be obtained from the Saha equation. For the physics of thermonuclear plasmas where $T > T_i$ the Saha equation is unimportant.

3.2 Plasma Equation of State

In many aspects a plasma behaves like a monatomic gas made up of n ions and nZ electrons. As for an ideal gas it has the equation of state

$$p = (1 + Z)\, nkT \,. \tag{3.4}$$

Similarly, the energy density of a monatomic gas is given by $\varepsilon = \frac{3}{2}nkT$, the energy density of a Z times ionized plasma is

$$\varepsilon = \frac{3\,(1+Z)}{2} nkT = \frac{3}{2} p \,. \tag{3.5}$$

Other useful relations are obtained by introducing the plasma density $\rho = nMA$, where M is the proton mass and A the atomic weight. With the expression for the density one has, with c_{v} the specific heat at constant volume,

$$\varepsilon = \rho c_{\text{v}} T \tag{3.6a}$$

$$c_{\text{v}} = \frac{3\,(1+Z)\,k}{2AM} = \frac{3\,(1+Z)}{2A} R \tag{3.6b}$$

where $R = k/M = 8.3 \times 10^7$ erg/g°K is the gas constant. The equation of state (3.4) can then be written as follows:

$$\frac{p}{\rho} = \frac{2}{3}\, c_{\text{v}} T. \tag{3.7}$$

For a high temperature plasma in thermodynamic equilibrium dominated by blackbody radiation, the energy density is

$$u = aT^4 \tag{3.8}$$

($a = 7.67 \times 10^{-15}$ erg/cm^3 °K^4). With the radiation pressure

$$p = \frac{u}{3} \tag{3.9}$$

hence

$$pT^{-4} = const. \tag{3.10}$$

For an ideal gas an adiabatic change leads to the relation

$$pT^{-\gamma/(\gamma-1)} = const. \tag{3.11}$$

where $\gamma = c_p/c_v$ is the specific heat ratio. For a plasma, as in a monatomic gas, $\gamma = \frac{5}{3}$, one has $pT^{-5/2} = const.$ This is not true for a radiation dominated plasma where $4 = \gamma/(\gamma - 1)$, or where $\gamma = \frac{4}{3}$.

Whether the plasma is radiation dominated or not depends on the optical mean free path, that is the absorption length of the photons. If the plasma dimensions are small compared to this length, the plasma is not in equilibrium with the blackbody radiation and can then be treated like an ideal monatomic gas with the specific heat ratio $\gamma = \frac{5}{3}$.

For an arbitrary number of degrees of freedom f,

$$\gamma = \frac{2+f}{f} . \tag{3.12}$$

For a monatomic gas (and degenerate Fermi gas) one has $\gamma = \frac{5}{3}$, but for blackbody radiation with $\gamma = \frac{4}{3}$ one has $f = 6$.

3.3 Microscopic Plasma Theory

Here we consider the motion of electrons and ions (the particles making up a plasma), in an electric (**E**) and magnetic (**H**) field. Using electrostatic cgs units the (nonrelativistic) equation of motion of a particle with charge Ze and mass m is

$$m\frac{d\mathbf{v}}{dt} = Ze\left(\mathbf{E} + \frac{1}{c}\mathbf{v} \times \mathbf{H}\right). \tag{3.13}$$

If $\mathbf{H} = 0$ or $\mathbf{v} \parallel \mathbf{H}$ one has

$$m\frac{d\mathbf{v}}{dt} = Ze\mathbf{E} \tag{3.14}$$

and hence (assuming $\mathbf{E} = const.$),

$$\left.\begin{aligned} \mathbf{v} &= \frac{Ze\mathbf{E}}{m} t \\ \mathbf{r} &= \frac{1}{2} \frac{Ze\mathbf{E}}{m} t^2 \end{aligned}\right\} \tag{3.15}$$

If $\mathbf{E} = 0$ one has

$$m \frac{d\mathbf{v}}{dt} = \frac{Ze}{c} \mathbf{v} \times \mathbf{H} \,. \tag{3.16}$$

Since $(\mathbf{v} \times \mathbf{H}) \perp \mathbf{v}$, $|\mathbf{v}|$ is constant, and from

$$\frac{m\mathrm{v}_\perp^2}{r} = \frac{Ze\mathrm{v}_\perp |\mathbf{H}|}{c} \tag{3.17}$$

one has (r_L Larmor radius)

$$r = r_L = \frac{m\mathrm{v}_\perp c}{ZeH} \,. \tag{3.18}$$

If $\mathrm{v}_\parallel = const.$ ($\mathrm{v}_\parallel \parallel \mathbf{H}$), the motion is a helix with a pitch angle α given by $\mathrm{tg}\alpha = \mathbf{v}_\parallel / \mathrm{v}_\perp$.

Putting $\mathbf{v}_\perp = r_L \omega_c$, one has ($\omega_c$ cyclotron frequency)

$$\omega_c = \frac{ZeH}{mc} \,. \tag{3.19}$$

If $\mathbf{E}$ and $\mathbf{H}$ are constant in space and time one puts

$$\mathbf{v} = \mathbf{v}' + c \, \frac{\mathbf{E} \times \mathbf{H}}{\mathrm{H}^2} \tag{3.20}$$

and substitutes (3.20) into (3.13):

$$\frac{d\mathbf{v}'}{dt} = \frac{Ze}{m} \left(\mathbf{E} + \frac{1}{c} (\mathbf{v}' \times \mathbf{H}) + \frac{1}{H^2} (\mathbf{E} \times \mathbf{H}) \times \mathbf{H} \right) \tag{3.21}$$

If $\mathbf{H} \cdot \mathbf{E} = 0$ ($\mathbf{H} \perp \mathbf{E}$), one has with $(\mathbf{E} \times \mathbf{H}) \times \mathbf{H} = -H^2 \mathbf{E}$

$$\frac{d\mathbf{v}'}{dt} = \frac{Ze}{mc} \, \mathbf{v}' \times \mathbf{H}, \tag{3.22}$$

describing a circular motion with the Larmor radius (3.18), superimposed on a drift motion

$$\mathbf{v}_D = c\,\frac{\mathbf{E}\times\mathbf{H}}{H^2}\,, \quad \mathrm{v}_D = c\,\frac{E}{H} \tag{3.23}$$

perpendicular to both H and E.

The magnetic moment of a gyrating particle is defined by

$$\mu = IS \tag{3.24}$$

where $I = Ze\nu_c = Ze\omega_c/2\pi$ is the electric current by the gyrating particle, and $S = \pi r_L^2$ the area enclosed by the current. One obtains

$$\mu = \pi r_L^2 \frac{Ze\omega_c}{2\pi} = \frac{1}{2}\frac{m\mathrm{v}_\perp^2 c}{H}\,. \tag{3.25}$$

Because (V voltage)

$$\begin{aligned} \frac{d}{dt}\left(\frac{1}{2}m\mathrm{v}_\perp^2\right) &= VI = \oint \mathbf{E}\cdot d\mathbf{s}\cdot\mathbf{I} \\ &= -\frac{1}{c}\int\frac{\partial H}{\partial t}\cdot d\mathbf{S}\cdot\mathbf{I} \\ &= \frac{\mu}{c}\frac{dH}{dt} \end{aligned} \tag{3.26}$$

it follows that

$$\frac{d}{dt}\left(\frac{\mu}{c}H\right) = \frac{d}{dt}\left(\frac{1}{2}m\mathrm{v}_\perp^2\right) \tag{3.27}$$

or that

$$\frac{d\mu}{dt} = 0 \tag{3.28}$$

and hence

$$\mu = const. \tag{3.29}$$

This result though is only approximately true, because in (3.26) it was assumed that the Larmor radius in $\oint \mathbf{E}\cdot d\mathbf{s}$ remains constant during one revolution. One therefore calls μ an "adiabatic" invariant, implying that H changes slowly on a time scale of one revolution for the gyrating particle.

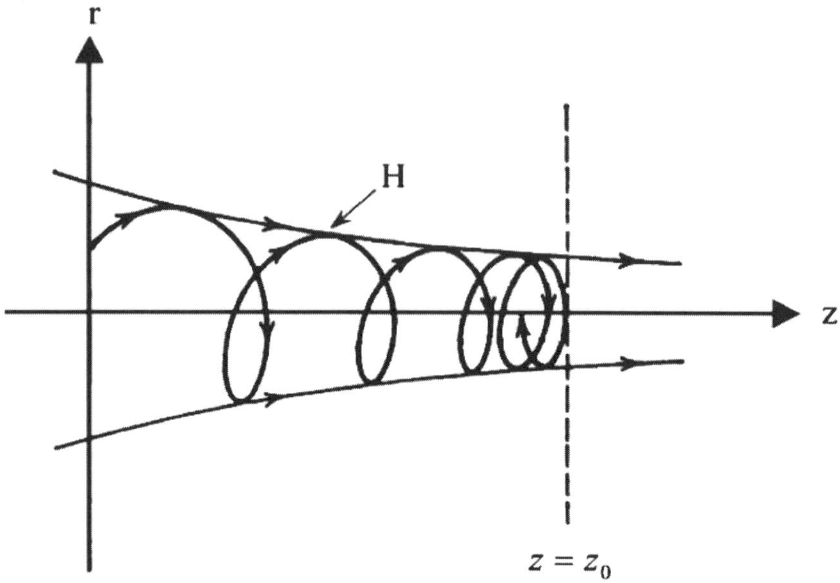

Figure 3.1: Magnetic mirror.

But μ is also an adiabatic invariant if H changes slowly in space. For this consider the motion of a charged particle into a magnetic mirror (see Fig. 3.1). In cylindrical coordinates, div $\mathbf{H} = 0$ is

$$\frac{1}{r}\frac{\partial}{\partial r}(rH_r) + \frac{\partial H}{\partial z} = 0\,. \tag{3.30}$$

For $\partial H/\partial z = const.$ one has

$$H_r = -\frac{r}{2}\frac{\partial H}{\partial z}\,. \tag{3.31}$$

Putting $r = r_L$ one can write

$$H_r = -\frac{r_L}{2}\nabla_{\parallel} H \tag{3.32}$$

and

$$m\,\frac{d\mathrm{v}_{\parallel}}{dt} = \frac{e}{c}\,\mathrm{v}_{\perp} H_r = -\frac{e}{c}\,\mathrm{v}_{\perp}\frac{r_L}{2}\,\nabla_{\parallel} H\,, \tag{3.33}$$

or

$$m\frac{d\text{v}_{\parallel}}{dt} = -\frac{\mu}{c}\nabla_{\parallel} H\,. \tag{3.34}$$

Multiplying with $\text{v}_{\parallel}$ this becomes

$$\frac{d}{dt}\left(\frac{1}{2}m\text{v}_{\parallel}^2\right) = -\frac{\mu}{c}\frac{\partial H}{\partial z}\text{v}_{\parallel} = -\frac{\mu}{c}\frac{\partial H}{\partial z}\frac{dz}{dt} = -\frac{\mu}{c}\frac{dH}{dt}. \tag{3.35}$$

With (3.25) and

$$\frac{d}{dt}\left(\frac{1}{2}m\text{v}_{\perp}^2 + \frac{1}{2}m\text{v}_{\parallel}^2\right) = 0 \tag{3.36}$$

one has

$$\frac{d}{dt}\left(\frac{1}{2}m\text{v}_{\parallel}^2\right) = -\frac{d}{dt}\left(\frac{1}{2}m\text{v}_{\perp}^2\right) = -\frac{d}{dt}\left(\frac{\mu}{c}H\right), \tag{3.37}$$

and with (3.35) again (as in adiabatic approximation),

$$\mu = const. \tag{3.38}$$

If θ is the angle $\mathbf{v}$ makes with the z-axis,

$$\frac{\text{v}_{\perp}}{\text{v}} = \sin\theta, \tag{3.39}$$

one has $\sin^2\theta = (\text{v}_{\perp}/\text{v})^2$ and with $\text{v}_{\perp}^2/H = const.$ (μ-invariance) that

$$\sin^2\theta = \frac{H}{H_0}\sin^2\theta_0, \tag{3.40}$$

where $\sin\theta_0 = \text{v}_{\perp}/\text{v}$ and H_0 are the initial values for a particle entering the mirror where $H > H_0$. Total reflection occurs for $\sin\theta = 1$, that is for $\theta = 90°$, and if H_m is the maximum mirror field, with $H_m > H_0$ all particles for which

$$\sin^2\theta_0 > \frac{H_0}{H_m} \tag{3.41}$$

holds are reflected. The number of particles "hitting" the mirror per second and solid angle $d\Omega = 2\pi \sin\theta_0 \, d\theta_0$ is proportional to $\cos\theta_0 \, d\Omega$ ($\mathrm{v}_\perp^{(0)}/\,\mathrm{v} = \cos\theta_0$), and the reflection coefficient accordingly is

$$R = \frac{\int_{\theta_0=\theta_1}^{\pi/2} \cos\theta_0 \; d\Omega}{\int_{\theta_0=0}^{\pi/2} \cos\theta_0 \; d\Omega} = 1 - \frac{H_0}{H_m}\,, \qquad \sin^2\theta_1 = \frac{H_0}{H_m}\,. \tag{3.42}$$

For particles trapped in between two mirrors separated by a distance L, one has the longitudinal adiabatic invariant

$$\mathrm{v}_{\shortparallel} L = const. \tag{3.43}$$

which can be viewed as follows: For two mirrors approaching each other, the relative velocity of these mirrors is

$$V = -\frac{dL}{dt} \tag{3.44}$$

but one also has for the change in the longitudinal particle velocity by the reflections

$$\frac{d\mathrm{v}_{\shortparallel}}{dt} = \frac{\mathrm{v}_{\shortparallel}}{2L}\,2V \tag{3.45}$$

or

$$\frac{d\mathrm{v}_{\shortparallel}}{dt} = -\frac{\mathrm{v}_{\shortparallel}}{L}\frac{dL}{dt} \tag{3.46}$$

and hence (3.43).

With the total number of particles in a magnetic flux tube constant, one has

$$nLr^2 = const. \tag{3.47}$$

and with div $\mathbf{H} = 0$,

$$Hr^2 = const. \tag{3.48}$$

If $H = const.$

$$L \propto \frac{H}{n} \propto \frac{1}{n}\,. \tag{3.49}$$

With $\mathrm{v}_{\shortparallel} L = const.$, and with

$$T_{\shortparallel} \propto \mathrm{v}_{\shortparallel}^2 \,, \tag{3.50}$$

one has for particles trapped between magnetic mirrors

$$T_{\shortparallel} \propto n^2 \,. \tag{3.51}$$

For particles in a magnetic flux tube where $H = H(t)$, one has $\mu = const.$, and hence

$$\mathrm{v}_{\perp}^2 \propto H \,. \tag{3.52}$$

For the number of particles one there has

$$nr^2 = const. \tag{3.53}$$

and because of (3.48),

$$H \propto n \tag{3.54}$$

and

$$\mathrm{v}_{\perp}^2 \propto n \tag{3.55}$$

Hence,

$$T_{\perp} \propto n \,. \tag{3.56}$$

Comparing (3.51) and (3.56) with the general formula for an adiabatic change of state,

$$T \propto n^{\gamma-1} = n^{2/f} \tag{3.57}$$

with f the number of degrees of freedom, one finds that for (3.51) the gas of particles behaves like a one-dimensional gas with $f = 1$, and for (3.56) like a two-dimensional gas with $f = 2$.

There is another kind of drift motion perpendicular to the magnetic field, which can become important in thermonuclear configurations. It occurs in the presence of a gradient of the magnetic field perpendicular to the direction of the field. With the constancy of the magnetic moment it results from the force

$$\mathbf{F}_{\perp} = -\frac{\mu}{c}\,\nabla_{\perp} H \,. \tag{3.58}$$

The effect of this force on a particle's trajectory can be obtained from (3.23) by replacing $\mathbf{E}$ with $-(\mu/cZe)\,\nabla_{\perp} H$. The result is

$$\mathbf{v}_{D} = \frac{\mu}{Ze}\,\frac{\mathbf{H}\times(\nabla_{\perp} H)}{H^2} \tag{3.59}$$

or with $\mu = m\mathrm{v}_{\perp}^2 c/2H$ and $r_{L} = m\mathrm{v}_{\perp} c/ZeH$,

$$\frac{\mathbf{v}_{D}}{\mathrm{v}_{\perp}} = \frac{r_{L}}{2}\,\frac{H\times(\nabla_{\perp} H)}{H^2}\,, \qquad \frac{\mathbf{v}_{D}}{\mathrm{v}_{\perp}} = \frac{r_{L}}{2}\,\frac{\nabla_{\perp} H}{H}\,. \tag{3.60}$$

3.4 Debye-Length

On a macroscopic scale a plasma is always almost electrically neutral. Even a small separation of the positive and negative charges would result in large electrostatic forces trying to restore charge neutrality. On a small scale though, charge neutrality can be violated by thermal motion.

If all the electronic charges per unit volume $n_e e$ are displaced by the distance x in one direction against the ionic background charges, the electric field E caused by this charge separation is determined by Poisson's equation,

$$\mathrm{div}\mathbf{E} = 4\pi n_e e \,. \tag{3.61}$$

It follows that

$$E = 4\pi n_e e x \tag{3.62}$$

with the potential

$$V = \int_0^x E\,dx = 2\pi n_e e x^2 \,. \tag{3.63}$$

If the energy $eV = 2\pi n_e e^2 x^2$ is supplied by the thermal motion along the x-direction one has to set $eV = \frac{1}{2}kT$ and obtains from (3.63)

$$x = \lambda_D = \sqrt{\frac{kT}{4\pi n_e e^2}} \,. \tag{3.64}$$

The distance λ_D is called the Debye length, with charge neutrality violated for distances smaller than λ_D. This means that in a plasma the electrostatic potential of an ion is screened for distances larger than λ_D. The ionic potential of a Z times charged ion is thus given by

$$V(r) = \frac{Ze}{r} e^{-r/\lambda_D} , \qquad \lambda_D = \sqrt{\frac{kT}{4\pi Z n e^2}} , \tag{3.65}$$

with n ions per unit volume, screened by nZ electrons. For the example $T \sim 10^8$ °K, $n \simeq 5 \times 10^{22}$ cm^{-3}, $Z = 1$, one has $\lambda_D \sim 10^{-6}$ cm. If the Debye length is compared with the average distance between the ions

$$d = n^{-1/3} , \tag{3.66}$$

one speaks of a plasma only when $\lambda_D > d$. If $\lambda_D < d$ one says the plasma is nonideal. For the given example $d \sim 10^{-7}$ cm, $\lambda_D \sim 10^{-6}$ cm the plasma is ideal. However, as can be seen from (3.65) for lower temperatures and higher densities the plasma can become nonideal.

3.5 Macroscopic Plasma Theory

Provided the plasma is ideal ($\lambda_D > n^{-1/3}$), it can be described by a two-fluid model for the ions and the electrons. These are the Euler equations for the ions and electrons with external forces and a mutual friction term. The index i stands for the ions and the index e for the electrons:

$$n_i m_i \left(\frac{\partial \mathbf{v}_i}{\partial t} + (\mathbf{v}_i \cdot \nabla) \mathbf{v}_i \right) = n_i Z e \left(\mathbf{E} + \frac{1}{c} \mathbf{v}_i \times \mathbf{H} \right) - \nabla p_i + \mathbf{P}_{ie} , \tag{3.67a}$$

$$m_e m_e \left(\frac{\partial \mathbf{v}_e}{\partial t} + (\mathbf{v}_e \cdot \nabla) \mathbf{v}_e \right) = -n_e e \left(\mathbf{E} + \frac{1}{c} \mathbf{v}_e \times \mathbf{H} \right) - \nabla p_e + \mathbf{P}_{ei} . \tag{3.67b}$$

From Newton's actio = reactio principle it follows for the mutual friction terms $\mathbf{P}_{ie}$ and $\mathbf{P}_{ei}$ that

$$\mathbf{P}_{ie} + \mathbf{P}_{ei} = 0 . \tag{3.67c}$$

Introducing

$\rho = n_i m_i + n_e m_e$	MASS DENSITY
$\mathbf{v} = \frac{1}{\rho}\left(n_i m_i \mathbf{v}_i + n_e m_e \mathbf{v}_e\right)$	CENTER OF MASS VELOCITY
$\mathbf{j} = e\left(n_i Z \mathbf{v}_i - n_e \mathbf{v}_e\right)$	ELECTRIC CURRENT DENSITY
$\rho_e = e\left(Z n_i - n_e\right)$	ELECTRIC CHARGE DENSITY

and assuming charge neutrality, i.e., $Zn_i = n_e$, one obtains by adding (3.67a) to (3.67b) and neglecting quadratic terms in $\mathbf{v}_e$ and $\mathbf{j}$:

$$\rho\left(\frac{\partial \mathbf{v}}{\partial t} + (\mathbf{v}\cdot\nabla)\,\mathbf{v}\right) = -\nabla p + \frac{1}{c}\,\mathbf{j}\times\mathbf{H} \tag{3.68}$$

where $p = p_e + p_i$. Multiplying (3.67a) with Zm_e and (3.67b) with m_i, and thereafter subtracting the second from the first, and as before neglecting quadratic terms in $\mathbf{v}_e$ and $\mathbf{j}$, one obtains:

$$\frac{m_i m_e}{Ze^2\rho}\,\frac{\partial \mathbf{j}}{\partial t} = \mathbf{E} + \frac{1}{c}\mathbf{v}\times\mathbf{H} - \frac{1}{\sigma}\mathbf{j} + \\ + \frac{1}{Ze\rho}\left(m_i\nabla p_e - Zm_e\nabla p_i - \frac{1}{c}\left(m_i - Zm_e\right)\mathbf{j}\times\mathbf{H}\right) \tag{3.69}$$

In (3.69) we have made the "collision ansatz," with σ still to be determined:

$$\mathbf{P}_{ei} = -\mathbf{P}_{ie} = \frac{1}{\sigma}\,en_e\mathbf{j}\,. \tag{3.70}$$

Equations (3.68) and (3.69) must be supplemented by:

1. the equation of state with $n_e = Zn_i$,

$$p = p_e + p_i = n_e kT + n_i kT = (1+Z)\,n_i kT\,, \tag{3.71}$$

2. the equation of continuity,

$$\frac{\partial \rho}{\partial t} + \nabla\cdot(\rho\mathbf{v}) = 0\,, \tag{3.72}$$

3. the energy equation (first and second law of thermodynamics), and

4. Maxwell's equations, which in electrostatic cgs units are

$$\begin{aligned}
\operatorname{div}\mathbf{E} &= 4\pi e\,(Zn_i - n_e) \\
\operatorname{div}\mathbf{H} &= 0 \\
-\frac{1}{c}\frac{\partial \mathbf{H}}{\partial t} &= \operatorname{curl}\mathbf{E} \\
\frac{4\pi}{c}\,\mathbf{j} + \frac{1}{c}\frac{\partial \mathbf{E}}{\partial t} &= \operatorname{curl}\mathbf{H}
\end{aligned} \tag{3.73}$$

For $\mathbf{v} = 0$, one has from (3.68)

$$\frac{1}{c}\,\mathbf{j}\times\mathbf{H} = \nabla p \tag{3.74}$$

furthermore, if $\partial\mathbf{E}/\partial t = 0$ one has $(4\pi/c)\,\mathbf{j} = \operatorname{curl}\mathbf{H}$ and thus

$$\operatorname{curl}(\mathbf{H}\times\operatorname{curl}\mathbf{H}) = 0\,. \tag{3.75}$$

If $\operatorname{curl}\mathbf{H} = 0$ then there are no currents flowing through the plasma and (3.75) is trivially satisfied, but it is also satisfied if $\mathbf{H}\times\operatorname{curl}\mathbf{H} = 0$. In this case, called a force-free magnetic field, the current and magnetic field lines are everywhere parallel.

For $\mathbf{v} = 0$, one obtains from (3.69), neglecting terms of the order m_e/m_i and expressing $\mathbf{j}\times\mathbf{H}$ by (3.74):

$$\frac{1}{\sigma}\,\mathbf{j} = \mathbf{E} - \frac{1}{en_e}\nabla p_i\,. \tag{3.76}$$

At constant pressure one has

$$\mathbf{j} = \sigma\mathbf{E}\,, \tag{3.77}$$

which shows that σ is the electrical conductivity and (3.69) the generalized Ohm's law.

With Maxwell's equation $(4\pi/c)\,\mathbf{j} = \operatorname{curl}\mathbf{H}$ one obtains for the resistive losses

$$\mathbf{j}\cdot\mathbf{E} = \frac{\mathbf{j}^2}{\sigma} = \left(\frac{1}{\sigma}\right)\left(\frac{c}{4\pi}\right)^2(\operatorname{curl}\mathbf{H})^2\,. \tag{3.78}$$

If $\mathbf{v} \neq 0$, one obtains from (3.69) (neglecting the $\partial \mathbf{j}/\partial t$ and pressure gradient terms):

$$\mathbf{j} = \sigma \left(\mathbf{E} + \frac{1}{c}\, \mathbf{v} \times \mathbf{H} \right) . \tag{3.79}$$

Eliminating $\mathbf{E}$ from (3.79) and Maxwell's equation $-(1/c)\, \partial \mathbf{H}/\partial t = \mathrm{curl}\, \mathbf{E}$, one has

$$-\frac{1}{c}\frac{\partial \mathbf{H}}{\partial t} = \frac{1}{\sigma}\, \mathrm{curl}\, \mathbf{j} - \frac{1}{c}\, \mathrm{curl}\, \mathbf{v} \times \mathbf{H} \tag{3.80}$$

and eliminating $\mathbf{j}$ from (3.80) and Maxwell's equation $(4\pi/c)\, \mathbf{j} = \mathrm{curl}\, \mathbf{H}$ one has

$$\frac{\partial \mathbf{H}}{\partial t} = \mathrm{curl}\, \mathbf{v} \times \mathbf{H} - \frac{c^2}{4\pi\sigma}\, \mathrm{curl}\, \mathrm{curl}\, \mathbf{H} . \tag{3.81}$$

With the vector identity $\mathrm{curl}\, \mathrm{curl} = \mathrm{grad}\, \mathrm{div} - \nabla^2$ and $\mathrm{div}\, \mathbf{H} = 0$, one finally obtains

$$\frac{\partial \mathbf{H}}{\partial t} = \mathrm{curl}\, \mathbf{v} \times \mathbf{H} + \frac{c^2}{4\pi\sigma}\, \nabla^2\, \mathbf{H} . \tag{3.82}$$

Putting $\mathbf{r} = R\xi$, $\mathbf{v} = U\mu$, $t = (R/U)\,\tau$, where R and U are a characteristic length and velocity of the problem considered, one can write for (3.82)

$$\frac{\partial \mathbf{H}}{\partial \tau} = \mathrm{curl}\, \mu \times \mathbf{H} + \frac{1}{Rem}\, \nabla^2\, \mathbf{H} \tag{3.83}$$

where the operations curl and ∇^2 are with regard to ξ, and where

$$Rem = \frac{4\pi\sigma U R}{c^2} \tag{3.84}$$

is the magnetic Reynolds number. If $Rem \gg 1$ one has

$$\frac{\partial \mathbf{H}}{\partial t} \simeq \mathrm{curl}\, \mathbf{v} \times \mathbf{H} \tag{3.85a}$$

and if $Rem \ll 1$

$$\frac{\partial \mathbf{H}}{\partial t} \simeq \frac{c^2}{4\pi\sigma} \nabla^2 \mathbf{H} \tag{3.85b}$$

the latter is exactly true for $\mathbf{v} = 0$.

From (3.85b) one obtains the characteristic diffusion time $t_0 = (R/U)\,\tau$ for a magnetic field to penetrate and to be dissipated in a conducting plasma

$$\mathbf{t}_0 = \frac{4\pi\sigma R^2}{c^2} \tag{3.86}$$

One can say that for $Rem \gg 1$ the plasma is dragged along by the magnetic field, and for $Rem \ll 1$ that the magnetic field is dissipated. For $Rem = 1$ one has a steady state case realized in a magnetohydrodynamic dynamo. This occurs in the liquid metallic core of the earth creating the earth's magnetic field.

For the special case where $\mathbf{H} \perp \mathbf{j}$ one has $(\mathbf{H}\cdot\nabla)\,\mathbf{H} = 0$, and because of $\mathbf{H} \times \operatorname{curl}\mathbf{H} = \nabla\,(\mathbf{H}^2/2) - (\mathbf{H}\cdot\nabla)\,\mathbf{H}$

$$\mathbf{H} \times \operatorname{curl}\mathbf{H} = \nabla\left(\mathbf{H}^2/2\right) \tag{3.87}$$

in this case then, one can introduce the concept of a magnetic pressure

$$p_H = \frac{\mathbf{H}^2}{8\pi} \tag{3.88}$$

whereby the magnetostatic equation (3.74) becomes

$$\nabla\,(p + p_H) = 0 \tag{3.89}$$

resp. $p + p_H = const.$ This last result can be applied to describe the pinch effect, where a surface current I flows along a plasma cylinder (Fig. 3.2). From

$$\frac{4\pi}{c}\mathbf{j} - \operatorname{curl}\mathbf{H} \tag{3.90}$$

follows

$$\int \frac{4\pi}{c}\mathbf{j}\cdot d\mathbf{S} = \int \operatorname{curl}\mathbf{H}\cdot d\mathbf{S} = \oint \mathrm{H}\cdot d\mathbf{s} \tag{3.91}$$

or

$$\frac{4\pi}{c}I = 2\pi r H \tag{3.92}$$

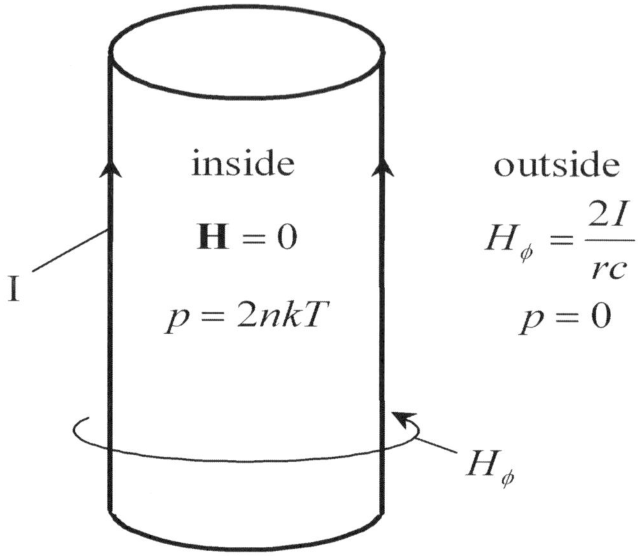

Figure 3.2: The pinch effect.

hence

$$H = \frac{2I}{rc} \, . \tag{3.93}$$

Inside the plasma cylinder one has (assuming $Z = 1$)

$$p = 2nkT \, , \quad p_H = 0 \tag{3.94}$$

and outside

$$p = 0 \, , \quad p_H = \frac{H^2}{8\pi} = \frac{I^2}{2\pi r^2 c^2} \, . \tag{3.95}$$

From $p + p_H = const.$, then follows at the plasma-vacuum boundary

$$(p + p_H)_{inside} = (p + p_H)_{outside} \tag{3.96}$$

or

$$2nkT = \frac{H^2}{8\pi} \tag{3.97}$$

or

$$2nkT = \frac{I^2}{2\pi r_0^2 c^2} \,. \tag{3.98}$$

With $N = \pi r_0^2 n$ the total number of ions or electrons per pinch length, one finds that (Bennett relation)

$$4NkT = \frac{I^2}{c^2} \,. \tag{3.99}$$

If I is measured in amperes (3.99) becomes

$$I^2 = 400NkT \,. \tag{3.100}$$

It should be noted, that a high temperature reached with a large current is entirely due to magnetic compression, since no resistive heating enters the Bennett equation.

One important parameter is the β-factor

$$\beta = \frac{p}{p_H} \,, \tag{3.101}$$

where p is the internal plasma pressure and p_H the externally acting magnetic pressure. According to (3.97) $\beta = 1$ for the pinch effect.

The pinch effect is an example of magnetic plasma confinement where a current flowing through the plasma sets up a magnetic field with its magnetic pressure $H^2/8\pi$ exerting the confining force on the plasma. But a plasma can also be confined by an externally applied magnetic field. As we will see later, both cases are of interest for thermonuclear microexplosions.

With regard to the microscopic working of magnetic plasma confinement the following comment can be made: As can be seen from (3.67a) and (3.67b) one roughly has $\mathrm{v}_e/\mathrm{v}_i \sim \sqrt{m_i/m_e} \gg 1$, which means that the electric current is mainly carried by the electrons. And with the ratio of the electron - to the ion - Larmor radius $r_{Le}/r_{Li} \sim (m_e \mathrm{v}_e)\,/\,(m_i \mathrm{v}_i) \sim \sqrt{m_e/m_i} \ll 1$, it is the electrons which are tightly bound to the magnetic lines of force, with the ions held to the electrons by electrostatic forces to sustain charge neutrality of the electron-ion plasma.

Finally we summarize the macroscopic plasma equations in the so-called magnetohydrodynamic (MHD) approximation, sufficient for most high density thermonuclear plasmas:

1. Ohm's law combined with Maxwell's equations

$$\left.\begin{aligned} &\operatorname{div}\mathbf{H} = 0 \\ &\frac{\partial \mathbf{H}}{\partial t} = \operatorname{curl}\mathbf{v}\times\mathbf{H} + \frac{c^2}{4\pi\sigma}\nabla^2\mathbf{H} \end{aligned}\right\} \tag{3.102}$$

2. Equation of motion combined with Maxwell's equations (ν kinematic viscosity)

$$\begin{aligned} \frac{\partial \mathbf{v}}{\partial t} + (\mathbf{v}\cdot\operatorname{grad})\,\mathbf{v} \;=\; & -\frac{1}{\rho}\operatorname{grad}p - \\ & -\frac{1}{4\pi\rho}\mathbf{H}\times\operatorname{curl}\mathbf{H} + \nu\nabla^2\mathbf{v} \end{aligned} \tag{3.103}$$

3. Equation of continuity

$$\frac{\partial \rho}{\partial t} + \operatorname{div}(\rho\mathbf{v}) = 0 \tag{3.104}$$

4. Energy equation (s specific entropy, κ heat conduction coefficient)

$$\begin{aligned} \rho T\left(\frac{\partial s}{\partial t} + \mathbf{v}\cdot\operatorname{grad}s\right) \;=\; & \operatorname{div}(\kappa\operatorname{grad}T) \\ & + \frac{c^2}{16\pi^2\sigma}(\operatorname{curl}\mathbf{H})^2 + \frac{\rho\nu}{2}(\operatorname{curl}\mathbf{v})^2 \end{aligned} \tag{3.105}$$

5. Equation of state

$$p = p(\rho, T)\ , \tag{3.106}$$

all together 10 equations, for the 10 unknowns $\mathbf{v}$, $\mathbf{H}$, ρ, p, s, T. In the MHD approximation the displacement current in Maxwell's equations is neglected and charge neutrality assumed.

3.6 Magnetohydrodynamics of Thermonuclear Plasmas

At the high temperatures of thermonuclear plasmas one can set the electrical conductivity infinitely large. Absent viscous dissipation and heat conduction the magnetohydrodynamic equations under these conditions are

$$\frac{\partial \mathbf{H}}{\partial t} = \operatorname{curl} \mathbf{v} \times \mathbf{H}\,, \quad \operatorname{div} \mathbf{H} = 0 \tag{3.107}$$

$$\frac{\partial \mathbf{v}}{\partial t} + (\mathbf{v} \cdot \nabla)\, \mathbf{v} = -\frac{1}{\rho}\, \nabla p - \frac{1}{4\pi\rho}\, \mathbf{H} \times \operatorname{curl} \mathbf{H} \tag{3.108}$$

$$\frac{\partial \rho}{\partial t} + \nabla \cdot (\rho \mathbf{v}) = 0 \tag{3.109}$$

$$p = p\,(\rho) \tag{3.110}$$

The latter is here the isentropic equation of state. With the Eulerian derivative

$$\frac{d}{dt} = \frac{\partial}{\partial t} + (\mathbf{v} \cdot \nabla)\,, \tag{3.111}$$

one has

$$\begin{aligned} \frac{d\mathbf{H}}{dt} &= \frac{\partial \mathbf{H}}{\partial t} + (\mathbf{v} \cdot \nabla)\, \mathbf{H} \\ &= \operatorname{curl} \mathbf{v} \times \mathbf{H} + (\mathbf{v} \cdot \nabla)\, \mathbf{H} \end{aligned} \tag{3.112}$$

With the vector identity

$$\operatorname{curl} \mathbf{v} \times \mathbf{H} = (\mathbf{H} \cdot \nabla)\, \mathbf{v} - (\mathbf{v} \cdot \nabla)\, \mathbf{H} + \mathbf{v} \operatorname{div} \mathbf{H} - \mathbf{H} \operatorname{div} \mathbf{v}$$

and with

$$\operatorname{div} \mathbf{H} = 0$$

one can write for (3.112)

$$\frac{d\mathbf{H}}{dt} = (\mathbf{H} \cdot \nabla)\, \mathbf{v} - \mathbf{H} \operatorname{div} \mathbf{v}\,. \tag{3.113}$$

Furthermore since

$$\begin{aligned} \frac{d\rho}{dt} &= \frac{\partial \rho}{\partial t} + \mathbf{v} \cdot \operatorname{grad} \rho \\ &= -\operatorname{div} \rho \mathbf{v} + \mathbf{v} \cdot \operatorname{grad} \rho \\ &= -\rho \operatorname{div} \mathbf{v} \end{aligned} \tag{3.114}$$

one has

$$\begin{aligned}\frac{d\mathbf{H}}{dt} &= (\mathbf{H}\cdot\nabla)\,\mathbf{v} - \mathbf{H}\mathrm{div}\,\mathbf{v} \\ &= (\mathbf{H}\cdot\nabla)\,\mathbf{v} + \frac{\mathbf{H}}{\rho}\frac{d\rho}{dt}\end{aligned} \tag{3.115}$$

or

$$\frac{d}{dt}\left(\frac{\mathbf{H}}{\rho}\right) = \left(\frac{\mathbf{H}}{\rho}\cdot\nabla\right)\mathbf{v}\,. \tag{3.116}$$

The physical interpretation of (3.116) is as follows: Consider the change of a fluid filament δ moving with the fluid,

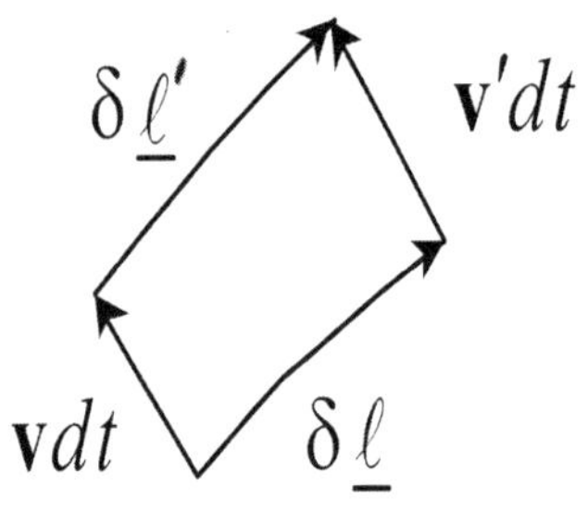

$$\delta\underline{\ell}' - \delta\underline{\ell} = \frac{d\,\delta\underline{\ell}}{dt}\cdot dt = (\mathbf{v}' - \mathbf{v})\,dt\,,$$

but with

$$\mathbf{v}' = \mathbf{v} + (\delta\underline{\ell}\cdot\nabla)\,\mathbf{v}\,,$$

it follows that

$$\frac{d\,\delta\underline{\ell}}{dt} = (\delta\underline{\ell}\cdot\nabla)\,\mathbf{v}$$

or that $\delta\underline{\ell}$ is "frozen" into the fluid.

According to (3.116) the same happens to H. A plasma into which a magnetic field is "frozen in" is called a magnetized plasma.

Next we consider small amplitude disturbances of a magnetized plasma. There are three modes: 1. Transverse Alfvén waves; 2. longitudinal sound waves along the lines of force; 3. longitudinal "mixed" magnetosonic waves perpendicular to the lines of force.

The modes are obtained from the linearized set of equations, where $\mathbf{h} \ll \mathbf{H}_0$ is the magnetic field disturbance imposed on $\mathbf{H}_0$, likewise $\rho \ll \rho_0$ the density disturbance imposed on ρ_0, and where all the small nonlinear terms are neglected:

$$\frac{\partial \mathbf{h}}{\partial t} = \operatorname{curl}(\mathbf{v} \times \mathbf{H}_0) \tag{3.117}$$

$$\frac{\partial \mathbf{v}}{\partial t} = -\frac{1}{\rho_0}\operatorname{grad} p - \frac{1}{4\pi\rho_0}\mathbf{H}_0 \times \operatorname{curl} \mathbf{h} \tag{3.118}$$

$$\frac{\partial \rho}{\partial t} + \rho_0 \operatorname{div} \mathbf{v} = 0 \tag{3.119}$$

With $\operatorname{grad} p = (\partial p/\partial \rho) \operatorname{grad} \rho = a^2 \operatorname{grad} \rho$ and $\mathbf{u} = \mathbf{H}_0/\sqrt{4\pi\rho_0}$, where a and u are the velocity of sound and Alfvén velocity, one obtains from (3.117 - 119), by elimination of ρ and $\mathbf{h}$, the wave equation:

$$\frac{\partial^2 \mathbf{v}}{\partial t^2} = a^2 \operatorname{grad} \operatorname{div} \mathbf{v} + \mathbf{u} \times \operatorname{curl} \operatorname{curl}(\mathbf{u} \times \mathbf{v}) \tag{3.120}$$

1. The transverse Alfvén wave for which $\operatorname{div} \mathbf{v} = 0$, propagates along the lines of force. It obeys the wave equation

   $$\frac{\partial^2 \mathbf{v}}{\partial t^2} = \mathbf{u} \times \operatorname{curl} \operatorname{curl}(\mathbf{u} \times \mathbf{v}) \tag{3.121}$$

   If $\mathbf{u}$ is directed along the x-axis and $\mathbf{v}$ perpendicular to $\mathbf{u}$ (3.121) becomes

   $$-\frac{1}{u^2}\frac{\partial^2 \mathrm{v}}{\partial t^2} + \frac{\partial^2 \mathrm{v}}{\partial x^2} = 0 \tag{3.122}$$

   describing a plane Alfvén wave propagating with the Alfvén speed $u = H_0/\sqrt{4\pi\rho_0}$.

2. The sonic wave where $\mathbf{v}$ is parallel to $\mathbf{H}_0$, with $\mathbf{u} \times \mathbf{v} = 0$, obeys the wave equation

   $$\frac{\partial^2 \mathbf{v}}{\partial t^2} = a^2 \operatorname{grad} \operatorname{div} \mathbf{v} \tag{3.123}$$

   or

   $$-\frac{1}{a^2}\frac{\partial^2 \mathrm{v}}{\partial t^2} + \frac{\partial^2 \mathrm{v}}{\partial x^2} = 0 \tag{3.124}$$

3. For the longitudinal wave propagating perpendicular to $\mathbf{H}_0$, we rewrite the wave equation (3.120) as follows

$$\begin{aligned}\frac{\partial^2 \mathbf{v}}{\partial t^2} &= a^2 \text{grad div } \mathbf{v} + \mathbf{u} \times \text{curl curl} (\mathbf{u} \times \mathbf{v}) \\ &= \left(a^2 + u^2\right) \text{grad div } \mathbf{v} - \mathbf{u} (\mathbf{u} \cdot \nabla) \text{ div } \mathbf{v} - \mathbf{u} \times (\mathbf{u} \cdot \nabla) \text{ curl } \mathbf{v} \,. \end{aligned} \tag{3.125}$$

If $\mathbf{u} \perp \mathbf{k}$ one has $\mathbf{u} \cdot \mathbf{k} = 0$ and hence $\mathbf{u} \cdot \nabla = 0$, whereby the wave equation becomes

$$\frac{\partial^2 \mathbf{v}}{\partial t^2} = \left(a^2 + u^2\right) \text{grad div } \mathbf{v} \tag{3.126}$$

with the phase velocity $a^* = \sqrt{a^2 + u^2}$. (For a wave propagating oblique to $\mathbf{H}_0$ the situation is more complex).

For an isentropic equation of state one has $p \propto \rho^\gamma$ (γ specific heat ratio), and $\partial p / \partial \rho = \gamma p / \rho$. The ratio of the velocity of sound to the Alfvén speed is there

$$\frac{a}{u} = \sqrt{\frac{\gamma \beta}{2}} \tag{3.127}$$

where $\beta = p/p_{_H}$, $p_{_H} = H^2/8\pi$. For a monatomic gas $a \approx u$ if $\beta \approx 1$.

3.7 Electrostatic and Electromagnetic Plasma Disturbances

The largest electrostatic plasma disturbance arises from a charge separation between the electrons and ions. Assuming $H = 0$, $\sigma = \infty$, $\nabla p_e = \nabla p_i = 0$, (3.69) becomes

$$\frac{m_i m_e}{Z e^2 \rho} \frac{\partial \mathbf{j}}{\partial t} = \mathbf{E} \,. \tag{3.128}$$

With $m_i \gg m_e$, one can assume that the ions remain at rest, and that $\mathbf{j} = -n_e e \mathrm{v} = -Z n_i e \mathrm{v}$, where v is the electron velocity. With $\rho \simeq n_i m_i$ and putting $m_e \equiv m$ one obtains from (3.69)

$$m \frac{\partial \mathbf{v}}{\partial t} = -eE \tag{3.129}$$

From Maxwell's equation

$$\frac{4\pi}{c}\,\mathbf{j} + \frac{1}{c}\frac{\partial \mathbf{E}}{\partial t} = 0 \tag{3.130}$$

with $\mathbf{j} = -n_e e\mathbf{v}$ one obtains

$$\frac{\partial \mathbf{E}}{\partial t} = 4\pi n_e e\mathbf{v}\,. \tag{3.131}$$

The elimination of $\mathbf{E}$ from (3.129) and (3.131) leads to

$$\ddot{\mathbf{v}} + \omega_p^2\,\mathbf{v} = 0 \tag{3.132}$$

where

$$\omega_p = \sqrt{\frac{4\pi n_e e^2}{m}} \tag{3.133}$$

is the electron plasma frequency. Comparing (3.133) with the Debye length (3.64) one sees that $\lambda_D\omega_p = \sqrt{kT/m}$ is the thermal electron velocity.

Next we derive the equation for an electromagnetic plasma wave, assuming as before that $H = 0$, $\sigma = \infty$. From Maxwell's equations,

$$\frac{4\pi}{c}\mathbf{j} + \frac{1}{c}\frac{\partial \mathbf{E}}{\partial t} = \operatorname{curl}\mathbf{H} \tag{3.134}$$

$$-\frac{1}{c}\frac{\partial \mathbf{H}}{\partial t} = \operatorname{curl}\mathbf{E} \tag{3.135}$$

one obtains by eliminating $\mathbf{H}$ and using $\operatorname{div}\mathbf{E} = 0$:

$$\begin{aligned}\frac{4\pi}{c^2}\frac{\partial \mathbf{j}}{\partial t} + \frac{1}{c^2}\frac{\partial^2 \mathbf{E}}{\partial t^2} &= -\operatorname{curl}\operatorname{curl}\mathbf{E}\\ &= \nabla^2\mathbf{E}\,.\end{aligned} \tag{3.136}$$

For $\sigma = \infty$ Ohm's law (3.69) ($n_e = Zn_i$, $\rho = n_i m_i$) is:

$$\frac{\partial \mathbf{j}}{\partial t} = \frac{\omega_p^2}{4\pi}\mathbf{E}\,. \tag{3.137}$$

Inserting (3.137) into (3.136) one has

$$-\frac{1}{c^2}\frac{\partial^2 \mathbf{E}}{\partial t^2} + \nabla^2\mathbf{E} = \frac{\omega_p^2}{c^2}\mathbf{E}\,. \tag{3.138}$$

Putting $\mathbf{E} = \mathrm{A}e^{i(kx-\omega t)}$ one obtains from (3.138)

$$\frac{\omega^2}{c^2} - k^2 = \frac{\omega_p^2}{c^2} \tag{3.139}$$

and from there the phase velocity

$$V = \frac{\omega}{k} = \frac{c}{\sqrt{1-\left(\omega_p^2/\omega^2\right)}}\,. \tag{3.140}$$

For V to be real $\omega > \omega_p$, with total reflection for $\omega < \omega_p$. For $\omega < \omega_p$ the k-value becomes imaginary, whereby e^{ikx} becomes $e^{-\kappa x}$ with the wave penetration depth into the plasma ($d = 1/\kappa$):

$$d = \frac{c}{\omega_p}\frac{1}{\sqrt{1-\left(\omega^2/\omega_p^2\right)}}\,. \tag{3.141}$$

Neglecting in Ohm's law (3.69) $\left(4\pi/\omega_p^2\right)\partial\mathbf{j}/\partial t$ against $(1/\sigma)\,\mathbf{j}$, one has

$$\frac{\partial \mathbf{j}}{\partial t} = \sigma\frac{\partial \mathbf{E}}{\partial t} \tag{3.142}$$

and after inserting into (3.136):

$$k^2 = \frac{\omega^2}{c^2} + i\,\frac{4\pi\sigma\omega}{c^2}\,. \tag{3.143}$$

For strong wave attenuation

$$k^2 = i\,\frac{4\pi\sigma\omega}{c^2} \tag{3.144}$$

hence

$$\kappa = \frac{1}{c}\sqrt{2\pi\sigma\omega} \tag{3.145}$$

which is the well-known skin effect.

Next we take for Ohm's law the better approximation:

$$\frac{4\pi}{\omega_p^2}\frac{\partial \mathbf{j}}{\partial t} = \mathbf{E} - \frac{1}{\sigma}\mathbf{j}\,. \tag{3.146}$$

With $\mathbf{j} \propto e^{-i\omega t}$ one has

$$\left(-\frac{4\pi i\omega}{\omega_p^2} + \frac{1}{\sigma}\right)\mathbf{j} = \mathbf{E}\,. \tag{3.147}$$

Inserting (3.147) into (3.134) one obtains

$$\left(\frac{kc}{\omega}\right)^2 = 1 - \left(\frac{\omega_p}{\omega}\right)^2 \frac{1}{1 + \left(i\omega_p^2/4\pi\omega\sigma\right)}\,. \tag{3.148}$$

If $\omega_p^2/4\pi\omega\sigma \ll 1$ one has approximately

$$\begin{aligned}\left(\frac{kc}{\omega}\right)^2 &\simeq 1 - \left(\frac{\omega_p}{\omega}\right)^2 \left(1 - \frac{i\omega_p^2}{4\pi\omega\sigma}\right) \\ &= 1 - \left(\frac{\omega_p}{\omega}\right)^2 + i\left(\frac{\omega_p}{\omega}\right)^2 \frac{\omega_p^2}{4\pi\omega\sigma}\,.\end{aligned} \tag{3.149}$$

With the imaginary part in (3.149) small compared to the real part one has

$$\frac{kc}{\omega} \simeq \left[1 - (\omega_p/\omega)^2\right]^{1/2}\left[1 - i\,\frac{(\omega_p/\omega)^2\left(\omega_p^2/8\pi\sigma\omega\right)}{1 - (\omega_p/\omega)^2}\right]. \tag{3.150}$$

The imaginary part of (3.150) leads to the attenuation factor

$$\kappa = \frac{1}{c}\frac{(\omega_p/\omega)^2\left(\omega_p^2/8\pi\sigma\omega\right)}{\left[1 - (\omega_p/\omega)^2\right]^{1/2}}\,. \tag{3.151}$$

In chapter 4.2, we show that σ is related to the electron-ion collision frequency ν by $\sigma = \omega_p^2/4\pi\nu$. This gives (3.151) an especially simple form:

$$\kappa = \frac{\nu}{2c}\,\frac{(\omega_p/\omega)^2}{\left[1 - (\omega_p/\omega)^2\right]^{1/2}}\,. \tag{3.152}$$

Finally, we give a numerical expression for the plasma frequency:

$$\nu_p = \frac{\omega_p}{2\pi} = 8.97 \times 10^3 n_e^{1/2}\,. \tag{3.153}$$

3.8 Magnetohydrodynamic Instabilities

In the context of thermonuclear microexplosions there are two magnetohydrodynamic instabilities of special importance. They occur in the pinch discharge described in chapter 3.5, and shown here in Fig. 3.3. The first of these instabilities, Fig. 3.3a, is the so-called sausage or $m = 0$ instability, while the second one, Fig. 3.3b, is the so-called kink or $m = 1$ instability. In both cases the instability arises from an unstable equilibrium of the magnetic and plasma pressure for a cylindrical pinch discharge column. In the $m = 0$ instability a localized constriction of the pinch discharge raises there the magnetic pressure over the plasma pressure, with the tendency to cut off the current, the source of the magnetic field. In the $m = 1$ instability, the magnetic pressure is increased at the concave side of a bent discharge channel, which tries to disrupt the current as in the $m = 0$ instability. The

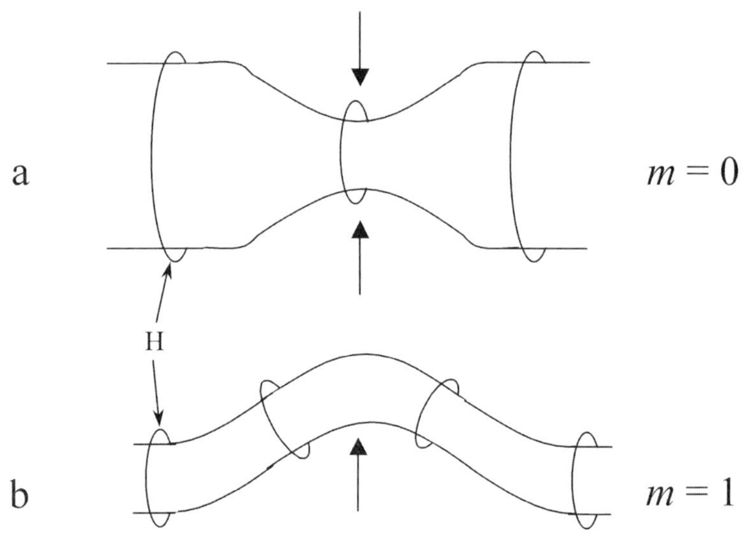

Figure 3.3: $m = 0$ sausage and $m = 1$ kink instability of pinch discharge.

velocity by which the instabilities grow is determined by the excess of the magnetic pressure $p_H = H^2/8\pi$ over the plasma pressure p. For $p_H >> p$ one has for this velocity

$$a = \sqrt{\frac{p_H}{\rho}} = \left(\frac{1}{\sqrt{2}}\right) \mathrm{v_A} \,, \tag{3.154}$$

where $\mathrm{v_A} = H/\sqrt{4\pi\rho}$ is the Alfvén velocity. The time needed to disrupt the pinch discharge channel of radius r by the $m = 0$ or $m = 1$ instability is thus of the order

$$\tau_{inst.} \simeq \frac{r}{\mathrm{v}_A} \,. \tag{3.155}$$

Both the $m = 0$ and $m = 1$ instabilities can be suppressed in three ways:

1. By entrapping an axial magnetic field H_z inside the pinch discharge channel, with a strength comparable to the strength of the azimuthal magnetic field H_ϕ of the pinch discharge:

$$H_z \gtrsim H_\phi \,. \tag{3.156}$$

2. By an axial shear flow $\mathrm{v}_z(r)$, with a stagnation pressure $\frac{1}{2}\rho \mathrm{v}_z^2$ larger than the magnetic pressure of the pinch discharge:

$$\frac{1}{2}\rho \mathrm{v}_z^2 \gtrsim \frac{H_\phi^2}{8\pi} \,. \tag{3.157}$$

3. By the rapid rotation of the discharge channel, making the centrifugal force equal to the gradient of the magnetic pressure force:

$$\frac{\rho \mathrm{v}_\phi^2}{r} = \nabla \left(\frac{H^2}{8\pi}\right) \tag{3.158}$$

3.9 Radiation Pressure

The radiation pressure on a fully ionized plasma is computed by the force density

$$\mathbf{f} = \frac{1}{c} \mathbf{j} \times \mathbf{H} \tag{3.159}$$

Expressing $\mathbf{j}$ by the Maxwell equation

$$\frac{4\pi}{c}\mathbf{j} = \text{curl}\,\mathbf{H} - \frac{1}{c}\frac{\partial \mathbf{E}}{\partial t} \tag{3.160}$$

and using the other Maxwell equation

$$\frac{1}{c}\frac{\partial \mathbf{H}}{\partial t} + \text{curl}\,\mathbf{E} = 0 \tag{3.161}$$

one obtains

$$\mathbf{f} = \frac{1}{4\pi}\left[(\text{curl}\,\mathbf{H}) \times \mathbf{H} - \frac{1}{c}\frac{\partial \mathbf{E}}{\partial t} \times \mathbf{H}\right]. \tag{3.162}$$

Next, we use the identity

$$-\frac{1}{c}\frac{\partial \mathbf{E}}{\partial t} \times \mathbf{H} = \frac{1}{c}\mathbf{E} \times \frac{\partial \mathbf{H}}{\partial t} - \frac{1}{c}\frac{\partial}{\partial t}(\mathbf{E} \times \mathbf{H}) \tag{3.163}$$

where we can omit the last term, since for an electromagnetic wave its time average vanishes. We therefore have

$$\mathbf{f} = \frac{1}{4\pi}\left[(\text{curl}\,\mathbf{H}) \times \mathbf{H} + (\text{curl}\,\mathbf{E}) \times \mathbf{E}\right]. \tag{3.164}$$

For an electromagnetic wave propagating along the x-axis, this becomes

$$\begin{aligned} f_x &= -\frac{1}{4\pi}\left[\frac{\partial H_z}{\partial x}H_z + \frac{\partial H_y}{\partial x}H_y + \frac{\partial E_z}{\partial x}E_z + \frac{\partial E_y}{\partial x}E_y\right] \\ &= -\frac{1}{8\pi}\frac{\partial}{\partial x}\left(\mathbf{H}^2 + \mathbf{E}^2\right). \end{aligned} \tag{3.165}$$

With the energy density of the electromagnetic wave,

$$u = \frac{\mathbf{H}^2 + \mathbf{E}^2}{8\pi} \tag{3.166}$$

(3.165) becomes

$$f_x = -\frac{\partial u}{\partial x} \tag{3.167}$$

and with the radiation energy flux density $S = uc$ this is

$$f_x = -\frac{1}{c}\frac{\partial S}{\partial x} \,. \tag{3.168}$$

If the frequency ω of the electromagnetic wave is less than the plasma frequency ω_p, the amplitude of the wave decreases rapidly with the plasma electrons accelerated to the kinetic energy

$$\begin{aligned} \varepsilon &= \frac{1}{n_e}\int_0^\infty f_x \, dx \\ &= -\frac{1}{n_e c}\int_0^\infty \frac{\partial S}{\partial x} \, dx \\ &= \frac{1}{n_e c} S = \frac{u}{n_e} \end{aligned} \tag{3.169}$$

where S is the radiation energy flux density at the vacuum-plasma interface at $x = 0$.

3.10 Equation of State for Cold Matter

For the problem of imploding shells, but also for the isentropic compression of cold DT, one needs the equation of state. For cold DT it is given by the Fermi equation of state (m electron mass)

$$\begin{aligned} p &= \frac{h^2}{5m}\left(\frac{3}{8\pi}\right)^{2/3} n^{5/3} \\ &\simeq 2.3 \times 10^{-27} n^{5/3} \ [\mathrm{dyn/cm^2}] \end{aligned} \tag{3.170}$$

behaving like a monatomic gas with $\gamma = 5/3$.

For high-Z materials, typical for imploding metallic shells, in particular for fissile shells, one can use the Thomas-Fermi equation of state. In its lowest approximation it is

$$\begin{aligned} p &= \frac{h^2}{5m}\left(\frac{3}{8\pi}\right)^{2/3} (nZ)^{5/3} \\ &\simeq 2.3 \times 10^{-27} \, (nZ)^{5/3} \, [\mathrm{dyn/cm^2}] \end{aligned} \tag{3.171}$$

which has the same dependence on n as a monatomic gas with $\gamma = 5/3$. For a given pressure one has $n \approx 10^{16} p^{3/5}/Z$. However, in accordance with (3.2), Z should be replaced by $Z^{2/2.42} \simeq Z^{0.83}$, to take into account the binding energy of the inner shell electrons, with the improved formula $n \approx 10^{16} p^{3/5}/Z^{0.83}$. Therefore, if for example DT is compressed $\sim 10^3$ fold, at a pressure of $\sim 10^{16}$ dyn/cm^2 = 10^{10} atm, the same pressure would compress $U^{235} \sim 23$ fold. This means that hydrogen is compressed about 43 times more than uranium.

In its lowest approximation (3.171), the Thomas-Fermi equation of state, is good only at very high pressures. At megabar pressures many materials can better be described by an equation of state of the form $p = A\rho^\gamma$ with $\gamma \approx 10$, but for higher pressures which can be reached with high explosives ($\approx$ 10 megabar), one rather has $\gamma \approx 4$ Therefore the function $\gamma = \gamma(p)$ begins with $\gamma = \infty$ for $p = 0$ (incompressible), for $p \to \infty$ asymptotically reaching $\gamma = 5/3$ (Fermi gas).

The isentropic compression energy from the volume V_0 to V is given by

$$E = -\int_{V_0}^{V} p\,dV \simeq \frac{pV}{\gamma - 1}, \qquad V_0 \gg V . \tag{3.172}$$

3.11 Bibliography for Chapter 3

L. Spitzer Jr., Physics of Fully Ionized Gases, Interscience Publishers, 1962.

W. B. Thompson, An Introduction to Plasma Physics, Pergamon Press, 1962.

J. G. Linhart, Plasma Physics, EURATOM, Brussels, 1969.

N. A. Krall and A. W. Trivelpiece, Principles of Plasma Physics, McGraw Hill Book Company, New York, 1973.

L. D. Landau and E. M. Lifshitz, Electrodynamics of Continuous Media, Pergamon Press, Oxford, 1960.

F. Winterberg, Beitr. Plasmaphys. **3**, 117 (1985).

T. D. Arber and D. F. Howell, Phys. Plasmas **3**, 554 (1996).

R. S. Bradley, High Pressure Physics and Chemistry, Academic Press, London, 1963.

Chapter 4

Collision Processes in Thermonuclear Plasmas

4.1 Collision Cross Sections and Mean Free Path

In a gas, where the forces between the molecules fall off rapidly, one can define a collision cross section σ putting $\sigma = \pi r_0^2$ where the radius r_0 of the molecules is equal to the range of their forces, and the mean free path is equal to $\lambda = 1/n\sigma = 1/n\pi r_0^2$. In a plasma where the forces between the electrons and ions are Coulomb forces, screened off at the Debye-length, such a simple expression for the collision cross section and mean free path cannot be given.

In a plasma a large deflection of a particle from its trajectory can result either from the close encounter with one particle, or from the cumulative effect of distant encounters with many particles.

In a close encounter between an electron and an ion the potential energy at the closest distance of approach b_0 is $E_{pot} = Ze^2/b_0$. By equating E_{pot} with $E_{kin} = \frac{3}{2}kT$, the particle kinetic energy, one can compute b_0:

$$b_0 = \frac{2Ze^2}{3kT} \tag{4.1}$$

and then define a collision cross section for close encounter collisions by putting

$$\sigma_c = \pi b_0^2 = \frac{4\pi}{9} \frac{Z^2 e^4}{(kT)^2} \ . \tag{4.2}$$

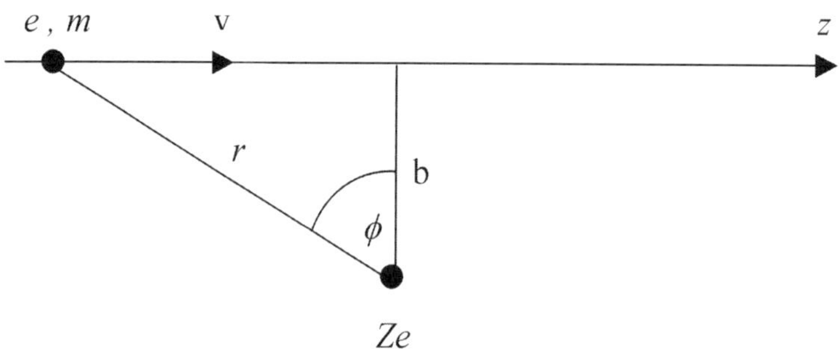

Figure 4.1: Distant collision between an electron and an ion.

To compute the cumulative effect of many distant encounters, we approximate the electron trajectory by a straight line (see Fig. 4.1). The electric field of the ion in the direction of the electron is $E = Ze/r^2$. Deflection of the electron trajectory is caused by the transverse component of the electric field $E_\perp = (Ze/r^2)\cos\phi$, with the momentum transfer perpendicular to the electron trajectory

$$\Delta p = \int_{-\infty}^{+\infty} E_\perp(t) \ dt \ . \tag{4.3}$$

With $b = r\cos\phi$, $z = b \ \tan\phi$, where b is called the impact parameter, and with $dt = (1/\mathrm{v}) \ dz$, where v is the electron velocity one has

$$\Delta p = \frac{Ze^2}{b\mathrm{v}} \int_{-\pi/2}^{+\pi/2} \cos\phi \, d\phi = \frac{2Ze^2}{b\mathrm{v}} \ . \tag{4.4}$$

The momentum transfer by many distant encounters is random and the mean square momentum change is equal to the number of the encounters times the square of the momentum change per encounter. As the electron moves the distance L, it collides with all the ions in a cylindrical shell of length L of radius b and thickness db. The contribution to $(\Delta p)^2$ from the collisions of the electron with the ions in this shell is

$$d\,(\Delta p)^2 = nL2\pi b\,db\frac{4Z^2e^4}{b^2\mathrm{v}^2} \tag{4.5}$$

where n is the number of density of the ions. The total change in $(\Delta p)^2$ is then given by

$$(\Delta p)^2 = \frac{8\pi nLZ^2e^4}{\mathrm{v}^2}\int_{b_{min}}^{b_{max}}\frac{db}{b} = \frac{8\pi nLZ^2e^4}{\mathrm{v}^2}\ln\Lambda \tag{4.6}$$

where $\Lambda = b_{max}/b_{min}$, and where $\ln\Lambda$ is called the Coulomb logarithm. We set b_{max} equal the Debye Length (3.65), and b_{min} equal b_0 (4.1). Hence

$$\Lambda = \frac{3}{4\sqrt{\pi}}\frac{1}{\sqrt{n}}\left(\frac{kT}{Ze^2}\right)^{3/2} . \tag{4.7}$$

A typical value for dense thermonuclear plasmas is $\ln\Lambda \approx 10$.

From (4.6) one can obtain a value for the mean free path λ for distant collisions by equating L with λ and $(\Delta p)^2$ with p^2. Since $p^2 = m^2\mathrm{v}^2$ one has

$$m^2\mathrm{v}^2 = \frac{8\pi n\lambda Z^2e^4}{\mathrm{v}^2}\ln\Lambda\ . \tag{4.8}$$

Defining for cumulative collisions a cross section $\sigma_d = 1/n\lambda$, and putting $m^2\mathrm{v}^4 = 4E_{kin}^2$, one obtains

$$\sigma_d = \frac{8\pi}{9}\frac{Z^2e^4\ln\Lambda}{(kT)^2} \tag{4.9a}$$

With the value $\ln\Lambda \approx 10$, one has $\sigma_d \approx 20\sigma_c$. The cumulative effect of distant collisions is thus more important, but this is true only for ideal plasmas. For a hydrogen plasma with kT expressed in keV one has

$$\sigma_d \approx \frac{10^{-18}}{(kT)^2}\ [\mathrm{cm}^2]\ . \tag{4.9b}$$

If for example $kT = 10$ keV ($\approx 10^8$ °K) one has $\sigma_d \approx 10^{-20}$ cm^2.

With the collision cross section inversely proportional to T^2, the mean free path goes as

$$\lambda = const.\frac{T^2}{n} \tag{4.10}$$

for a hydrogen plasma (kT in keV)

$$\lambda \approx \frac{10^{18}\,(kT)^2}{n} \tag{4.11}$$

4.2 Electrical Conductivity

The equation of motion for an electron under an applied electric field, and a frictional force caused by collisions, is

$$\frac{d\mathrm{v}}{dt} = \frac{e}{m}\,E - \nu\mathrm{v} \tag{4.12}$$

where the collision frequency ν is given by $\nu = n\sigma_d\mathrm{v}_{th}$, with $\mathrm{v}_{th} = (3kT/m)^{1/2}$ the thermal electron velocity. For a steady state electron current one has $d\mathrm{v}/dt = 0$ and

$$\mathrm{v} = \frac{eE}{m\nu}\,. \tag{4.13}$$

Putting $j = n_e\mathrm{v} = Zne\mathrm{v}$ and $j = \sigma E$, where σ is the electrical conductivity, one has

$$\sigma = \frac{n_e e^2}{m\nu} = \frac{Zne^2}{m\nu} = \frac{(3kT)^{3/2}}{8\pi m^{1/2} Z e^2 \ln\Lambda}\,. \tag{4.14}$$

A more rigorous derivation with the Boltzmann equation gives

$$\sigma = \frac{2\,(2kT)^{3/2}}{\pi^{3/2} m^{1/2} Z e^2 \ln\Lambda} \tag{4.15a}$$

which is about 5 times larger. However, if electron-electron collisions are taken into account, the conductivity is only about 1/2 as large. If T is in °K one has (in cgs units)

$$\sigma = 1.38 \times 10^8\,\frac{T^{3/2}}{Z\ln\Lambda}\;\;[\mathrm{s}^{-1}] \tag{4.15b}$$

4.3 Heat Conduction

According to elementary gas kinetic theory the heat conduction coefficient for a monatomic gas is

$$\kappa = \frac{1}{2} n_e \lambda k \mathrm{v}_{th} = \frac{1}{2} Z n \lambda k \mathrm{v}_{th} \tag{4.16}$$

where v_{th} is the thermal electron velocity, with the heat transported mainly by the electrons. The heat flux vector $\mathbf{Q}$ obeys the Fourier law $\mathbf{Q} = -\kappa \operatorname{grad} T$. With $\lambda = 1/n\sigma_d$ and $\mathrm{v}_{th} = (3kT/m)^{1/2}$ one has

$$\kappa = \frac{1}{2} \frac{Zk}{\sigma_d} \left(\frac{3kT}{m} \right)^{1/2} = \frac{3^{5/2} k \, (kT)^{5/2}}{16\pi m^{1/2} Z e^4 \ln \Lambda} \, . \tag{4.17a}$$

A lengthy derivation with the Boltzmann equation gives

$$\kappa \simeq 20 \left(\frac{2}{\pi} \right)^{3/2} \frac{k \, (kT)^{5/2}}{m^{1/2} Z e^4 \ln \Lambda} \tag{4.17b}$$

which is about 30 times larger. However, it has to be reduced by two factors, by 0.42 and by 0.22 hence by ~ 0.1, whereby (4.17b) is only about 3 times larger. If T is expressed in °K one has

$$\kappa = 2 \times 10^{-5} \frac{T^{5/2}}{Z \ln \Lambda} \left[\mathrm{erg/s\,^\circ Kcm} \right] \tag{4.18}$$

4.4 Viscosity

The elementary gas kinetic expression for the viscosity is

$$\eta = \frac{1}{3} \frac{M_i \mathrm{v}^i_{th}}{\sigma^i_d} \tag{4.19}$$

where $\mathrm{v}^i_{th} = (3kT/M_i)^{1/2}$ is the ion thermal velocity with $M_i = AM$ the ion mass (A atomic weight, M hydrogen mass), and σ^i_d the ion-ion collision cross section. With the force of the ion-ion interaction proportional to $Z^2 e^2$, rather than Ze^2 as for an electron-ion collision, one has $\sigma^i_d = Z^2 \sigma_d$. One then obtains from (4.19)

$$\eta = \frac{3^{3/2}}{8\pi} \frac{A^{1/2} M^{1/2} (kT)^{5/2}}{Z^4 e^4 \ln \Lambda} \, . \tag{4.20a}$$

The value obtained with the Boltzmann equation is

$$\eta = 0.406 \, \frac{A^{1/2} M^{1/2} (kT)^{5/2}}{Z^4 e^4 \ln \Lambda} \tag{4.20b}$$

which is approximately twice as large.

4.5 Energy Gain of Cold Ions by Hot Electrons

If a cold ion of energy E is put into a hot plasma of temperature T, resp. if $E \ll \frac{3}{2}kT$, the plasma electrons "heat" the ion. The energy gain for the ion can be written as follows:

$$\frac{dE}{dt} = n_e \sigma_d \mathrm{v}_{th} \left(\frac{3}{2} kT \right) \left(\frac{m}{M_i} \right) \tag{4.21}$$

where $\mathrm{v}_{th} = (3kT/m)^{1/2}$ is the electron thermal velocity and $m/M_i = m/AM$ the electron-ion mass ratio. The factor $n_e \mathrm{v}_{th}$ is the flux of electrons colliding with the ion, raising its energy in each collision by $\frac{3}{2}kT\,(m/M_i)$. One must here distinguish between the charge Z_i of the ion, which is "heated up", and the ions with the charge Ze of the hot plasma. With $n_e = Zn$ and σ_d given by (4.9), one has

$$\frac{dE}{dt} = \frac{4\pi}{\sqrt{3}} \frac{Z Z_i^2 \, n e^4 \ln \Lambda}{A \sqrt{mkT}} \left(\frac{m}{M} \right) . \tag{4.22a}$$

With the Boltzmann equation one obtains

$$\frac{dE}{dt} = 4\sqrt{2\pi} \, \frac{Z Z_i^2 \, n e^4 \ln \Lambda}{A \sqrt{mkT}} \left(\frac{m}{M} \right) \tag{4.22b}$$

larger by the factor $\sqrt{6/\pi} \simeq 1.4$. As can be seen the heating becomes less efficient with increased plasma temperature, going down in proportion to $1/\sqrt{T}$.

A better approximation is given by

$$\frac{dE}{dt} = 4\sqrt{2\pi} \, \frac{Z Z_i^2 \, n e^4 \ln \Lambda}{A \sqrt{mkT}} \left(\frac{m}{M} \right) \left(1 - \frac{E}{(3/2)kT} \right) \tag{4.23}$$

which takes into account that for $E = \frac{3}{2}kT$ one must have $dE/dt = 0$.

4.6 Energy Loss of Hot Ions by Cold Electrons

If $E \gg \frac{3}{2}kT$, one obtains from (4.23)

$$\frac{1}{E}\frac{dE}{dt} = -\frac{8\sqrt{2\pi}}{3\sqrt{m}}\frac{ZZ_i^2\, ne^4 \ln\Lambda}{A\,(kT)^{3/2}}\left(\frac{m}{M}\right) . \tag{4.24}$$

With $E = (M_i/2)\mathrm{v}_i^2$, and $\mathrm{v}_i = dx/dt$ the ion velocity, one computes from (4.24) the stopping range of a hot ion in a "cooler" plasma

$$\begin{aligned}\lambda_0 &= -\left[\frac{1}{E}\frac{dE}{dx}\right]^{-1} = -\left[\frac{1}{E}\frac{dE}{dt}\frac{dt}{dx}\right]^{-1} = -\mathrm{v}_i\left[\frac{1}{E}\frac{dE}{dt}\right]^{-1} \\ &= \frac{3}{8\sqrt{\pi}}\frac{A^{1/2}E^{1/2}\,(kT)^{3/2}}{ZZ_i^2\, ne^4 \ln\Lambda}\left(\frac{M}{m}\right)^{1/2}\end{aligned} \tag{4.25}$$

The range has here the form $\lambda_0 = const.\, T^{3/2}/n$, rather than T^2/n as it is in the case of the mean free path.

4.7 Transport Coefficients in the Presence of a Strong Magnetic Field

If the ion cyclotron frequency $\omega_i = ZeH/AMc$ is large compared to the ion-electron collision frequency $\nu_{ie} = \mathrm{v}_{th}^i n_e \sigma_d = (3kT/MA)^{1/2}\, Zn\sigma_d$, the ion trajectories in a plane perpendicular to the magnetic field can be viewed as circles, which after each ion-ion collision taking place in the time $\nu_{ii} = \mathrm{v}_{th}^i n\sigma_d^i = (3kT/MA)^{1/2}\, Z^2 n\sigma_d$, are displaced by the ion Larmor radius. With $\mathrm{v}_\perp^i = (2kT/MA)^{1/2}$ the thermal ion velocity perpendicular to H, one has

$$r_L = \frac{MA\mathrm{v}_\perp^i c}{ZeH} = \frac{(2MAkT)^{1/2}\, c}{ZeH} . \tag{4.26}$$

Since the electrons have a much smaller Larmor radius, heat conduction by ions dominates. To compute the heat conduction coefficient $\kappa_\perp$ perpendicular to a strong magnetic field, one has to substitute into (4.16) $n_e \to n$, $\lambda \to r_L$, and $\mathrm{v}_{th}^i = r_L\nu_{ii}$. One thus has

$$\kappa_\perp = \frac{1}{2}nkr_L^2\nu_{ii} \tag{4.27}$$

and hence

$$\kappa_\perp = \frac{8\pi}{3^{3/2}} \frac{n^2 k \left(MA\right)^{1/2} Z^2 e^2 c^2 \ln \Lambda}{H^2 \left(kT\right)^{1/2}} \,. \tag{4.28a}$$

The correct expression obtained with transport theory is

$$\kappa_\perp = \frac{8\sqrt{\pi}}{3} \frac{n^2 k \left(MA\right)^{1/2} Z^2 e^2 c^2 \ln \Lambda}{H^2 \left(kT\right)^{1/2}} \,. \tag{4.28b}$$

From (4.19) one can likewise obtain a value for the viscosity $\eta_\perp$ perpendicular to a strong magnetic field by making the substitution $\sigma_d^i \to 1/nr_L$, $\mathrm{v}_{th}^i \to r_L \nu_{ii} = r_L \mathrm{v}_{th}^i n \sigma_d^i = r_L \mathrm{v}_{th}^i Z^2 n \sigma_d$. One has

$$\eta_\perp = \frac{1}{3} MAnr_L^2 \nu_{ii} = \frac{16\pi}{9\sqrt{3}} \frac{c^2 \left(MA\right)^{3/2} n^2 Z^2 e^2 \ln \Lambda}{H^2 \left(kT\right)^{1/2}} \,. \tag{4.29a}$$

The correct value obtained with transport theory is

$$\eta_\perp = \frac{2\sqrt{\pi}}{5} \frac{c^2 \left(MA\right)^{3/2} n^2 Z^2 e^2 \ln \Lambda}{H^2 \left(kT\right)^{1/2}} \,. \tag{4.29b}$$

The heat conduction perpendicular to a strong magnetic field with the heat conduction coefficient proportional to $1/H^2$ is not what is observed, except in a "quiescent" plasma. What is observed is a $1/H$ dependence. This can be understood to result from electric fields at distances smaller than the Debye length. These microfields are obtained by (3.62) putting $x = \lambda_D$:

$$E = \sqrt{4\pi n_e kT} \,. \tag{4.30}$$

In a quiescent plasma these microfields are spherically symmetric around the ions, but not in a slightly turbulent plasma, with a turbulence at the microscale of a Debye length. In the presence of a magnetic field there will then be a drift motion perpendicular to E and H, given by (3.23). With (4.30) this drift motion is

$$\mathrm{v}_D = c \frac{\sqrt{4\pi n_e kT}}{H} \,. \tag{4.31}$$

A mass diffusion current is given by

$$\mathbf{J} = -D \,\mathrm{grad}\, n \tag{4.32}$$

where $D = \frac{1}{3}\lambda \mathrm{v}^i_{th}$ is the diffusion coefficient in the absence of a magnetic field. In the presence of a magnetic field one has to put $\lambda \to r_L$, and $\mathrm{v}^i_{th} \to r_L \nu_{ie}$ where $\nu_{ie} = \mathrm{v}^i_{th}/\lambda$, hence

$$D_\perp = \frac{1}{3}\frac{r_L^2 \mathrm{v}^i_{th}}{\lambda} \,. \tag{4.33}$$

In case the diffusion is dominated by the drift motion (4.31) one should rather set $\lambda \to \lambda_D$ and $\mathrm{v}^i_{th} \to \mathrm{v}_D$, hence

$$D_B = \frac{1}{3}\lambda_D \mathrm{v}_D = \frac{c}{3}\frac{kT}{eH} \,. \tag{4.34}$$

This is the result obtained by Bohm (except that the factor is there 1/16 instead of 1/3). With $\mathrm{v}_\perp = (2kT/MA)^{1/2}$ and r_L this can be also written as follows ($\mathrm{v}^i_\perp \to (2/3)^{1/2}\mathrm{v}^i_{th}$)

$$D_B = \frac{1}{6} Z r_L \mathrm{v}_\perp = \frac{1}{3\sqrt{6}} Z r_L \mathrm{v}^i_{th} \,. \tag{4.35}$$

The ratio

$$\frac{D_B}{D_\perp} = \frac{1}{\sqrt{6}}\frac{Z\lambda}{r_L} \tag{4.36}$$

expresses the increase of Bohm over "classical" diffusion. With Bohm diffusion the heat conduction coefficient is increased by multiplying (4.27) with (4.36):

$$\kappa_B = \frac{1}{2\sqrt{6}} nkr_L Z \mathrm{v}^i_{th} = \frac{1}{2}\frac{c}{e}\frac{nk^2T}{H} \,. \tag{4.37}$$

Finally, we mention the thermomagnetic Nernst effect. In the presence of a temperature gradient and magnetic field there is a thermomagnetic current directed perpendicular to the temperature gradient and magnetic field. It is given by

$$\mathbf{j}_N = \frac{3kn_e c}{2H^2}\mathbf{H} \times \mathrm{grad}\, T \,. \tag{4.38}$$

4.8 Collective Collision — The Two-Stream Instability

Because of the long-range forces between the charged particles of a plasma, there can be collective "collision effects", absent in a gas of neutral particles with short-range forces. The most important of these collective collision effects is the two-stream instability. It occurs if a beam of charged particles, rather than a single particle, interacts with a background plasma. Of particular importance is the collective interaction of an electron beam with a plasma. It is of interest for thermonuclear ignition concepts with intense relativistic electron beams.

The plasma and electron beam are assumed to have an electron number density n_0 and εn_0. The velocity of the electron beam is v_0, and the beam is perturbed by n_1, v_1 from εn_0 and v_0. The plasma electrons are perturbed by n_2, v_2 from εn_0 and $\mathrm{v} = 0$. The linearized equations of motion and continuity for the electron beam (moving in the x-direction) are

$$\frac{\partial \mathrm{v}_1}{\partial t} + \mathrm{v}_0 \frac{\partial \mathrm{v}_1}{\partial x} = -\frac{e}{m} E \tag{4.39a}$$

$$\frac{\partial n_1}{\partial t} + \varepsilon n_0 \frac{\partial \mathrm{v}_1}{\partial x} + \mathrm{v}_0 \frac{\partial n_1}{\partial x} = 0 \tag{4.39b}$$

and for the plasma electrons

$$\frac{\partial \mathrm{v}_2}{\partial t} = -\frac{e}{m} E \tag{4.40a}$$

$$\frac{\partial n_2}{\partial t} + n_0 \frac{\partial \mathrm{v}_2}{\partial x} = 0 \tag{4.40b}$$

to be supplemented by Maxwell's equation

$$\frac{\partial E}{\partial t} = -4\pi j \tag{4.41a}$$

where

$$j = -e\left(n_1 \mathrm{v}_0 + \varepsilon n_0 \mathrm{v}_1 + n_0 \mathrm{v}_2\right) . \tag{4.41b}$$

Setting

$$\mathrm{v}_1 \, , \; \mathrm{v}_2 \, , \; n_1 \, , \; n_2 \, , \; E = const.\, e^{i(kx-\omega t)} \tag{4.42}$$

one obtains from (4.41a) and (4.41b)

$$E = -\frac{4\pi e}{i\omega}\left(n_1 \mathrm{v}_0 + \varepsilon n_0 \mathrm{v}_1 + n_0 \mathrm{v}_2\right) \tag{4.43}$$

and from (4.39a) and (4.40a):

$$-\,i\omega \mathrm{v}_1 + ik_0 \mathrm{v}_1 = -\frac{e}{m}E \tag{4.44a}$$

$$-\,i\omega \mathrm{v}_2 = -\frac{e}{m}E \tag{4.44b}$$

from which E can be eliminated:

$$\mathrm{v}_2 = -\frac{-\omega + k\mathrm{v}_0}{\omega}\,\mathrm{v}_1\ . \tag{4.45}$$

From (4.39b) one has

$$-\,i\omega n_1 + ik\varepsilon n_0 \mathrm{v}_1 + ik\mathrm{v}_0 n_1 = 0 \tag{4.46a}$$

hence

$$n_1 = -\frac{k\varepsilon n_0 \mathrm{v}_1}{-\omega + k\mathrm{v}_0}\ . \tag{4.46b}$$

From (4.43), (4.44a), (4.45), and (4.46b) one obtains the dispersion relation:

$$1 - \frac{\omega_p^2}{\omega^2} - \frac{\varepsilon\omega_p^2}{\left(\omega - k\mathrm{v}_0\right)^2} = 0 \tag{4.47}$$

where $\omega_p^2 = 4\pi n_0 e^2/m$. Assuming that $\varepsilon \ll 1$, and putting $\omega = \omega_p + \Delta\omega$, $\Delta\omega \ll \omega_p$, further putting $k\mathrm{v}_0 \approx \omega_p$, one has

$$1 - \frac{\omega_p^2}{\left(\omega_p + \Delta\omega\right)^2} - \frac{\varepsilon\omega_p^2}{\left(\Delta\omega\right)^2} = 0\ . \tag{4.48}$$

With $\Delta\omega/\omega_p = x$ this is

$$1 - \frac{1}{\left(1+x\right)^2} - \frac{\varepsilon}{x^2} = 0\ . \tag{4.49}$$

For $x \ll 1$ this can be approximated by setting $(1-x)^2 \simeq 1-2x$. One then has

$$x^3 = \frac{\varepsilon}{2} \tag{4.50}$$

or with $\xi = \left(2^{1/3}/\varepsilon^{1/3}\right) x$

$$\xi^3 = 1 \tag{4.51}$$

with the solutions

$$\left.\begin{aligned} \xi_1 &= e^{2\pi i/3} = \cos(120)^\circ + i\sin(120)^\circ \\ \xi_2 &= e^{4\pi i/3} = \cos(240)^\circ + i\sin(240)^\circ \\ \xi_3 &= e^{2\pi i} = 1 \end{aligned}\right\} \tag{4.52}$$

For a growing, hence unstable, wave one has

$$e^{-i\omega t} = e^{-i(\omega_p+\Delta\omega)t} = e^{-i\omega_p t} e^{-i\Delta\omega_p x t} \tag{4.53}$$

with the imaginary part of x of the form iA, or where

$$e^{-i\omega_p x t} = e^{\omega_p A t} \tag{4.54}$$

realized for ξ_1. One there has

$$\operatorname{Im} x = \varepsilon^{1/3} 2^{-1/3} \operatorname{Im} \xi_1 = \varepsilon^{1/3} 2^{-1/3} \frac{\sqrt{3}}{2} i = iA \tag{4.55}$$

hence

$$e^{\omega_p A t} = \exp\left[\left(\frac{3^{1/2}}{2^{4/3}}\right) \varepsilon^{1/3} \omega_p t\right] \simeq \exp\left[0.7 \varepsilon^{1/3} \omega_p t\right] . \tag{4.56}$$

The collective stopping range

$$\lambda_c = \left(\frac{2^{4/3}}{3^{1/2}}\right) \frac{c}{\varepsilon^{1/3}\omega_p} \simeq 1.4 \, \frac{c}{\varepsilon^{1/3}\omega_p} . \tag{4.57a}$$

For relativistic electron beams one has to set $\varepsilon^{1/3} \to \varepsilon^{1/3}/\gamma$ because of the longitudinal electron mass, $m_{\|} = \gamma^3 m$. For relativistic beams the fastest growing mode is actually inclined by about 45° against the direction of the beam. With the transverse mass given by $m_{\perp} = \gamma m$, one would have to there set $\varepsilon \to \varepsilon/\gamma$. In liquid hydrogen the range of MeV electrons can be quite small and much less than 1 cm. By comparison, the range of single energetic electrons in cold matter is orders of magnitude larger and given by the approximate formula

$$\lambda^* \simeq (1/\rho)\left[0.543E_0 - 0.16\right]\ [\mathrm{cm}] \tag{4.57b}$$

where ρ is the density of the cold matter and E_0 the electron energy in MeV. Accordingly, the range for single MeV electrons in liquid hydrogen is larger than 1 cm.

4.9 Plasma Radiation

If a plasma is in thermodynamic equilibrium it emits radiation from its surface at the rate (Stefan-Boltzmann law)

$$\phi = \sigma T^4\ ,\ \sigma = 5.75 \times 10^{-5}\ \frac{\mathrm{erg}}{\mathrm{cm^2 s}\ {}^\circ\mathrm{K}^4}\ . \tag{4.58}$$

This law can be understood in a somewhat different way. In thermodynamic equilibrium the radiation is the blackbody radiation with an energy density

$$\varepsilon_r = aT^4\ ,\ a = 7.67 \times 10^{-15} \frac{\mathrm{erg}}{\mathrm{cm^3}\ {}^\circ\mathrm{K}^4} \tag{4.59}$$

With (4.59), (4.58) can be written as follows:

$$\phi = \frac{c}{4}\varepsilon_r = \frac{ac}{4}T^4\ . \tag{4.60}$$

The Stefan-Boltzmann law therefore implies that photons are emitted from a surface element of the hot body, each with the velocity of light but in all possible directions, resulting in the factor (1/4).

The Stefan-Boltzmann law is true as long as the optical mean free path $\lambda_{opt} = 1/n\sigma_{opt}$, where σ_{opt} is the optical cross section (for visible or invisible light) of the plasma ions, is small compared to the linear plasma dimensions.

For a completely ionized plasma σ_{opt} has no simple meaning. There the radiation is bremsstrahlung, emitted from electrons making collisions with ions. To compute the bremsstrahlung rate we refer to Fig. 4.1. The ions are assumed to be at rest, with the electrons accelerated by the electric field of the ions. According to Larmor's formula, the energy radiated by the electrons as a result of this acceleration is

$$\frac{dw}{dt} = \frac{2}{3}\frac{e^2}{c^3}\bar{\dot{\mathrm{v}}}^2 \tag{4.61}$$

with the acceleration given by

$$\dot{\mathrm{v}} = \frac{Ze^2}{mr^2} \, . \tag{4.62}$$

One therefore has

$$w = \frac{2}{3}\frac{Z^2e^6}{m^2c^3}\int_{-\infty}^{+\infty}\frac{dt}{r^4} \, . \tag{4.63}$$

Putting $dt = (1/\mathrm{v})\, dz$, $r = b/\cos\phi$, $z = b\, \tan\phi$, one has

$$w = \frac{2}{3}\frac{Z^2e^6}{m^2c^3\mathrm{v}b^3}\int_{-\pi/2}^{+\pi/2}\cos^2\phi\, d\phi = \frac{\pi}{3}\frac{Z^2e^6}{m^2c^3\mathrm{v}b^3} \, . \tag{4.64}$$

The total bremsstrahlung loss W is obtained by integrating over the contribution from all electrons $n_e = Zn$ making a collision with an ion. With the electron flux equal to $n_e\mathrm{v} = Zn\mathrm{v}$ one has

$$W = \int_{b_{min}}^{\infty} wZn\mathrm{v}\, 2\pi b\, db = \frac{2\pi^2}{3}\frac{Z^3e^6n}{m^2c^3}\int_{b_{min}}^{\infty}\frac{db}{b^2} = \frac{2\pi^2}{3}\frac{Z^3e^6}{m^2c^3}\frac{n}{b_{min}} \, . \tag{4.65}$$

Actually, we should have integrated to $b_{max} = \lambda_D$ rather than $b_{max} = \infty$, but the error made is insignificant. For b_{min} we set $b_{min} = \hbar/m\mathrm{v}$, the De Broglie wave length of an electron. We thus have ($h = 2\pi\hbar$)

$$W = \frac{4\pi^3}{3}\frac{Z^3e^6}{mc^3h}n\mathrm{v} \, . \tag{4.66}$$

To obtain the bremsstrahlung losses resulting from the collisions with all ions per cm^3, (4.66) has to be multiplied with n. Putting $\text{v} = \sqrt{3kT/m}$ one finally has

$$\varepsilon_r = \frac{4\pi^3}{3}\frac{Z^3e^6k^{1/2}}{m^{3/2}c^3h}n^2T^{1/2} = 2.1\times 10^{-27}Z^3n^2T^{1/2}\left[\frac{\text{erg}}{\text{cm}^3\ \text{s}}\right]. \tag{4.67a}$$

A more correct (but rather complicated) calculation gives

$$\varepsilon_r = 1.42\times 10^{-27}Z^3n^2T^{1/2}\left[\frac{\text{erg}}{\text{cm}^3\ \text{s}}\right] \tag{4.67b}$$

With the expression for ε_r we can compute the effective optical cross section. If we apply the Stefan-Boltzmann law to an infinite slab of thickness d, the radiation emitted per unit area is

$$\phi' = \phi\left[1 - e^{-n\sigma_{opt}d}\right] \tag{4.68}$$

where $\phi = \sigma T^4$. If $n\sigma_{opt}d \gg 1$ or $d \gg \lambda_{opt}$ one has $\phi' = \phi$. If $n\sigma_{opt}d \ll 1$ we can expand the square bracket in (4.68) and obtain

$$\phi' = \phi n\sigma_{opt}d = \sigma T^4 n\sigma_{opt}d \tag{4.69}$$

for the radiative energy emitted per unit area from a slab of thickness d. The energy radiated per unit volume therefore is ϕ'/d which has to be equal to ε_r. From this it follows that

$$\sigma_{opt} = 2.5\times 10^{-23}Z^3nT^{-3.5}\,. \tag{4.70}$$

It seems odd that a cross section should be proportional to the density n. In the theory of stellar structure it is common to replace σ_{opt} by an opacity coefficient

$$\kappa = 7.23\times 10^{24}\rho T^{-3.5}\sum_i \frac{w_iZ_i^2}{A_i}\frac{g}{t} \tag{4.71}$$

where w_i are the relative fractions of the elements of charge Z_i and atomic number A_i in the radiating plasma, with g the Gaunt and t the guilliotine factor. For a fully ionized hydrogen plasma $g/t \sim 1$. With κ the optical path length is $\lambda_{opt} = (\kappa\rho)^{-1}$, and $\sigma_{opt} = const.\kappa = const.\rho T^{-3.5}$.

For not fully ionized plasmas, as they occur in nuclear fission explosions but also inside stars, one can still work with (4.71), but there the factor g/t becomes important.

With the definition of a photon path length λ_{opt}, one can compute the radiation flux inside a plasma with a diffusion equation where

$$j_r = -\frac{\lambda_{opt} c}{3} \nabla \left(aT^4\right) . \tag{4.72}$$

With the help of (4.23) and (4.67b), one can show that the electron temperature T_e lags behind the ion temperature T_i. The reason is that primarily electrons lose energy by bremsstrahlung, with the ions losing energy to the electrons by inelastic ion-electron collisions. For a plasma with n ions of charge $Z = Z_i$, kinetic energy $E = \frac{3}{2}kT_i$, the energy loss according to (4.23) is

$$n\frac{dE}{dt} = 4\sqrt{2\pi}\frac{Z^3 n^2 e^4 \ln \Lambda}{A\sqrt{mkT}} \left(\frac{m}{M}\right)\left(\frac{T_i}{T_e} - 1\right) . \tag{4.73}$$

This expression must be equated with the bremsstrahlung losses (4.67b), setting $T = T_e$ (cgs units):

$$\varepsilon_r = \alpha Z^3 n^2 \sqrt{T_e}\,, \ \alpha = 1.42 \times 10^{-27} \text{ [cgs]} . \tag{4.74}$$

One has

$$\left.\begin{aligned} T_e^2 &= \left(\frac{B}{A}\right)(T_i - T_e) \\ B &= \frac{4\sqrt{2\pi}e^4 \ln \Lambda}{\sqrt{mk}\alpha\,(M/m)} \simeq 5.8 \times 10^9 \text{ [cgs]} \end{aligned}\right\} \tag{4.75}$$

Solving for T_e one finds

$$T_e = -\left(\frac{B}{2A}\right) + \sqrt{\left(\frac{B}{2A}\right)^2 + \left(\frac{B}{A}\right)T_i} . \tag{4.76}$$

With the correction factor

$$\gamma = \sqrt{\frac{T_e}{T_i}} \tag{4.77}$$

the bremsstrahlung losses are

$$\varepsilon_r = \gamma \alpha Z^3 n^2 \sqrt{T} \, . \tag{4.78}$$

4.10 Radiative Plasma Cooling and Collapse

If the radiative energy loss of a plasma exceeds the heating rate its temperature falls, and if it is confined by forces acting on the plasma it can be compressed to high densities. We have to distinguish two cases:

1. The plasma is magnetically confined by currents flowing through the plasma.

2. The plasma is confined by externally applied magnetic fields.

The first case is important for high current pinch discharges, while the second can be applied for the bunching to high densities of plasmas accelerated to large velocities.

In the first case where the radiative losses shall exceed resistive heating one has

$$\frac{j^2}{\phi} < \varepsilon_r \tag{4.79}$$

where σ and ε_r are given by (4.15) and (4.67b). Putting $\sigma = aT^{3/2}/Z$, $\varepsilon_r = bZ^3n^2T^{1/2}$ with a, b numerical constants, one has from (4.79):

$$j < \sqrt{ab} nTZ \, . \tag{4.80}$$

In applying (4.80) to the linear pinch discharge we have

$$I = \pi r^2 j \tag{4.81}$$

whereby (4.80) becomes

$$I < \sqrt{ab} Z \pi r^2 n T \, . \tag{4.82}$$

Furthermore, from

$$\frac{H^2}{8\pi} = (Z+1)\, nkT \tag{4.83}$$

and

$$H = \frac{2I}{rc} \tag{4.84}$$

one obtains by elimination of H

$$\pi r^2 n T = \frac{I^2}{2kc^2\,(Z+1)} \tag{4.85}$$

which by inserting into (4.82) and solving for I results in

$$I > \frac{2kc^2}{\sqrt{ab}} \frac{Z+1}{Z} \, . \tag{4.86}$$

The critical current

$$I_{PB} = \frac{2kc^2}{\sqrt{ab}} \frac{Z+1}{Z} \tag{4.87}$$

called Pease-Braginskii current, is for a hydrogen plasma $\sim 1.6 \times 10^6$ ampere. If the current is larger than the Pease-Braginskii current, the pinch shrinks until it becomes optically opaque. In realistic situations though the collapse may not be fast enough to take place in a time shorter than the time for instabilities to disrupt the pinch.

The Pease-Braginskii current (4.87) was computed for a fully ionized plasma. If the plasma is not completely ionized and contains high Z atoms, the radiation rate can be much higher and the critical current for collapse much smaller. The highly focused heavy metal vacuum sparks may for this reason be a manifestation of the Pease-Braginskii collapse.

The large radiation loss rate of a high Z plasma can also be used for the axial bunching of radially confined high velocity plasmas. This is an application of the second case listed above. The idea is explained in Fig. 4.2a. A high Z plasma jet with a variable injection velocity $\mathrm{v}(t)$ is shot into a magnetic solenoid. If the variable injection velocity is

$$\mathrm{v}\,(t) = \frac{L}{(t_m - t)} \tag{4.88}$$

where L is the length of the solenoid, all the particles making up the jet arrive simultaneously at the end of the solenoid. In general this is not possible because the density, and with it the pressure, in an axially compressed jet will rise. But it is possible if the jet is cooled by radiative energy losses. If $\mathrm{v}_0 = L/t_m$ is the initial injection velocity, one can write for (4.88)

$$\mathrm{v}(t) = \mathrm{v}_0 \left(1 - \frac{t}{t_m}\right)^{-1} \tag{4.89}$$

and for the density in the jet

$$n = n_0 \left(1 - \frac{t}{t_m}\right)^{-1} . \tag{4.90}$$

The energy equation is

$$T\frac{\partial s}{\partial t} = \frac{\partial q}{\partial t} \tag{4.91}$$

where $\partial q/\partial t$ is the heat added or removed from the jet per unit time and mass. With heat removed by radiation one has

$$\rho\frac{\partial q}{\partial t} = -b\, f\,(Z)\, n^2 T^{1/2} \; . \tag{4.92}$$

For a fully ionized plasma $f(Z) = Z^3$, but for a high Z, not fully ionized plasma, $f(Z)$ can be much larger.

Let us assume a singly ionized high Z plasma at a constant temperature of the order $\sim 10^5$ °K, with a plasma pressure $p = 2nkT$. From the thermodynamic relation $T\,ds = c_\mathrm{v}\,dt + pd(1/\rho)$ it then follows that $T\,ds = -\left(p/\rho^2\right) d\rho$ and hence from (4.91), (4.92) that

$$\frac{1}{n^2}\frac{\partial n}{\partial t} = \frac{b\, f(Z)}{2kT^{1/2}} \, . \tag{4.93}$$

With (4.90) one then finds that

$$\frac{1}{n^2}\frac{\partial n}{\partial t} = \frac{1}{n_0 t_m} \tag{4.94}$$

and

$$\frac{2k}{bn_0 t_m} = \frac{f(Z)}{T^{1/2}} \tag{4.95}$$

or

$$\frac{2k}{bn_0}\frac{\mathrm{v}_0}{L} = \frac{f(Z)}{T^{1/2}} \tag{4.96}$$

or

$$\frac{2k}{6n_{max}}\frac{\mathrm{v}_{max}}{L} = \frac{f(Z)}{T^{1/2}} \ . \tag{4.97}$$

Let us assume that $n_{max} = 10^{18}$ cm^{-3}, which for $T = 10^5$ °K results in the plasma pressure $p_{max} = 2n_{max}kT \simeq 3 \times 10^7$ dyn/cm^2. A solenoidal magnetic field $H \simeq 2.6 \times 10^4$ G would thus be sufficient to confine the jet. We finally obtain from (4.97)

$$L \simeq 6 \times 10^{-5}\,\frac{\mathrm{v}_{max}}{f(Z)} \ . \tag{4.98}$$

For the example $\mathrm{v}_{max} \simeq 10^9$ cm/s, $f(Z) \simeq 30$, one finds that $L \simeq 30$ meter.

Not only can radiation cooling be used for the axial bunching of the plasma jet, but also for radial compression by projecting the jet into a magnetic mirror (Fig. 4.2b), and finally into a collapsing pinch discharge (Fig. 4.2c).

4.11 Stopping Cross Section of Ions in Cold Matter

According to (4.25) the range of "hot" ions in a "cold" plasma goes in proportion to $T^{3/2}$, where T is the plasma temperature. But according to (2.70)

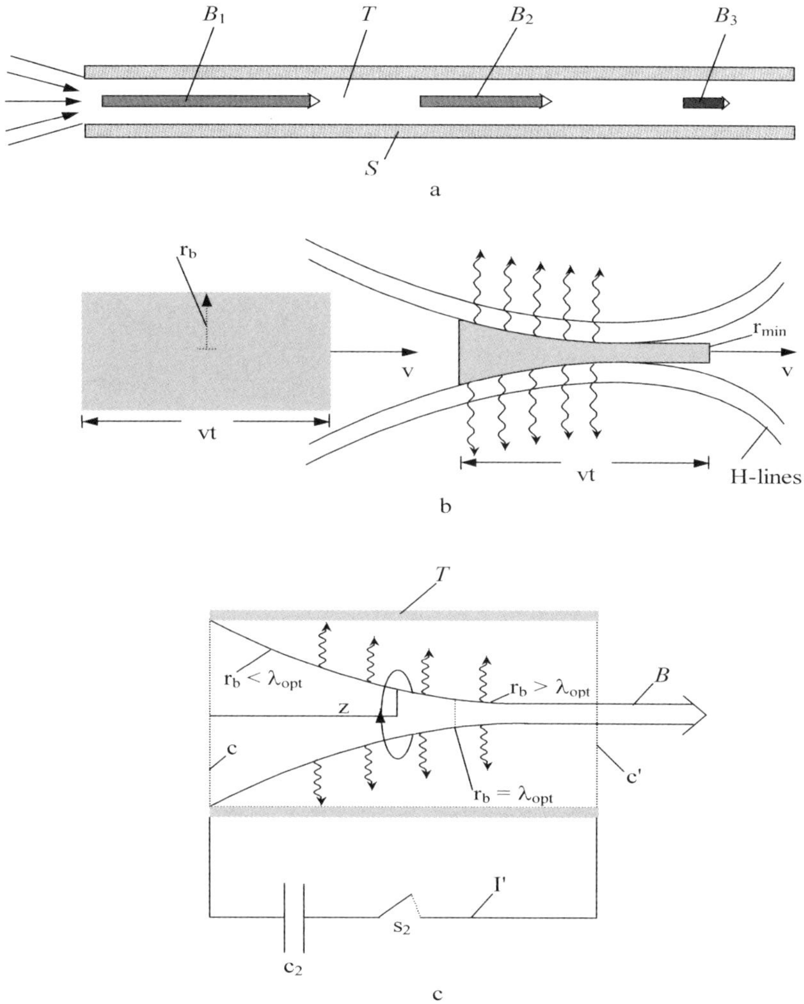

Figure 4.2: Radiative cooling and compression of intense plasma jets.

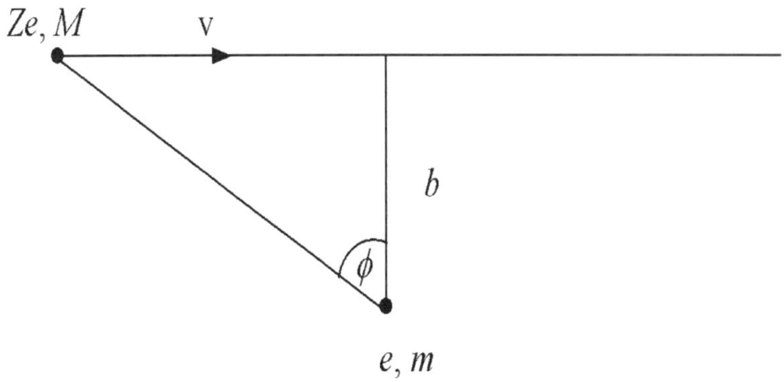

Figure 4.3: Stopping of fast ions in cold matter.

the range $1/n\sigma_s$ goes in proportion to E_0^2, where E_0 is the kinetic ion energy, not in proportion to $E_0^{3/2}$. And the mean free path in a plasma, (4.10), goes in proportion to T^2. To understand the difference and similarity of these range formulas, we go to Fig. 4.3.

We have here a fast energetic ion projectile moving through a cold electron gas. If the atomic number density of the target plasma is n, there will be nZ electrons per cm^3. The ion moving with the velocity v has the mass M and charge eZ_i. As for the computation of the mean free path in chapter 4.1, we have

$$\Delta p = \frac{Z_i e^2}{b\mathrm{v}} \int_{-\pi/2}^{+\pi/2} \cos\phi \, d\phi = \frac{2Z_i e^2}{b\mathrm{v}} \tag{4.99}$$

and

$$d\left(\Delta p\right)^2 = ZnL2\pi b \, db \frac{4Z_i^2 e^4}{b^2 \mathrm{v}^2} \tag{4.100}$$

hence

$$\begin{aligned}(\Delta p)^2 &= \frac{8\pi n L Z Z_i^2 e^4}{\mathrm{v}^2}\int\frac{db}{b}\\ &= \frac{8\pi n L Z Z_i^2 e^4}{\mathrm{v}^2}\ln\Lambda\end{aligned} \tag{4.101}$$

with the energy transferred to the electrons equal to

$$\Delta E = \frac{(\Delta p)^2}{2m}\ . \tag{4.102}$$

Further, with $\mathrm{v}^2 = 2E_0/M$, for $\Delta E = E_0$ and $L = 1/n\sigma_s$ one finds that

$$\sigma_s = \frac{2\pi M Z Z_i^2 e^4}{m E_0^2}\ln\Lambda \tag{4.103a}$$

or with $M = A_i M_H$ (M_H proton mass)

$$\sigma_s = \frac{2\pi\,(M_H/m)\,A_i Z Z_i^2 e^4}{E_0^2}\ln\Lambda\ . \tag{4.103b}$$

Comparing (4.103) with (2.70) one has to take into account that to define a cross section $\sigma_s = 1/nL$, $n_e = nZ$ applies to (4.3), and $n_e = n$ to (2.70).

The reason for the difference in the range formula (4.25), where the slowing down (i.e. stopping) cross section goes as $T^{-3/2}$ instead of T^{-2}, is the factor $(1 - E/(3/2)kT)$ which for $E \gg \frac{3}{2}kT$ becomes $-E/(3/2)kT$.

4.12 Magnetic Bremsstrahlung

One particular kind of radiation loss, called magnetic bremsstrahlung, occurs if electrons are in a strong magnetic field. In the nonrelativistic limit these losses can be obtained from Larmor's formula (4.61), putting $\dot{\mathrm{v}} = \mathrm{v}^2/r$ and $r = m\mathrm{v}c/eH$:

$$\frac{dw}{dt} = \frac{2}{3}\frac{e^4\mathrm{v}^2H^2}{m^2c^5}\ . \tag{4.104}$$

In the relativistic case (4.104) has to be multiplied by the factor $\gamma^2 = (w/mc^2)^2$. Thus for v $\to c$

$$\frac{dw}{dt} = \frac{2}{3}\frac{e^4H^2}{m^2c^3}\left(\frac{w}{mc^2}\right)^2 . \tag{4.105}$$

With $dw/dt = (dw/dx)\,(dx/dt) = c\,dw/dx$, one obtains for the stopping range of electrons by magnetic bremsstrahlung losses

$$\lambda_e \sim \frac{(mc^2)^4}{e^4H^2}\frac{1}{w} . \tag{4.106}$$

4.13 Radiation Losses Near a Wall

In several inertial confinement fusion concepts, the hot plasma is held together by cool solid walls, with a large temperature gradient in the vicinity of the wall. For a uniform plasma pressure one has $p = 2nkT = const.$ (hydrogen plasma), hence $n = const./T$. Accordingly, the bremsstrahlung losses (4.67) are proportional to $T^{-3/2}$, and they become very large near a cold wall. However, because of the thermomagnetic Nernst effect the plasma pressure distribution is far from uniform. This is especially true for wall confined magnetized plasmas where the magnetic field is parallel to the wall surface.

With the temperature gradient from the cold wall into the hot plasma, the thermomagnetic current for a hydrogen plasma, according to (4.38), is

$$\mathbf{j}_N = \frac{3knc}{2H^2}\,\mathbf{H} \times \nabla T \tag{4.107}$$

with the magnetic force density on the plasma

$$\mathbf{f} = \frac{1}{c}\,\mathbf{j}_N \times \mathbf{H} = \frac{3}{2}\frac{nk}{H^2}\,(\mathbf{H} \times \nabla T) \times \mathbf{H} \tag{4.108}$$

or with ∇T perpendicular and $\mathbf{H}$ parallel to the wall

$$\mathbf{f} = \frac{3}{2}\,nk\nabla T . \tag{4.109}$$

The magnetohydrodynamic equilibrium condition

$$\nabla p = \mathbf{f} \tag{4.110}$$

with $p = 2nkT$, $\nabla p = 2nk\nabla T + 2kT\nabla n$, becomes

$$2nk\nabla T + 2kT\nabla n = \frac{3}{2}nk\nabla T \tag{4.111}$$

which upon integration yields

$$Tn^4 = const. \tag{4.112}$$

With $n^2 = const./T^{1/2}$ the bremsstrahlung losses (4.67) are here independent of T and therefore constant across the entire plasma.

In a cartesian x, y, z coordinate system with the cold wall at $z = 0$, and the magnetic field $\mathbf{H}$ into the x-direction, the Nernst current density $\mathbf{j}$ is in the y-direction, and is

$$\mathbf{j}_y = -\frac{3knc}{2H}\frac{dT}{dz} . \tag{4.113}$$

From Maxwell's equation $4\pi\mathbf{j}/c = \mathrm{curl}\,\mathbf{H}$ one has

$$\mathbf{j}_y = \frac{c}{4\pi}\frac{dH}{dz} . \tag{4.114}$$

Eliminating $\mathbf{j}_y$ from (4.113) and (4.114) one obtains

$$2H\frac{dH}{dz} = -12\pi kn\frac{dT}{dz} . \tag{4.115}$$

If in the plasma far away from the wall $n = n_0$, $T = T_0$, one has from (4.112)

$$n = \frac{n_0 T_0^{1/4}}{T^{1/4}} \tag{4.116}$$

inserting this into (4.115) one finds

$$dH^2 = -\frac{12\pi kn_0 T_0^{1/4}}{T^{1/4}}\,dT . \tag{4.117}$$

With the boundary condition $H = H_0$ at $z = 0$ and $T = T_0$ at $z = \infty$, integration of (4.117) yields

$$\frac{H_0^2}{8\pi} = 2n_0 kT_0 . \tag{4.118}$$

The meaning of (4.118) is that the magnetic pressure $H_0^2/8\pi$ exerted on the plasma from the wall surface at $z = 0$, balances the plasma pressure $2n_0kT_0$ at $z = \infty$.

4.14 Integrated Heat Conduction Losses of a Magnetized Plasma Cylinder

As an important example we will compute the heat conduction losses of a magnetized plasma cylinder (magnetic field directed along the axis of the cylinder) with a hot core of temperature T_0 and radius r_0, within which thermonuclear reactions take place to sustain the temperature T_0. The heat conduction losses per unit length of the column are

$$Q = -2\pi r \kappa_\perp \frac{dT}{dr} \tag{4.119}$$

which by integration yields

$$Q = \frac{2\pi}{\ln\left(r_1/r_0\right)} \int_0^{T_0} \kappa_\perp \, dT \, . \tag{4.120}$$

Here we wish to have a better expression for $\kappa_\perp$ than the one given by (4.28b) which is valid only for strong magnetic fields, more precisely for $\omega_i t_i \gg 1$, where t_i is the ion collision time ($t_i = 1/\nu$, ν collision frequency) and ω_i the ion cyclotron frequency. A more accurate value for $\kappa_\perp$ is

$$\kappa_\perp = \frac{3}{4\sqrt{\pi}} \frac{k\left(kT\right)^{5/2}}{\sqrt{M}e^4 \ln\Lambda} \frac{2.645 + 2\left(\omega_i t_i\right)^2}{0.677 + 2.70\left(\omega_i t_i\right)^2 + \left(\omega_i t_i\right)^4} \tag{4.121}$$

where

$$\omega_i t_i = \frac{eH}{Mc} \frac{3\sqrt{M}\left(kT\right)^{3/2}}{4\sqrt{\pi}e^4 n_i \ln\Lambda} \, . \tag{4.122}$$

If the pressure is constant, the ion density varies as T^{-1} with $\omega_i t_i$ proportional to $T^{5/2}$. Inserting (4.121) and (4.122) into (4.120) one can write

$$Q = bKT_1^{7/2} \tag{4.123}$$

where $T_1 = 4.0 \times 10^4 \left(p/H\right)^{2/5}$ is approximately the temperature where $\omega_i t_i \approx 1$. Further $K = 0.53 \times 10^{-14}$ W/cm°K$^{7/2}$ and

$$b = \frac{2\pi}{\ln\left(r_1/r_0\right)} \int_0^\infty \frac{x^{5/2}\left(1 + 0.756x^5\right)}{1 + 3.99x^5 + 1.48x^{10}} \, dx \, . \tag{4.124}$$

If $r_1/r_0 \approx 10$ one finds that $b \approx 3$. For $H = 10^4$ G and $H = 10^5$ G one has

$H = 10^4$ G	$H = 10^5$ G
$T_1 = 4 \times 10^5$ °K	$T_1 = 1.6 \times 10^5$ °K
$Q = 600$ kW/cm	$Q = 24$ kW/cm

4.15 Electron Run-Away

The expression for the electric plasma conductivity was obtained with a simplified diffusion model, which is valid only under the assumption that the electric field is sufficiently small to keep the potential energy gained along a mean free path small compared to kT. With the mean free path proportional to T^2, it is clear that this assumption must break down for sufficiently large temperatures. If this happens the plasma electrons are continuously accelerated, ultimately forming a relativistic electron beam inside the plasma. However, for the electric field to be able to act on the electrons, the radius of the plasma column must be smaller than the penetration of the electric field into the plasma.

The condition for electron run-away is

$$eE\lambda \gg \frac{3}{2}kT \tag{4.125}$$

where $\lambda = 1/n\sigma_c$ is the mean free path with σ_c given by (4.2). One thus obtains from (4.125)

$$2.2 \times 10^9 \frac{TE\ [\text{Volt/cm}]}{Z^2 n} \gg 1. \tag{4.126}$$

The penetration depth is given by

$$\delta = \frac{c}{\omega_p} \tag{4.127}$$

where for relativistic plasma electrons with the longitudinal mass $\gamma^3 m$ along the direction of the electric field one has $\omega_p = \sqrt{4\pi n e^2/\gamma^3 m}$.

4.16 Bibliography for Chapter 4

R. F. Post, Review of Modern Physics **28**, 338 (1956).

L. Spitzer, Jr. Physics of Fully Ionized Gasses, Interscience Publishers, 1962.

R. S. Pease, Proc. Phys. Soc. London Ser. **B70**, 11 (1957).

S. I. Braginskii, Zh. Eksp. Teor. Fiz. **33**, 645 (1957).

F. Winterberg, Phys. Rev. Lett. **37**, 713 (1976).

F. Winterberg, Z. f. Naturforsch. **32a**, 840 (1977).

F. Winterberg, J. Plasma Physics **21**, 301 (1979).

J. D. Jackson, Classical Electrodynamics, John Wiley & Sons, New York, 1962, p. 429 ff.

L. D. Landau and E. M. Lifshitz, The Classical Theory of Fields, Pergamon Press, Oxford 1971, p. 197 ff.

H. Alfvén, C. G. Fälthammar, R. B. Johansson, E. A. Smärs, B. Wilner and E. Witalis. Nuclear Fusion, 1962 Supplement Part I.

Chapter 5

Shock and Compression Waves

5.1 Shock Waves

A shock wave consists of a discontinuity, where p, ρ, T and s jump in passing through the discontinuity. In simplifying the following analysis we assume that the shock is plane, which means that the shock propagates perpendicular to its front of discontinuity. We also assume that the shock is in an ideal gas where one has the following relations

$$c_s^2 = \gamma \frac{p}{\rho} \tag{5.1}$$

with c_s the isentropic sound velocity,

$$\varepsilon = c_{\mathrm{v}} T = \frac{p}{\rho\,(\gamma - 1)} = \frac{c_s^2}{\gamma\,(\gamma - 1)} \tag{5.2}$$

the internal energy per unit mass. Further

$$w = c_p T = \varepsilon + \frac{p}{\rho} = \frac{\gamma p}{\rho\,(\gamma - 1)} = \frac{c_s^2}{\gamma - 1} \tag{5.3}$$

the enthalpy per unit mass, and finally

$$s = c_\nu \log\left(\frac{p}{\rho^\gamma}\right) = c_p \log\left(\frac{p^{1/\gamma}}{\rho}\right) \tag{5.4}$$

the entropy per unit mass. For a fully ionized plasma one has $\gamma = \frac{5}{3}$. To obtain the "jump conditions" of a shock, one goes into a reference system at rest with respect to the front of the shock. In this reference system the fluid in front of the shock intersects the front with the velocity v_1, leaving the front with the smaller velocity v_2. As seen from an observer at rest with respect to the fluid in front of the shock, the front moves with the velocity v_1, and the fluid behind the discontinuity with the velocity

$$\mathrm{v} = \mathrm{v}_1 - \mathrm{v}_2 \,. \tag{5.5}$$

In passing through the shock front, the conservation of mass, momentum and energy lead to three equations (Rankine-Hugoniot equations)

$$\left.\begin{aligned} \rho_1 \mathrm{v}_1 &= \rho_2 \mathrm{v}_2 \\ p_1 + \rho_1 \mathrm{v}_1^2 &= p_2 + \rho_2 \mathrm{v}_2^2 \\ c_p T_1 + \frac{\mathrm{v}_1^2}{2} &= c_p T_2 + \frac{\mathrm{v}_2^2}{2} \end{aligned}\right\} \tag{5.6}$$

where the index 1 refers to the quantities in front, and the index 2 to those behind the front, both measured in the reference system at rest with respect to the front. There are three equations for the six unknowns ρ_1, v_1, p_1, ρ_2, v_2, p_2 since $c_p T_1$, and $c_p T_2$ can be expressed with (5.3) by p_1, ρ_1, and p_2, ρ_2. However, since we are only interested in the ratios ρ_2/ρ_1, $\mathrm{v}_2/\mathrm{v}_1$, p_2/p_1, the number of unknowns is reduced to three, with the jump condition T_2/T_1 obtained with the help of (5.3).

One finds

$$\left.\begin{aligned} \frac{\rho_2}{\rho_1} &= \frac{\mathrm{v}_1}{\mathrm{v}_2} = \frac{(\gamma+1)\,M_1^2}{\left[(\gamma-1)\,M_1^2+2\right]} \\ \frac{p_2}{p_1} &= \frac{2\gamma M_1^2}{\gamma+1} - \frac{\gamma-1}{\gamma+1} \\ \frac{T_2}{T_1} &= \frac{\left[2\gamma M_1^2-(\gamma-1)\right]\left[(\gamma-1)\,M_1^2+2\right]}{(\gamma+1)^2\,M_1^2} \,. \end{aligned}\right\} \tag{5.7}$$

In these equations

$$M_1 = \frac{\mathrm{v}_1}{c_{s1}} \tag{5.8}$$

is the Mach number of the shock front. The Mach number behind the shock front M_2 is in terms of M_1 given by

$$M_2^2 = \frac{2 + (\gamma - 1)\, M_1^2}{2\gamma M_1^2 - (\gamma - 1)} \; . \tag{5.9}$$

In the limit of very strong shock waves, which are typical for thermonuclear processes, one has

$$\left.\begin{aligned} \frac{\rho_2}{\rho_1} &= \frac{\mathrm{v}_1}{\mathrm{v}_2} = \frac{(\gamma+1)}{(\gamma-1)} \\ \frac{p_2}{p_1} &= \frac{2\gamma M_1^2}{(\gamma+1)} \\ \frac{T_2}{T_1} &= \frac{2\gamma M_1^2\,(\gamma-1)}{(\gamma+1)^2} \end{aligned}\right\} \tag{5.10}$$

further

$$M_2 = \left[\frac{(\gamma-1)}{2\gamma}\right]^{1/2} \; . \tag{5.11}$$

for $\gamma = 5/3$, $M_2 = 1/\sqrt{5} < 1$.

From (5.5) and (5.10) one obtains

$$\mathrm{v}_1 = \frac{\gamma+1}{2}\,\mathrm{v} \geq \mathrm{v} \; . \tag{5.12}$$

For $\gamma = 5/3$ one has $\mathrm{v}_1 = (4/3)\mathrm{v}$. This means that the shock "snowplows" the fluid in front of it. With the "snowplow" moving with the velocity v, the front of the "plowed" fluid moves with a velocity larger than v.

With the help of eq. (5.1), (5.2) and (5.12) one can write

$$\left.\begin{aligned} p_2 &= \left[\frac{2}{(\gamma+1)}\right]\rho_1 \mathrm{v}_1^2 = \frac{1}{2}\,(\gamma+1)\,\rho_1 \mathrm{v}^2 \\ T_2 &= \left[\frac{2}{(\gamma+1)^2}\right]\frac{\mathrm{v}_1^2}{c_\mathrm{v}} = \frac{1}{2}\frac{\mathrm{v}^2}{c_\mathrm{v}} \end{aligned}\right\} \tag{5.13}$$

which shows that both pressure and temperature behind the shock front are, by order of magnitude, equal the stagnation pressure $\rho \mathrm{v}^2$, and stagnation temperature $\mathrm{v}^2/c_\mathrm{v}$, of a fluid with the velocity v coming to rest.

The thickness δ of the shock front can be estimated from the viscous force per unit area in the front

$$f = \eta \frac{d\text{v}}{dx} \sim \eta \frac{\text{v}}{\delta} \tag{5.14}$$

to be set equal to the stagnation pressure ρv^2 in the front; hence

$$\delta \sim \frac{\eta}{\rho\text{v}} . \tag{5.15}$$

With $\eta \sim M_i \text{v}^i_{th}/\sigma^i_d \sim \rho\text{v}^i_{th}/\lambda$, and with $\text{v}^i_{th} \sim \text{v}$ according to (5.13), it follows that

$$\delta \sim \lambda . \tag{5.16}$$

In the presence of a strong magnetic field, with the magnetic lines of force parallel to the shock front, one has to replace the mean free path λ with the ion Larmor radius r_L, and one there has

$$\delta_\perp \sim r_L . \tag{5.17}$$

5.2 Von Neumann Artificial Viscosity

The problem of a shock wave in a plasma is more complicated. Not only must a detailed analysis take into account both the electrons and ions, at the high temperatures of thermonuclear plasmas radiation can become important in ways not included in the above presented simple analysis. Further complications arise if there are thermonuclear reactions. Because all of this is important for the design of thermonuclear weapons, large computer simulation programs have been developed to deal with these problems.

One important concept used in these computations is the Von Neumann artificial viscosity. Because there is a sharp discontinuity across the shock front, this discontinuity must be made smooth to permit a numerical finite mesh integration through the front. This problem can be treated with Von Neumann's artificial viscosity. It is similar to Prandtl's turbulent eddy viscosity

$$\eta_t = \rho \ell^2 \left(\frac{\partial \text{v}}{\partial x} \right) \tag{5.18}$$

where ℓ is a mixing length for turbulent eddies in a boundary layer near a wall. One puts $\ell = c\Delta x$, $c \sim 1$, where Δx is the distance from the wall, with the fluid flow along the wall into the y-direction. Putting $\partial \mathrm{v}/\partial x \sim \mathrm{v}/\Delta x$ one then has for the eddy viscosity

$$\eta_t \sim \rho \Delta x \cdot \mathrm{v} \, . \tag{5.19}$$

It has the same form as the molecular viscosity $\eta = \frac{1}{3}\rho\lambda \mathrm{v}_{th}$, setting $\lambda \sim \Delta x$, $\mathrm{v} \sim \mathrm{v}_{th}$. In Prandtl's theory the eddy viscosity describes the turbulent shear stress of the flow along the wall in the y-direction:

$$\sigma_{xy} = \eta_t \frac{\partial \mathrm{v}}{\partial x} = \rho \left(\Delta x \frac{\partial \mathrm{v}}{\partial x} \right)^2 \tag{5.20}$$

with the shear force density in the y-direction

$$f_y = \frac{\partial \sigma_{xy}}{\partial x} = \frac{\partial}{\partial x} \left[\rho \left(\Delta x \frac{\partial \mathrm{v}}{\partial x} \right)^2 \right] . \tag{5.21}$$

With the eddy viscosity describing the nonlinear term $\rho \, (\mathbf{v} \cdot \mathrm{grad}) \, \mathbf{v}$ in Euler's equation, this term is replaced by (5.21) and put on the r.h.s. of Euler's equation.

In Von Neumann's artificial viscosity one has

$$\sigma_{xx} = \eta_a \frac{\partial \mathrm{v}}{\partial x} \tag{5.22}$$

where in analogy to Prandtl's η_t

$$\eta_a = b\rho \, (\Delta x)^2 \frac{\partial \mathrm{v}}{\partial x} \tag{5.23}$$

with b a numerical constant of order unity. The viscous force density by this artificial viscosity η_a is

$$f_x = \frac{\partial \sigma_{xx}}{\partial x} = b \frac{\partial}{\partial x} \left[\rho \left(\Delta x \frac{\partial \mathrm{v}}{\partial x} \right)^2 \right] \tag{5.24}$$

which has to be put on the r.h.s. of the equation of motion (3.103), replacing the nonlinear term $\rho \, (\mathbf{v} \cdot \mathrm{grad}) \, \mathbf{v}$ on the l.h.s.

In the theory by Von Neumann Δx is the mesh distance for the numerical computer integration. It is obvious that Δx must be smaller than the shock thickness, which gives a lower limit for b. Inserting η_a given by (5.23) into (5.15) with $\partial \mathrm{v}/\partial x \sim \mathrm{v}/\delta$ one finds that

$$\delta = \sqrt{b}\Delta x \,. \tag{5.25}$$

In computer calculations choosing $b \geq 2$ is already sufficient to make $\delta > \Delta x$ as required.

5.3 Convergent Shock Waves

The pressure in a cylindrical convergent sound wave rises as $\approx 1/\sqrt{r}$ and in a convergent spherical sound wave as $1/r$. In a convergent cylindrical and spherical shock wave the rise is less, because, unlike a sound wave, a shock wave is nonisentropic with part of the energy dissipated into heat.

The rise in pressure and temperature for a convergent cylindrical and spherical shock wave are

$$p,\ T \propto r^{-\kappa} \tag{5.26a}$$

$$p,\ T \propto r^{-2\kappa} \tag{5.26b}$$

where κ is obtained by a similarity solution. For strong shock waves a good approximation for κ is

$$\kappa^{-1} = \frac{1}{2} + \frac{1}{\gamma} + \left[\frac{\gamma}{2\,(\gamma - 1)}\right]^{1/2} . \tag{5.27}$$

If $\gamma = 5/3$ one obtains $\kappa \simeq 0.45$.

If the pressure and temperature at the initial radius R of a convergent shock wave are p_0 and T_0, the pressure and temperature rise in a cylindrical convergent shock wave is

$$\frac{p}{p_0} = \frac{T}{T_0} = \left(\frac{R}{r}\right)^{0.45} \tag{5.28}$$

and in a spherical convergent shock wave it is

$$\frac{p}{p_0} = \frac{T}{T_0} = \left(\frac{R}{r}\right)^{0.9} \tag{5.29}$$

not too much different from a convergent cylindrical and spherical sound wave. But unlike a sound wave, the density behind a shock wave is increased by the factor $(\gamma+1)/(\gamma-1)$, given by (5.10), for $\gamma = 5/3$ by the factor 4. The shock wave is reflected at the center of convergence, whereby the density is increased a second time by the same factor, hence all together by the factor $\left[(\gamma+1)/(\gamma-1)\right]^2$, for $\gamma = 5/3$ by the factor 16. Finally in a cylindrical convergent shock wave there is a further increase by the factor $\sqrt{2}$ due to the convergence of a cylindrical compared to a plane wave, and in a spherical wave by the factor 2. Therefore, the total increase in density for a cylindrical wave in the center of convergence is $\sqrt{2}\left[(\gamma+1)/(\gamma-1)\right]^2$ for $\gamma = 5/3$ equal to ≈ 23, and for a spherical wave it is $2\left[(\gamma+1)/(\gamma-1)\right]^2$, for $\gamma = 5/3$ equal to 32. Through isentropic compression alone, the temperature in a cylindrical shock would rise by the factor $\left[\sqrt{2}(\gamma+1)/(\gamma-1)\right]^{\gamma-1}$ and for a spherical shock by the factor $\left[2(\gamma+1)/(\gamma-1)\right]^{\gamma-1}$, for $\gamma = 5/3$ equal to $4/2^{1/3}$, resp. equal to 4.

According to (5.28) and (5.29), the temperature in a cylindrical or spherical convergent shock wave should go to infinity. In reality though the rise in temperature is limited, with no further rise possible if the distance of the shock front from the center of convergence becomes equal to a mean free path. As an example, we take a convergent shock in liquid DT, where the particle number density is $n = 5 \times 10^{22}$ cm^{-3}, and where the ignition temperature for the DT thermonuclear reaction is ~ 10 keV $\simeq 10^8$ °K. According to (4.11) the mean free path at this density and temperature is $\lambda \sim 10^{-2}$ cm. Therefore, at an initial temperature of $\sim 10^4$ °K, an initial radius of $R \sim 10^2$ cm would be needed to reach $T \sim 10^8$ °K at $r \sim 10^{-2}$ cm with a spherical convergent shock wave.

To ignite a thermonuclear burn wave at the center of the convergent shock wave requires that there $\rho r \gtrsim 1$ g/cm^2. At $\rho \sim 30\rho_0$, ($\rho_0 \sim 0.1$ g/cm^3 for liquid DT), $r \geq 0.3$ cm would then rather be needed as the distance where the ignition temperature is reached, requiring an initial radius of $R \sim 30$ meter. Therefore, to make possible the ignition of a thermonuclear burn in much smaller assemblies, one has either to start with a much higher temperature at the initial radius R, or to precompress the DT to higher than liquid densities, or a combination of both.

5.4 Isentropic Compression Waves

The compression by a shock wave is nonisentropic and accompanied by a rise in temperature. The rise in temperature is less if the compression is isentropic. Therefore, to reach the highest densities the compression must be isentropic. Whether it is the isentropic compression of a plane, cylindrical or spherical assembly, the compression process transmits at any given temperature a disturbance from the surface of the assembly to its center. The velocity by which this disturbance is transmitted is the isentropic sound velocity.

$$c_s = \left(\frac{\gamma RT}{A}\right)^{1/2} \tag{5.30}$$

(R is here the universal gas constant). If the compression velocity is set equal to the isentropic sound velocity during the entire compression process, then all the disturbances launched from the surface of the assembly arrive simultaneously at its center. Shock waves only occur if the sound waves intersect. If the compression velocity is set equal to the isentropic sound velocity, this intersection takes place only at the center where all the waves coalesce.

With the isentropic equation of state

$$\frac{T}{T_0} = \left(\frac{n}{n_0}\right)^{\gamma-1} \tag{5.31}$$

where n_0 is the initial particle number density at the temperature T_0, the isentropic equation of state for a plane ($s = 1$), cylindrical ($s = 2$) and spherical assembly ($s = 3$) is

$$\frac{T}{T_0} = \left(\frac{r_0}{r}\right)^{s(\gamma-1)} \tag{5.32}$$

where r_0 is the initial radius of a plane, cylindrical or spherical assembly.

Equating the compression (resp. implosion) velocity with the isentropic sound velocity one has

$$\frac{dr}{dt} = -\left(\frac{\gamma RT}{A}\right)^{1/2} = -c_0\left(\frac{T}{T_0}\right)^{1/2} \tag{5.33}$$

where $c_0 = (\gamma RT_0/A)^{1/2}$ is the initial sound velocity. Inserting (5.32) into (5.33) one has

$$\frac{dr}{dt} = -c_0 \left(\frac{r_0}{r}\right)^{s(\gamma-1)/2} . \tag{5.34}$$

Integration of (5.34) yields

$$r(t) = r_0 \left(1 - \frac{t}{t_0}\right)^m \tag{5.35}$$

where

$$\left.\begin{aligned} m &= \frac{2}{s(\gamma-1)+2} \\ t_0 &= \frac{mr_0}{c_0} \\ r_0 &= r(0) \end{aligned}\right\}$$

For $s = 1$, one has $m = 2/(\gamma+1)$, for $s = 2$, $m = 1/\gamma$, and for $s = 3$, $m = 2/(3\gamma-1)$.

For the implosion velocity one has

$$\frac{dr}{dt} = -c_0 \left(1 - \frac{t}{t_0}\right)^{-q} \tag{5.36}$$

where

$$q = \frac{s(\gamma-1)}{s(\gamma-1)+2}$$

for $s = 1$, $q = (\gamma-1)/(\gamma+1)$, for $s = 2$, $q = (\gamma-1)/\gamma$, and for $s = 3$, $q = 3(\gamma-1)/(3\gamma-1)$.

The boundary pressure is

$$\frac{p}{p_0} = \left(\frac{\mathrm{n}}{\mathrm{n}_0}\right)^\gamma = \left(\frac{r_0}{r}\right)^{s\gamma} = \left(1 - \frac{t}{t_0}\right)^{-\ell} \tag{5.37}$$

where

$$\ell = ms\gamma = \frac{2s\gamma}{s(\gamma-1)+2} .$$

For $s = 1$, $\ell = 2\gamma/(\gamma+1)$; for $s = 2$, $\ell = 2$; for $s = 3$, $\ell = 6\gamma/(3\gamma+1)$.

The power needed for the compression is

$$\left.\begin{array}{llll} s=1: & P = pc_s, & P_0 = p_0 c_0 & \mathrm{erg/cm^2\,s} \\ s=2: & P = 2\pi r p c_s, & P_0 = 2\pi r_0 p_0 c_0 & \mathrm{erg/cm\,s} \\ s=3: & P = 4\pi r^2 p c_s, & P_0 = 4\pi r_0^2 p_0 c_0 & \mathrm{erg/s} \end{array}\right\} \quad (5.38)$$

With $c_s/c_0 = (r_0/r)^{s(\gamma-1)/2}$ one then has

$$\frac{P}{P_0} = \left(\frac{r_0}{r}\right)^u \quad (5.39)$$

where

$$u = (3s/2)(\gamma-1) + 1$$

or

$$\frac{P}{P_0} = \left(1 - \frac{t}{t_0}\right)^{-w} \quad (5.40)$$

where

$$w = \frac{3s(\gamma-1)+2}{s(\gamma-1)+2} \quad (5.41)$$

For $s = 1$ this is $w = (3\gamma-1)/(\gamma+1)$; for $s = 2$, $w = (3\gamma-2)/\gamma$; and for $s = 3$, $w = (9\gamma-7)/(3\gamma-1)$. For $\gamma = 5/3$ we obtain: $w = 3/2$ $(s = 1)$, $w = 9/5$ $(s = 2)$, $w = 2$ $(s = 3)$.

5.5 Implosion of Compressible Shells

In addition to convergent shock waves, imploding shells play an important role for both fission and fusion explosives. For convergent shock waves the acoustic approximation already gave a fairly good result for the rise in pressure in its dependence of the distance from the center of convergence, but a similar approximation for the implosion of an incompressible shell is not nearly as good.

For an imploding incompressible shell, the increase in the velocity at the inner wall results from the fattening of the shell during its implosion as shown

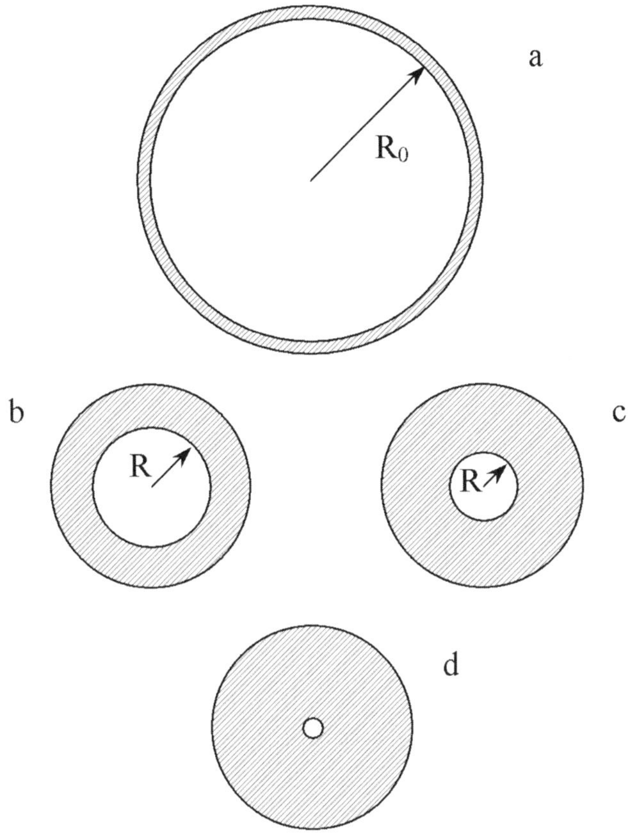

Figure 5.1: The imploding shell of initial radius $R = R_0$ in its initial configuration (a), and three consecutive stages, (b), (c) and (d), during its implosion.

in Fig. 5.1. This problem is similar to the classical cavitation problem by Rayleigh for an incompressible fluid. There the inner wall velocity at the distance R from the center of the cavity collapse is given by

$$\left.\begin{aligned} \mathrm{v} &\cong \frac{\mathrm{v}_0}{\sqrt{2}}\left(\frac{R_0}{R}\right)\left[\log\left(\frac{R_0}{R}\right)\right]^{-1/2} && \text{cylinder} \\ \mathrm{v} &\cong \frac{\mathrm{v}_0}{\sqrt{3}}\left(\frac{R_0}{R}\right)^{3/2} && \text{sphere} \end{aligned}\right\} \tag{5.42}$$

To obtain the rise in the inner wall velocity for a compressible shell, we assume that the shell material can be described as a frictionless compressible fluid, an assumption justified at high pressures.

The Euler and continuity equation for an inviscid compressible fluid are

$$\left.\begin{aligned} &\frac{\partial \mathbf{v}}{\partial t} + (\mathbf{v}\cdot\nabla)\,\mathbf{v} = -\frac{1}{\rho}\nabla p \\ &\frac{\partial \rho}{\partial t} + \nabla\cdot\rho\mathbf{v} = 0\,. \end{aligned}\right\} \tag{5.43}$$

Here we set $s = 1$ for cylindrical and $s = 2$ for spherical symmetry. For an imploding cylindrical or spherical shell one obtains from (5.43)

$$\frac{\partial \mathrm{v}}{\partial t} + \mathrm{v}\frac{\partial \mathrm{v}}{\partial r} + \frac{1}{\rho}\frac{\partial p}{\partial r} = 0 \tag{5.44a}$$

$$\frac{\partial \rho}{\partial t} + \mathrm{v}\frac{\partial \rho}{\partial r} + \rho\left(\frac{\partial \mathrm{v}}{\partial r} + s\frac{\mathrm{v}}{r}\right) = 0 \tag{5.44b}$$

where v is the radial fluid velocity and r the radial coordinate. Furthermore, the equation of state for the compressible shell material shall be given by ($A = const.$)

$$p = A\rho^{\gamma} \tag{5.45}$$

hence

$$c^2 = \frac{dp}{d\rho} = A\rho^{\gamma-1} \tag{5.46}$$

Whereby (5.44a) and (5.44b) can be written as follows

$$\frac{\partial \mathrm{v}}{\partial t} + \mathrm{v}\frac{\partial \mathrm{v}}{\partial r} + \frac{1}{\gamma - 1}\frac{\partial c^2}{\partial r} = 0 \tag{5.47a}$$

$$\frac{\partial c^2}{\partial t} + \mathrm{v}\frac{\partial c^2}{\partial r} + (\gamma - 1)\, c^2 \left(\frac{\partial \mathrm{v}}{\partial r} + s\frac{\mathrm{v}}{r}\right) = 0\,. \tag{5.47b}$$

To solve these two coupled nonlinear partial differential equations one sets

$$R(t) = (-\alpha t)^n \ , \ \alpha = const. \tag{5.48}$$

where $R(t)$ is the radius of the inner surface of the collapsing shell as a function of time. For $t < 0$ the radius decreases reaching $R = 0$ at $t = 0$.

One then introduces the similarity variable

$$\zeta = -\left(\frac{R}{r}\right)^{1/n} = \frac{\alpha t}{r^{1/n}}\,. \tag{5.49}$$

Comparing (5.49) with (5.48) shows that at the inner wall surface $\zeta = 1$, and that for the r-axis $\zeta = 0$.

From (5.48) one obtains for the velocity of the inner wall

$$\dot{R} = -n\alpha R^{1-1/n}\,. \tag{5.50}$$

The problem is now reduced to determine the number n, the so-called homology exponent. To obtain it one looks for solutions of the form

$$\mathrm{v} = -n\alpha r^{(1-1/n)}\, F(\zeta) \tag{5.51a}$$

$$c^2 = n^2\alpha^2 r^{(2-2/n)}\, G(\zeta)\,. \tag{5.51b}$$

Inserting (5.51a) and (5.51b) into (5.47a) and (5.47b), the dependence on r drops out, and one has two ordinary differential equations:

$$(\gamma - 1)(1 + \zeta F) F' + \zeta G' + (1 - n)\left[(\gamma - 1) F^2 + 2G\right] = 0 \tag{5.52a}$$

$$\begin{aligned}(\gamma - 1)\,\zeta G F' + (1 + \zeta F)\, G' + \\ + \left[(1 - n)(\gamma + 1) - s(\gamma - 1)\, n\right] F G = 0\end{aligned} \tag{5.52b}$$

where $F' \equiv dF/d\zeta$, $G' \equiv dG/d\zeta$. From (5.50), (5.51a) and (5.51b) it follows that at the wall where $\zeta = -1$, one has $F = 1$. Further, for $\zeta = -1$, one must have $G = 0$ with the pressure at the wall surface equal to zero and with it $c^2 = 0$. With the different set of variables

$$\left.\begin{aligned} x &= \ln(-\zeta) \\ y &= -\zeta F \\ z &= \zeta^2 G \end{aligned}\right\} \tag{5.53}$$

where at the inner wall surface $x = \ln(1) = 0$, (5.52a) and (5.52b) take the form of three coupled ordinary differential equations

$$\begin{aligned} dx : \ dy : \ dz = &\left[(y-1)^2 - z\right] \\ &: \left[y(y-1)(ny-1) - (s+1)nyz + \frac{2(1-n)}{\gamma-1}z\right] \\ &: \left[2z\begin{Bmatrix} -nz + \dfrac{ny^2}{2}\left[(2-s)+s\gamma\right] \\ +\dfrac{y}{2}\left[\gamma - 3 - ((s+1)\gamma + 1 - s)n\right] + 1 \end{Bmatrix}\right] \end{aligned} \tag{5.54}$$

Of these three ordinary differential equations only two are independent. One of them contains two variables only and can be separated from the other two. It is the differential equation:

$$\begin{aligned} \frac{dy}{dz} = &\left[y(y-1)(ny-1) - (s+1)nyz + \frac{2(1-n)}{\gamma-1}z\right] \\ &: \left[2z\begin{Bmatrix} -nz + \dfrac{ny^2}{2}\left[(2-s)+s\gamma\right] \\ +\dfrac{y}{2}\left[\gamma - 3 - ((s+1)\gamma + 1 - s)n\right] + 1 \end{Bmatrix}\right] \end{aligned} \tag{5.55}$$

With this equation a value of n can be determined from the condition that the solution is regular in passing through a singular point. A differential equation of the form $dy/dx = f(x)/g(x)$ is singular if both $f(x) = g(x) = 0$. For the regularity of the solution only one singular point is of importance. In our case it is located on the parabola

$$z = (y-1)^2 \tag{5.56}$$

where according to (5.54) both dx/dy and dx/dz vanish. On the singular point one has $d\zeta/dF = d\zeta/dG = 0$, which means that F and G are not single valued functions of ζ for an integral curve passing through this point. There is, however, one particular integral curve for a specific value of n where F and G are single valued, there the integral curve in the y-z plane does not have a turning point in crossing the parabola (5.56). It is this integral curve which determines the value of n.

A lower value of n can be obtained by inserting (5.56) into the denominator of (5.55) to be set equal to zero:

$$s(\gamma-1)ny^2 - \Big[(s+1)(\gamma-1)n - (\gamma-1) + 2(1-n)\Big]y + 2(1-n) = 0\,. \tag{5.57}$$

This equation has a real solution if

$$\Big[(s+1)(\gamma-1)n - (\gamma-1) + 2(1-n)\Big]^2 \geq 8s(\gamma-1)n(1-n) \tag{5.58}$$

In the limit where the r.h.s. of (5.58) is set equal to the l.h.s. one obtains a lowest value for n. For a cylindrical shell implosion ($s = 1$) it is

$$n_{min} = \frac{\gamma^3 - 3\gamma + 4 + \Big[\left(\gamma^2 - 3\gamma + 4\right)^2 - (\gamma-3)^2\left(\gamma^2 - 2\gamma + 2\right)\Big]^{1/2}}{2\gamma^2 - 4\gamma + 4} \tag{5.59a}$$

and for a spherical shell implosion ($s = 2$) it is

$$n_{min} = \frac{3\gamma^2 - 6\gamma + 7 + \Big[\left(3\gamma^2 - 6\gamma + 7\right)^2 - (\gamma-3)^2\left(9\gamma^2 - 14\gamma + 9\right)\Big]^{1/2}}{9\gamma^2 - 14\gamma + 9} \tag{5.59b}$$

In the limit $\gamma \to \infty$, corresponding to an incompressible shell one has

$$n_{min} = \begin{cases} \dfrac{1}{2} & (\mathrm{s} = 1,\ \text{cylinder}) \\ \dfrac{1}{3} & (\mathrm{s} = 2,\ \text{sphere}) \end{cases} \tag{5.60}$$

According to (5.50) this leads there to

$$\left.\begin{aligned} \dot{R} &= const./R \quad \text{(cylinder)} \\ \dot{R} &= const./R^2 \quad \text{(sphere)} \end{aligned}\right\} \tag{5.61}$$

This result differs from the cavitation solution (5.42), but is exactly what one would expect from the implosion of an incompressible shell. With the inner and outer shell radius set equal to R and R_0, one has for a cylindrical and spherical shell

$$\left.\begin{aligned} R_0^2 - R^2 &= const. \quad \text{(cylindrical)} \\ R_0^3 - R^3 &= const. \quad \text{(spherical)} \end{aligned}\right\} \tag{5.62}$$

which upon differentiation leads to

$$\left.\begin{aligned} \dot{R} &= \dot{R}_0 \left(\frac{R_0}{R}\right) \quad \text{(cylindrical)} \\ \dot{R} &= \dot{R}_0 \left(\frac{R_0}{R}\right)^2 \quad \text{(spherical)} \end{aligned}\right\} \tag{5.63}$$

where $\dot{R}_0 = const.$ is the externally exerted constant velocity onto the shell. This result is the same as (5.61)

To obtain the exact value of n, the differential equation (5.55) must be integrated numerically from $\zeta = -1$ to $\zeta = 0$, that is from $F = 1$, $G = 0$, to $F = G = 0$, or from $y = 1$, $z = 0$ to $y = z = 0$. The integration can be done with the Runge-Kutta method, where one can put $n = n_{min}$ for the first trial run, thereafter increasing n in small steps until the integral curve cuts the parabola (5.56) without a turning point.

With $m = 1/n - 1$ the shell implosion velocity for $R \to 0$ is there given by

$$\mathrm{v} = const.\ R^{-m} \tag{5.64}$$

In table 5.1 the values of n and m, obtained by the numerical integration of (5.55) are given for different values of γ, and in figures 5.2 and 5.3 are plotted together with n_{min} for the cylindrical and spherical shell implosion. For megabar pressures one has $\gamma \simeq 10$, $m = 0.45$ for cylindrical and $m = 0.92$ for a spherical shell. Hence there

$$\left.\begin{aligned} \mathrm{v} &= const.\ R^{-0.45} \quad \text{(cylindrical)} \\ \mathrm{v} &= const.\ R^{-0.92} \quad \text{(spherical)} \end{aligned}\right\} \tag{5.65}$$

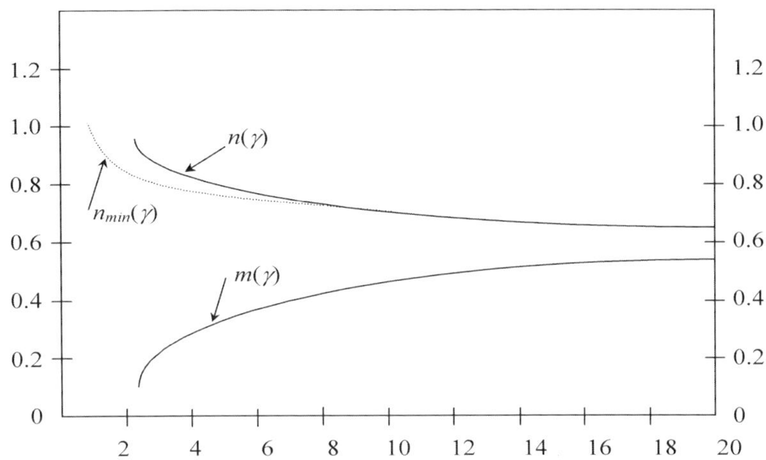

Figure 5.2: The functions $n(\gamma)$, $n_{min}(\gamma)$ and $m(\gamma)$ for the cylindrical shell implosion.

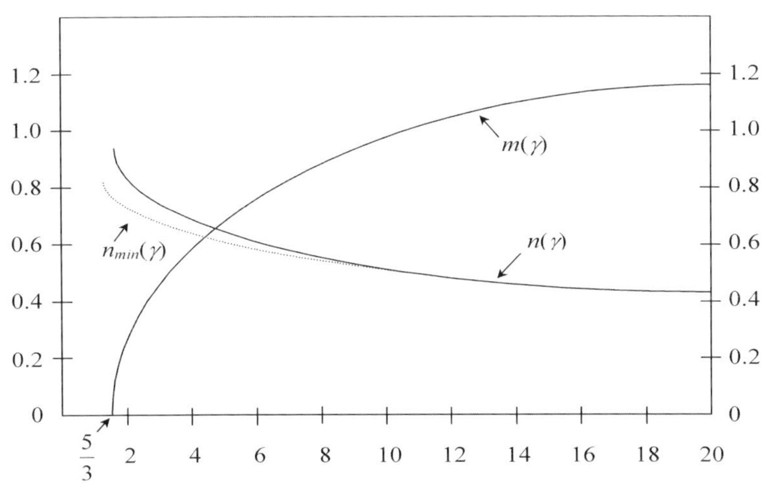

Figure 5.3: The functions $n(\gamma)$, $n_{min}(\gamma)$ and $m(\gamma)$ for the spherical shell implosion.

Cylindrical			Spherical	
γ	n	m	n	m
5/3			0.92	0.087
2			0.835	0.198
3	0.88	0.14	0.71	0.409
4	0.82	0.22	0.64	0.563
5	0.775	0.29	0.60	0.667
6	0.75	0.34	0.574	0.742
7	0.73	0.375	0.5574	0.804
8	0.71	0.41	0.5407	0.8495
9	0.70	0.43	0.5294	0.8889
10	0.69	0.45	0.5198	0.9238
11	0.68	0.47	0.5115	0.9549
12	0.67	0.49	0.5043	0.9830
13	0.67	0.50	0.4979	1.0085
14	0.66	0.51	0.4922	1.0318
15	0.655	0.53	0.4870	1.0533
16	0.65	0.54	0.4824	1.0732
17	0.65	0.54	0.4781	1.0916
18	0.64	0.56	0.4742	1.1089
19	0.64	0.56	0.4706	1.125
20	0.64	0.56	0.4673	1.1402

Table 5.1: Parameters for cylindrical and spherical shell implosions.

For higher pressures the value of γ goes rapidly down, approaching the limit $\gamma = 5/3$, valid for a cold Fermi gas (but also for a fully ionized plasma). There for a spherical shell $m = 0.087$. At these high pressures the flattening of the shell is less and does not lead to a substantial increase of the inner wall velocity. Higher implosion velocities can, of course, be reached by increasing the pressure on the outer shell surface, for example by ablating the surface with a powerful laser of particle beam, thereby increasing $\dot{R}_0$.

For an incompressible shell, the density of the shell material remains of course constant during the implosion, but not for a compressible shell.

There the density is computed by the continuity equation

$$R^s \rho \mathrm{v} = const. \tag{5.66}$$

therefore

$$\frac{\rho}{\rho_0} = \left(\frac{R_0}{R}\right)^s \frac{\mathrm{v}_0}{\mathrm{v}} = \left(\frac{R_0}{R}\right)^{s-m} . \tag{5.67}$$

For the example $\gamma = 10$ one finds

$$\frac{\rho}{\rho_0} = \begin{cases} \left(\dfrac{R_0}{R}\right)^{0.55} & \text{(cylindrical)} \\ \left(\dfrac{R_0}{R}\right)^{1.08} & \text{(spherical)} \end{cases} \tag{5.68}$$

5.6 Multishell Implosions

According to (5.13) the rise in pressure and temperature in a plane shock wave is proportional to the square of the shock wave velocity. With the pressure and temperature in a convergent spherical shock wave going in proportion to $r^{-2\kappa}$ where κ is given by (5.27) (for $\gamma = 5/3$, $\kappa = 0.45$), it thus follows that the shock wave velocity rises in proportion to $r^{-\kappa}$, for $\gamma = 5/3$ in proportion to $r^{-0.45}$, and in the limit $\gamma \to \infty$ where $\kappa^{-1} = 1.2$, in proportion to $r^{-0.8}$. By comparison, the velocity of an imploding spherical shell, in the same limit $\gamma \to \infty$, goes in proportion to r^{-2}.

A convergent shock wave can be viewed as the implosion of many concentric shells, with a buffering gas between the shells to soften the impact of the shells onto each other. It therefore seems plausible that by removing the buffering gas the rise in velocity will be larger if compared with the velocity rise in a convergent shock, but less if compared with the implosion of one shell, at least if the shell is incompressible.

A velocity amplification between a series of colliding bodies occurs in elastic collisions of bodies with a decreasing mass. If the mass ratio $m_n/m_{n-1} = \alpha$ between the colliding bodies with the mass m_{n-1} and m_n is constant, and if the velocity of the first body with mass m_0 is v_0, then, following a collision with the body of mass m_{n-1}, the nth body m_n acquires the velocity

$$\mathrm{v}_n = \left[\frac{2}{1+\alpha}\right]^n \mathrm{v}_0 . \tag{5.69}$$

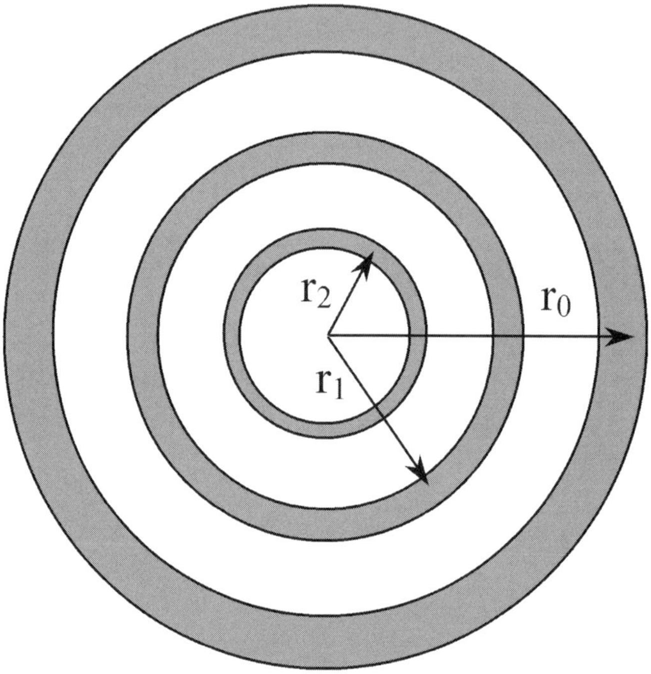

Figure 5.4: Raising the implosion velocity through the subsequent collision of several concentric shells of decreasing mass.

If the colliding bodies are concentric spherical shells of decreasing mass with the decreasing shell radius r_n and shell thickness Δr_n, one has (see Fig. 5.4)

$$m_n = 4\pi \rho r_n^2 \Delta r_n \tag{5.70}$$

where ρ is the density of the shell material. We further assume that the assembly of concentric shells is selfsimilar whereby

$$\Delta r_n = \varepsilon r_n \; , \; \varepsilon = const. < 1 \; . \tag{5.71}$$

We thus have

$$m_n = 4\pi \varepsilon r_n^3 \tag{5.72}$$

and

$$\frac{m_{n+1}}{m_n} = \left(\frac{r_{n+1}}{r_n}\right)^3 \tag{5.73}$$

or

$$r_{n+1} = \alpha^{1/3} r_n \; . \tag{5.74}$$

We then put

$$r_n = r_0 f(n) \; , \; f(0) = 1 \; . \tag{5.75}$$

Inserting (5.75) into (5.74) one obtains the functional equation for $f(n)$

$$f(n+1) = \alpha^{1/3} f(n) \tag{5.76}$$

with $f(0)$ it has the solution

$$f(n) = \alpha^{n/3} \tag{5.77}$$

and we have

$$r_n = r_0 \alpha^{n/3} \; . \tag{5.78}$$

Eliminating the parameter n from (5.69) and (5.78) one obtains

$$\left.\begin{aligned} \frac{\mathrm{v}_n}{\mathrm{v}_0} &= \left(\frac{r_0}{r_n}\right)^a \\ a &= -\frac{\log\left[2/\left(1+\alpha\right)\right]^3}{\log\alpha} \end{aligned}\right\} \tag{5.79}$$

$$\mathrm{v} = \frac{const.}{r^a} \tag{5.80}$$

As an example we take $\alpha = 1/2$ where $a = 1.25$. For $\alpha = 0.25$ we find $a \simeq 1$ and for $\alpha = 0.125$, $a \simeq 0.83$.

We can still improve this model by taking into account inelastic collision losses between the shells. There (5.69) is replaced by

$$\mathrm{v}_n = \left[g\left(\alpha\,,\,\eta\right)\right]^2 \mathrm{v}_0 \tag{5.81}$$

where

$$g\left(\alpha\,,\,\eta\right) = \frac{1}{1+\alpha}\left[1+\sqrt{1-\frac{1+\alpha}{\alpha}\,\eta}\right] \tag{5.82}$$

$$\eta = \frac{\varepsilon_t^n}{\left(\dfrac{m_n}{2}\right)\mathrm{v}_n^2} = const. \tag{5.83a}$$

where ε_t^n is the heat-dissipated kinetic energy of the shell m_n with a kinetic energy $(m_n/2)\,\mathrm{v}_n^2$, assuming that the fraction $\varepsilon_t^n/\,(m_n/2)\,\mathrm{v}_n^2$ is constant. In a similar way as for the completely elastic collision we have here

$$\begin{aligned} \frac{\mathrm{v}_n}{\mathrm{v}_0} &= \left(\frac{r_0}{r_n}\right)^a \\ a &= -\frac{\log g^3\left(\alpha\,,\,\eta\right)}{\log\alpha}\,,\quad g\left(\alpha\,,\eta\right) = \frac{2}{1+\alpha} \end{aligned} \tag{5.83b}$$

As an example we take $\alpha = 1/2$, and $\eta = 0.1$ and find $a \simeq 0.9$. For $\alpha = 0.25$ we find $a \simeq 0.7$.

5.7 Rayleigh-Taylor Instability

For imploding shells the Rayleigh-Taylor instability is explained in Fig. 5.5. An imploding shell, initially exactly of spherical or cylindrical symmetry, does not stay that way, but is deformed, for example as shown. The way this instability scales can be determined in a reference system at rest with respect to the inner surface of the shell. Near the inner shell surface the pressure vanishes, and the equation of motion at the surface is

$$\frac{\partial \mathbf{v}}{\partial t} = \mathbf{a} \tag{5.84}$$

where $\mathbf{a}$ is the inward acceleration of the shell surface. For curl $\mathbf{v} = 0$ the velocity can be derived from a velocity potential

$$\mathbf{v} = -\frac{\partial \phi}{\partial \mathbf{r}} \,. \tag{5.85}$$

Inserting (5.85) into (5.84) one has after integration along $\mathbf{r}$:

$$\frac{\partial \phi}{\partial t} + \mathbf{a} \cdot \mathbf{r} + const. = 0 \tag{5.86}$$

and by differentiation with regard to t:

$$\frac{\partial^2 \phi}{\partial t^2} + \mathbf{a} \cdot \frac{\partial \phi}{\partial \mathbf{r}} = 0 \,. \tag{5.87}$$

If $\mathbf{s}$ is a surface coordinate perpendicular to $\mathbf{r}$, and if $\mathrm{v}_s = -\partial\phi/\partial s$, div $\mathbf{v}$ $= 0$ results in

$$\frac{\partial^2 \phi}{\partial r^2} + \frac{\partial^2 \phi}{\partial s^2} = 0 \,. \tag{5.88}$$

With

$$\phi = e^{i(\mathbf{k}\cdot\mathbf{s} - \omega t)} \, f(\mathbf{r}) \tag{5.89}$$

describing a surface wave, one obtains from (5.88)

$$\frac{d^2 f}{d\mathbf{r}^2} - k^2 f = 0 \tag{5.90}$$

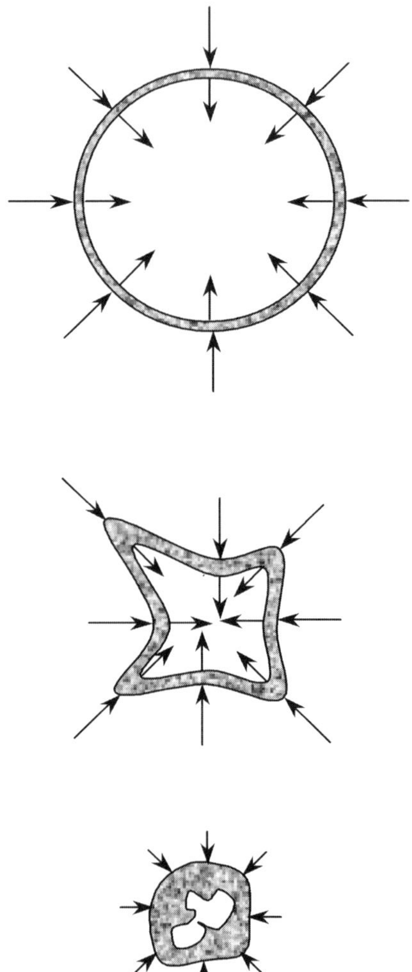

Figure 5.5: Rayleigh-Taylor instability of an imploding shell.

with the solution

$$f(r) = const.\, e^{\mathbf{k}\cdot\mathbf{r}} \tag{5.91}$$

and hence

$$\phi = const.\, e^{\mathbf{k}\cdot\mathbf{r}} e^{i(\mathbf{k}\cdot\mathbf{s}-\omega t)} \,. \tag{5.92}$$

Finally, inserting (5.92) into (5.87) one finds that

$$\omega^2 = -\mathbf{a}\cdot\mathbf{k}\,. \tag{5.93}$$

Because ω is imaginary, the wave amplitude grows exponentially in time as

$$A = A_0 e^{\sigma t} \tag{5.94}$$

where

$$\sigma = \sqrt{\mathbf{a}\cdot\mathbf{k}}\,. \tag{5.95}$$

If a gas or plasma with a density $\rho_p < \rho$ is inside the imploding shell, a similar calculation gives

$$\sigma^* = \sqrt{\frac{(\rho - \rho_p)}{(\rho + \rho_p)}}\,\sigma\,. \tag{5.96}$$

This means that a gas or plasma inside the imploding shell, while reducing the implosion velocity, also reduces the growth rate of the Rayleigh-Taylor instability.

An imploding shell sustains its symmetry as long as $A \ll R$, where R is the inner shell radius. If v is the implosion velocity and R_0 the initial shell radius, the implosion time is of the order $t = \sqrt{2R_0/a}$. The largest, most serious deformation has a wave number of the order $k \sim 1/R$, and the largest amplitude is $A \sim R$. Inserting these values into (5.94) results in

$$A_0 \sim Re^{-(2R_0/R)^{1/2}}\,. \tag{5.97}$$

If for example $R_0/R = 10$, one finds that $A_0 \sim 10^{-2}R$. This means that an initial nonspherical deformation of the shell must be less than 1% of the final implosion radius R. This is a severe demand for the implosion symmetry.

As (5.96) shows, the instability is less severe with a buffer gas inside the shell, and the same can be expected from a multishell configuration. In the limit of an infinite number of concentric shells one has the situation realized in a convergent shock wave, which is known to be quite stable, of importance for thermonuclear ignition schemes making use of the convergent shock wave solution. For long cylindrical shells the Rayleigh-Taylor instability can be overcome by rapid rotation, with the centrifugal force balancing the inertial force, used in certain schemes proposed to ignite thermonuclear reactions with a fast z-pinch.

We would like to add that non-spherical (ellipsoidal) implosions can also offer significant advantages. Even for pure fission explosions, the implosion of a prolate assembly seems to be advantageous as well. For the reasons why non-spherical implosions are of interest for fusion targets, we refer the reader to reference (Winterberg 1977).

5.8 Conical Implosion

One configuration of particular interest is the conical implosion explained in Fig. 5.6. As shown (Fig. 5.6a), the cone is imploded with the velocity V_0 perpendicular to the surface of the cone. The vertex point A of the imploding cone moves to the right with the velocity

$$V_1 = \frac{V_0}{\sin\alpha} \tag{5.98}$$

where α is 1/2 the opening angle of the cone. In the course of the implosion a "slug" of mass M_s moves to the left with the velocity V_s. To conserve momentum a jet of mass M_j moves to the right with the velocity V.

The streamlines in a reference system at rest with respect to the vertex point A, (shown in Fig. 5.6b) bifurcate at A with the outer streamlines merging into the slug. The inner streamlines, after making a sharp $360° - \alpha$ degree turn at the vertex point, move to the right. In the reference system at rest with reference to A, (see Fig. 5.6a), the velocity along the streamlines is everywhere equal to

$$V_2 = V_1 \cos\alpha = \frac{V_0}{\tan\alpha}\,. \tag{5.99}$$

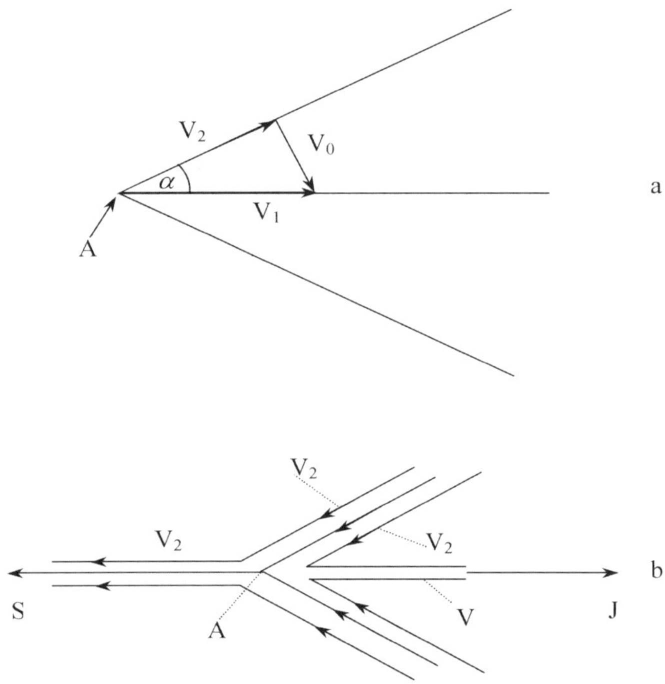

Figure 5.6: Conical implosion.

In a laboratory system on the other hand one has

$$\left.\begin{aligned} \mathrm{V} &= \mathrm{V}_1 + \mathrm{V}_2 \\ \mathrm{V}_s &= \mathrm{V}_1 - \mathrm{V}_2 \end{aligned}\right\} \tag{5.100}$$

The total mass of slug and jet is

$$M = M_s + M_J \; . \tag{5.101}$$

With linear momentum conservation along the cone axis requiring that

$$M\mathrm{V}_2 \cos\alpha = M_s \mathrm{V}_2 - M_j \mathrm{V}_2 \tag{5.102}$$

one obtains from (5.101) and (5.102)

$$\left.\begin{aligned} M_J &= \frac{M}{2}(1 - \cos\alpha) \\ M_s &= \frac{M}{2}(1 + \cos\alpha) \end{aligned}\right\} \tag{5.103}$$

And from (5.98)-(5.100)

$$\left.\begin{aligned} \mathrm{V} &= \frac{V_0}{\sin\alpha}(1 + \cos\alpha) \\ \mathrm{V_s} &= \frac{V_0}{\sin\alpha}(1 - \cos\alpha) \end{aligned}\right\} \tag{5.104}$$

From (5.103) and (5.104) one has

$$M_J \mathrm{V} = M_s \mathrm{V}_s = \frac{1}{2} M \mathrm{V}_0 \sin\alpha \; . \tag{5.105}$$

For $\alpha \to 0$ this becomes

$$\left.\begin{aligned} \mathrm{V} &\to \frac{2\mathrm{V}_0}{\alpha} \to \infty \\ \mathrm{V_s} &\to \frac{\mathrm{V}_0}{2}\alpha \to 0 \end{aligned}\right\} \tag{5.106}$$

$$M_J \mathrm{V} = M_s \mathrm{V}_s \to \frac{M\mathrm{V}_0}{2}\alpha \to 0 \; . \tag{5.107}$$

5.9 Bibliography for Chapter 5

L. D. Landau and E. M. Lifshitz, Fluid Mechanics, Pergamon Press, London 1959.

G. Guderley, Luftfahrtforschung, **19**, 302 (1942).

R. F. Chisnell, Journal of Fluid Mechanics, **2**, 286 (1957).

G. Birkhoff, D. P. MacDougall, E. M. Pugh and G. Taylor, Journal of Applied Physics, **19**, 563 (1948).

F. Winterberg, J. Plasma Physics, **18**, 473 (1977).

Chapter 6

Thermonuclear Ignition and Burn

6.1 Ignition of Thermonuclear Reactions

In a self-sustained thermonuclear burn the energy released by thermonuclear reactions exceeds the energy losses by expansion, heat conduction and radiation, and it must be the goal to keep these losses down. If the thermonuclear plasma is confined by some force field, the losses by expansion can be eliminated. The gravitational field of a burning plasma can confine the plasma by itself, but because of the weakness of the gravitational field only for masses of astronomical dimensions. A second example for the confinement by self-forces is the confinement by a magnetic field where electric currents flow through the plasma. However, unlike a gravitational field, there is no stable self-confining configuration. A third possibility is the confinement by inertial forces, but there the forces act only for a very short time. The time though can be increased by imploding the plasma inside a low temperature "tamp", realized through hypervelocity impact. There the problems are the heat conduction losses from the hot plasma into the tamp, but they can be reduced by strong magnetic fields. One is essentially then left with radiation losses. Because at the temperatures of more than 10 keV, the plasma radiation is in the x-ray regime and cannot be confined (for example by mirrors). Nevertheless, the wall confinement of (several 100 eV) soft x-ray blackbody radiation plays an important role in thermonuclear explosive devices. For small thermonuclear assemblies the escape of fusion products, not having dissipated their kinetic energy in the assembly, must be considered. There are two kinds of fusion products: charged particles and neutrons. In small thermonuclear assemblies neutrons escape unattenuated and do not contribute to the energy balance of the burning thermonuclear plasma. This is unfortunately true for the DT thermonuclear reaction, which has the largest cross section, but where 80% of the energy released goes into neutrons.

The stopping range of the charged fusion products can be substantially reduced by strong magnetic fields, which is of importance for small thermonuclear explosive devices.

6.2 Ignition Temperature for Black Body Radiation Losses

We consider first a plasma sphere of radius r and uniform temperature T in thermodynamic equilibrium, ignoring expansion and heat conduction losses. The thermonuclear reaction rate is given by (2.64) and the radiation loss rate by (4.58). For a plasma, like the DT plasma composed of two reacting nuclei with equal number density $n_1 = n_2 = n/2$, the energy balance equation is

$$4\pi r^2 \sigma T^4 = \left(\frac{4\pi}{3} r^3\right)\left(\frac{n^2}{4}\right)\langle\sigma \mathrm{v}\rangle\varepsilon_0 \tag{6.1}$$

from which a critical radius is derived above which makes a self-sustaining reaction possible. This radius is:

$$r = \frac{12\sigma T^4}{\varepsilon_0 n^2 \langle\sigma \mathrm{v}\rangle} . \tag{6.2}$$

With the expression for $\langle\sigma \mathrm{v}\rangle$ given by (2.59), (6.2) can be brought into the form

$$\left.\begin{aligned} r &= \left(\frac{a}{n^2}\right)\frac{e^x}{x^{14}} \\ a &= 12\sigma\frac{k_2^{12}}{\varepsilon_0 k_1} \\ x &= k_2 T^{-1/3} \end{aligned}\right\} \tag{6.3}$$

The critical radius (6.2) has a minimum at the optimal ignition temperature, where e^x/x^{14} has a minimum. This minimum is at $x_0 = 14$, and one finds

$$\left.\begin{aligned} T_{opt} &= \left(\frac{k_2}{14}\right)^3 \\ r_{min} &= 0.72 \times 10^{-10}\frac{a}{n^2} \end{aligned}\right\} \tag{6.4}$$

As an example we take the DT thermonuclear reaction with the values of k_1, k_2, and ε_0 given in table 2.1. We assume that the radius r is large enough to stop the neutrons. (If only the charged products are stopped, with 20% of the total energy released, ε_0 must be multiplied by 0.2). The result is

$$\left.\begin{aligned} T_{opt} &= 3.3 \times 10^7 \ {}^\circ\mathrm{K} \\ r_{min} &= \frac{0.84 \times 10^{49}}{n^2} \end{aligned}\right\} \tag{6.5}$$

For liquid or solid DT with $n = 5 \times 10^{22}$ cm^{-3} we obtain $r_{min} = 34$ m. However, if the plasma density is $\sim 10^3$ times solid density, one has $r_{min} = 3.4 \times 10^{-3}$ cm. This shows that, for highly compressed assemblies, blackbody radiation is tolerable. This might be of importance for reactions containing higher Z elements, like the HB^{11} thermonuclear reaction.

In (6.5) the ignition temperature was computed for the minimum radius of the plasma sphere. If both the radius and density are given, the ignition temperature follows from the solution of the equation

$$e^x = \left(\frac{n^2 r}{a}\right) x^{14} . \tag{6.6}$$

With the total plasma mass $m - (4\pi/3)r^3 nMA$ this can be written as follows:

$$e^x = \left(\frac{3}{4\pi}\right)^2 \left(\frac{m}{MA}\right)^2 \left(\frac{x^{14}}{ar^5}\right) . \tag{6.7}$$

This equation has two roots for a given value of r. To obtain the minimum temperature one must take the larger root. The value of r where both roots become equal is the same as the minimum radius given by (6.5). It thus follows that the ignition temperature can become arbitrarily small only if r becomes sufficiently large. This is so because an increase in r reduces the surface to volume ratio. For $r \to \infty$ the ignition temperature would therefore in principle go to zero. This seems to contradict the requirement that for thermonuclear reactions large particle energies are required. However, even at low temperatures a Maxwell distribution has a few energetic particles which can make a fusion reaction. A situation of this kind exists in the formation of a star from cold matter. There though, the assumption of a constant temperature within the entire gas sphere is invalid. Nevertheless, a thermonuclear burn process can there start from a comparatively low temperature.

6.3 Ignition Temperature for Optically Thin Plasmas

A plasma is optically thin if the optical path length $\lambda_{opt} = 1/n\sigma_{opt}$ is large compared to the plasma dimensions. This is especially true for hydrogen plasmas of thermonuclear explosive devices, where the optical cross section is given by (4.70). At solid state densities with $n = 5 \times 10^{22}$ cm^{-3} and for $Z = 1$, one has $\sigma_{opt} \approx 2 \times T^{-3.5}$. For the optimal ignition temperature of a DT plasma in thermodynamic equilibrium with its radiation, the blackbody temperature is $T_{opt} = 3.3 \times 10^7$ °K, and one finds that there $\sigma_{opt} \approx 2 \times 10^{-26}$ cm^2, and hence $\lambda_{opt} \approx 10^3$ cm, larger than even the largest thermonuclear explosive devices.

For optically thin plasmas the radiation losses are by bremsstrahlung which are given by (4.67b). Unlike the blackbody radiation losses they are proportional to the volume of the plasma, not its surface. With these radiation losses the energy balance equation becomes ($\alpha = 1.42 \times 10^{-27}$ cgs units)

$$\alpha n^2 T^{1/2} = \frac{n^2}{4}\langle\sigma \mathrm{v}\rangle\varepsilon_0 \,. \tag{6.8}$$

With (2.59) and the last of (6.3), one can bring this into the form

$$\left.\begin{aligned} e^x &= bx^{7/2} \\ b &= \frac{\varepsilon_0 k_1}{4\alpha k_2^{3/2}} \end{aligned}\right\} \tag{6.9}$$

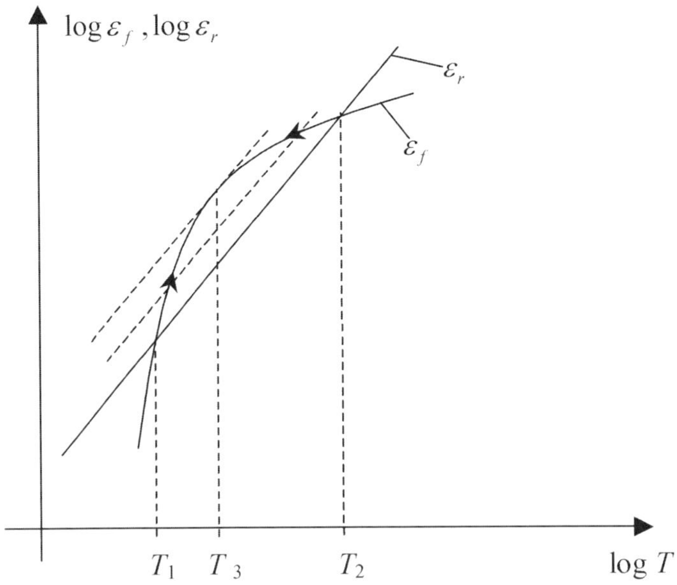

Figure 6.1: Ignition (T_1) and extinction (T_2) temperature of a thermonuclear reaction with bremsstrahlung losses. The dotted lines are for ε_r increasing with the addition of higher-Z fusion products to the burning plasma.

As before, there are two roots for x, with the larger root determining the ignition temperature. The smaller root gives the temperature above which the reaction is extinguished. The two roots can be understood as follows: The $\langle \sigma \mathrm{v} \rangle$ curve as a function of T goes through a maximum at which the overlapping of the cross section curve and Maxwellian velocity distribution is optimized. With the bremsstrahlung loss curve having no maximum, there are two points of intersection with the $\langle \sigma \mathrm{v} \rangle$ curve, the lower giving the ignition and the upper the extinction point (see Fig. 6.1). In the temperature interval $T_1 < T < T_2$ one has $\varepsilon_f > \varepsilon_r$ and there burn is possible.

After its ignition at T_1, the energy released in the burning plasma raises its temperature to T_2. The higher Z values of the charged fusion products will gradually increase the bremsstrahlung losses, whereby the ε_r curve moves upward with T_1 going up and T_2 going down until $T_1 = T_2 = T_3$. There the burn is brought to a halt. The temperature T_3 is also near the optimal burn temperature where $\varepsilon_f/\varepsilon_r$ reaches a maximum. It is determined by the maximum of the function

$$g(x) = bx^{7/2}e^{-x} \tag{6.10}$$

positioned at $x = 7/2$, with the optimal burn temperature

$$T_{opt} = \left(\frac{2}{7}k_2\right)^3 = \left(\frac{4}{7}\right)^3 T_0 \,. \tag{6.11}$$

It is near where $\langle\sigma\mathrm{v}\rangle$ reaches its maximum. For T_{opt} given by (6.11) the maximum ratio of the thermonuclear energy production over the bremsstrahlung losses is

$$g_{max} = g\left(\frac{7}{2}\right) = 2.4b \,. \tag{6.12}$$

Applying this result to the DT reaction under the assumption that all neutrons escape unattenuated (putting $\varepsilon_0 = 0.2\varepsilon_0$), one finds that $T_{opt} \simeq 1.5 \times 10^8$ °K and $g_{max} \simeq 280$, with the ignition temperature $T_1 \simeq 5 \times 10^7$ °K. For the DD reaction one has $g_{max} \simeq 2.8$, at $T_{opt} = 7 \times 10^8$ °K and $T_1 = 3 \times 10^8$ °K.

Finally, we treat the HB^{11} reaction. There we have to divide the constant b with Z^3_{eff} for this reaction. First, the average of Z^2 is $\bar{Z}^2 = 13$. Then n has to be multiplied with the number of electrons $n_e = \bar{Z}n$ where $\bar{Z} = 3$. One thus has $Z^3_{\mathit{eff}} = \bar{Z}^2\bar{Z} = 78$. One then obtains $b = 0.165$ and $g_{max} = 2.4b = 0.4$. Therefore no ignition of the HB^{11} reaction seems possible. Taking the corrected bremsstrahlung loss formula (4.78), and computing from (4.76) and (4.77) γ for the HB^{11} plasma, one finds $\gamma \simeq 0.66$ and that $g_{max} = 0.6$, still too small for ignition.

For the DD reaction $\gamma = 0.9$ and the corrected value is $g_{max} = 3.1$.

6.4 Ignition Temperature of Small Thermonuclear Assemblies

For small thermonuclear assemblies as they are typical for microexplosions, not all of the kinetic fusion product energy is dissipated within the assembly.

This alters the ignition condition. The amount of the kinetic fusion product energy dissipated is determined by the range equation (4.25) which can be written as follows

$$\lambda_0 = \frac{a\,(kT)^{3/2}}{n} \tag{6.13}$$

where a is a constant which depends on the energy, mass and charge of the fusion reaction product. For the He^4 fusion products of the DT reaction one has $a = 2.5 \times 10^{34}$ cm^{-2}erg$^{-3/2}$.

Introducing a macroscopic stopping power cross section defined by $\Sigma_0 = 1/\lambda_0$, and hence a microscopic stopping power cross section $\sigma_0 = \Sigma_0/n = (kT)^{-3/2}/a$, one can compute the probability for a charged fusion product to be stopped within the assembly. If the charged fusion product is released at the position $\mathbf{r}_1$, the probability that it will undergo a "stopping collision" at position $\mathbf{r}_2$ in the volume $d\mathbf{r}_2$ is

$$dP = \frac{\Sigma_0 \exp\left[-\Sigma_0\,|\mathbf{r}_1 - \mathbf{r}_2|\,\right]}{4\pi\,|\mathbf{r}_1 - \mathbf{r}_2|^2} d\mathbf{r}_2\ . \tag{6.14}$$

Averaging over the volume of the entire assembly one obtains the probability for the charged fusion product to have dissipated its kinetic energy within the assembly:

$$P = \frac{\Sigma_0}{4\pi} \frac{\int_{\mathbf{r}_1}\int_{\mathbf{r}_2} \dfrac{\exp\left[-\Sigma_0\,|\mathbf{r}_1 - \mathbf{r}_2|\right]}{|\mathbf{r}_1 - \mathbf{r}_2|^2}\, d\mathbf{r}_1\,\mathbf{r}_2}{\int_{\mathbf{r}_1} d\mathbf{r}_1} \tag{6.15}$$

For a sphere with radius r one has $\int_{\mathbf{r}_1} d\mathbf{r}_1 = (4\pi/3)r^3$. The integral in the numerator of (6.15) was obtained in a closed form by Dirac for a similar problem in neutron physics. With the abbreviation

$$\rho = r\Sigma_0 = \frac{rn}{a}\,(kT)^{-3/2} \tag{6.16}$$

one obtains

$$P(\rho) = 1 - \frac{3}{4\rho^3}\left[\rho^2 - \frac{1}{2} + \left(\frac{1}{2} + \rho\right)\exp\left(-2\rho\right)\right]. \tag{6.17}$$

The ignition condition, replacing (6.9), now is

$$\left.\begin{aligned} e^x &= bx^{7/2}P\left[\left(\frac{x}{x_0}\right)^{9/2}\right] \\ b &= \frac{\varepsilon_0 k_1}{4\alpha k_2^{3/2}} \\ x_0 &= k_2 k^{1/3}\left(\frac{a}{nr}\right)^{2/9} \end{aligned}\right\} \tag{6.18}$$

The ignition temperature is obtained by solving (6.18) for x, with the ignition temperature $T = (k_2/x)^3$. The ignition temperature as a function of temperature for the DT thermonuclear reaction is shown in Fig. 6.2a.

With only a fraction of the kinetic fusion product energy dissipated in the assembly, there is a minimum ignition energy with a temperature larger than the ignition temperature for an infinite assembly. The ignition energy

$$E_{ign} = \left(\frac{4\pi}{3}r^3\right) 3nkT_{ign} \tag{6.19}$$

has a sharp minimum for a particular value of r. In Fig. 6.2b $E_{ign}(r)$ is plotted for a DT sphere of solid density. There, the minimum of E_{ign} is at $r = 0.2$ cm where $E_{ign} \simeq 10^7$ Joule. The ignition temperature for $r = 0.2$ cm is $T_{ign} = 1.7 \times 10^8$ °K, larger than the minimum ignition temperature $T_{ign} = T_1 = 5 \times 10^7$ °K for $r \to \infty$. For $r = 0.2$ cm, $n = 5 \times 10^{22}$ cm^{-3} and $T = 1.7 \times 10^8$ °K one has $\rho \approx 0.1$, and $P(\rho) \simeq (3/4)\rho$, valid for $\rho \ll 1$, hence $P(\rho) \simeq 0.075$. Therefore, only 7.5% of the α-particle energy contributes to self-heating of the assembly.

Finally, we would like to know how the ignition energy depends on the density. According to (6.18) the ignition temperature remains unchanged if x_0 remains unchanged. To keep x_0 unchanged the product nr must be kept constant. If $nr = const.$, the ignition energy scales as

$$E_{ign} = E_{ign}^0 \left(\frac{n_0}{n}\right)^2 \tag{6.20}$$

where E_{ign}^0 is the ignition energy for solid state density $n_0 = 5 \times 10^{22}$ cm^{-3}.

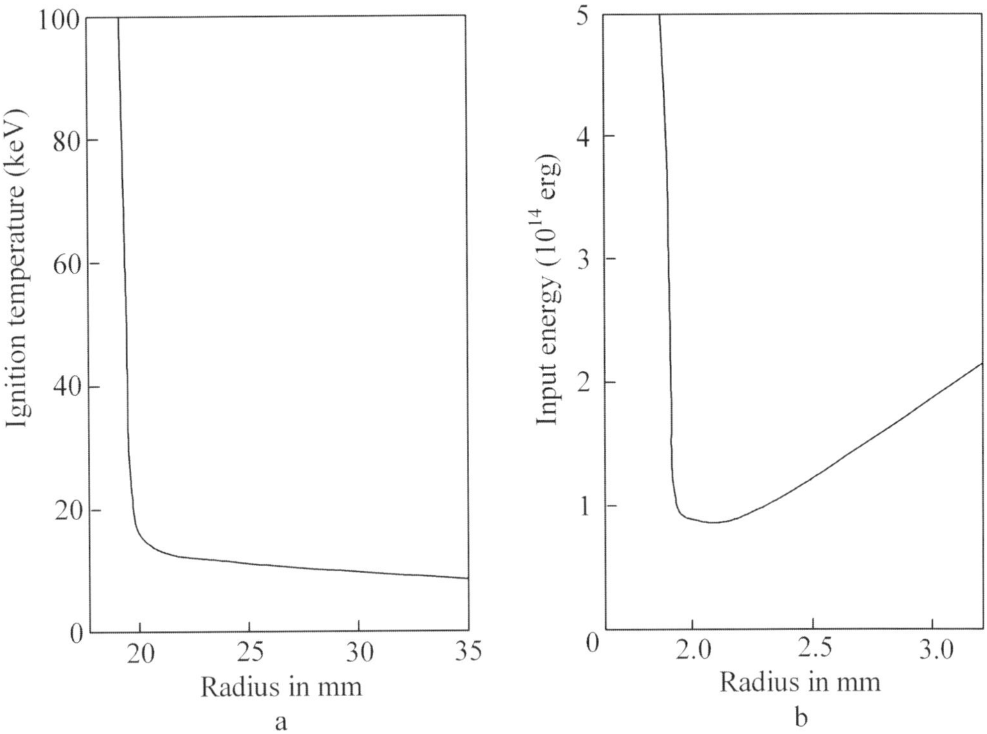

Figure 6.2: Ignition temperature and minimum ignition energy.

6.5 Ignition in the Presence of a Strong Magnetic Field

In the presence of a strong magnetic field the charged fusion products are forced into circular motion, effectively reducing their range. In addition, the topological properties of magnetic fields prefer a cylindrical symmetry for thermonuclear assemblies making use of the range-reducing effect by a strong magnetic field.

If the Larmor radius of a charged fusion product is small compared to the stopping power range (6.13), we can make the substitution $\lambda_0 \to 2r_L$ where r_L is the Larmor radius of the fusion product of charge Ze, mass MA and kinetic energy E:

$$\left.\begin{aligned} r_L &= \frac{\alpha}{H} \\ \alpha &= \frac{c}{e}\frac{(2MAE)^{1/2}}{Z} \end{aligned}\right\} \tag{6.21}$$

In one especially important case the magnetic field is generated by an axial current I flowing through a plasma column of radius r. This is the configuration of the pinch effect, and one there has

$$H = H_\phi = 0.2\frac{I}{r} \tag{6.22}$$

where I is measured in ampere. To confine the fusion products within the discharge channel of radius r, one should have $r_L \ll r$. Therefore I must be above a critical current I_c:

$$\left.\begin{aligned} I &> I_c \\ I_c &= 5\alpha \end{aligned}\right\} \tag{6.23}$$

Computer calculations indicate that a current $\sim 10I_c$ is actually needed to confine the charged fusion products within the channel, which becomes a necessary condition for thermonuclear burn. In table 6.1 the critical currents together with the values of α are put together for a number of important thermonuclear reactions. If $T > T_{ign}$ and $I \gtrsim 50\alpha$, thermonuclear burn takes place in the discharge channel.

Reaction	Fusion Product	Energy [MeV]	α [G cm]	I_c [A]
DT	He^4	3.6	2.7×10^5	1.35×10^6
DD	He^3	0.8	1.12×10^5	5.6×10^5
DD	T	1.0	2.5×10^5	1.25×10^6
DD	H	3.0	2.5×10^5	1.25×10^6
DHe^3	H	14.65	5.56×10^5	3.84×10^6
DHe^3	He^4	3.66	2.78×10^5	1.39×10^6
HB^{11}	He^4	2.93	2.84×10^5	1.24×10^6

Table 6.1: Critical ignition currents for thermonuclear reactions.

6.6 Self-Heating Following Ignition

Following ignition, the fusion products in dissipating their energy in the thermonuclear assembly raise the temperature of the latter, and with $\langle\sigma \mathrm{v}\rangle$ rising with temperature until its maximum is reached, the thermonuclear reaction is accelerated. To calculate this effect, we first repeat the estimates made in Chapter 1 with greater accuracy.

The input energy is

$$E_{in} = \left(\frac{4\pi}{3}r^3\right) 3nkT \tag{6.24}$$

and the output energy

$$E_{out} = \left(\frac{4\pi}{3}r^3\right)\left(\frac{n^2}{4}\right)\langle\sigma \mathrm{v}\rangle\varepsilon_0\tau \tag{6.25}$$

(for the DD reaction $n^2/4$ has to be replaced by $n^2/2$). Hence, one has the energy multiplication factor without self-heating

$$F = \frac{E_{out}}{E_{in}} = \frac{\langle\sigma \mathrm{v}\rangle\varepsilon_0 n\tau}{kT} \, . \tag{6.26}$$

A better estimate can be made as follows: After ignition to the temperature T, the spherical assembly disintegrates by a rarefaction wave, moving from the surface to the center. Therefore, only the central part burns for the time $t = r/c_s$ where c_s is the isentropic sound velocity which is equal to the velocity of the rarefaction wave. The outer layers of the sphere burn for a time less than $t = r/c_s$. An average burn time is given by

$$\bar{t} = \frac{3}{4\pi r^3 c_s} \int_0^r (r - r')\, 4\pi r'^2 \, dr' = \frac{r}{4c_s} \, . \tag{6.27}$$

Limiting the following calculation to the DT reaction, one has (M hydrogen atom mass):

$$c_s^2 = \frac{4kT}{3M} \tag{6.28}$$

and equating τ with $\bar{t}$

$$F = \frac{1}{32\sqrt{3}} \frac{\langle \sigma \mathrm{v} \rangle \varepsilon_0 \sqrt{M}}{(kT)^{3/2}} nr \tag{6.29}$$

or

$$F = \frac{\sqrt{3}}{240} \frac{\langle \sigma \mathrm{v} \rangle \varepsilon_0}{\sqrt{M}\,(kT)^{3/2}} \rho r \, . \tag{6.30}$$

Inserting the value $kT = 8 \times 10^{-8}$ erg, $\langle \sigma \mathrm{v} \rangle \simeq 10^{-15}$ cm^3/s, $\varepsilon_0 = 2.82 \times 10^{-5}$ erg, one finds that

$$F \simeq 7\rho r \ [\mathrm{g/cm^2}] \, . \tag{6.31}$$

For $F \geq 1$:

$$\rho r \geq 0.14 \ [\mathrm{g/cm^2}] \tag{6.32}$$

about the same as eq. (1.15) in Chapter 1.

To obtain a more accurate estimate, we have to compute the fuel depletion during burn. For the DT reaction it is given by

$$\frac{1}{2}\frac{dn}{dt} = -\frac{n^2}{4} \langle \sigma \mathrm{v} \rangle \, . \tag{6.33}$$

If n is the initial and n_f the final particle number density of unburned fuel, integration of (6.33) yields

$$\frac{1}{n_f} - \frac{1}{n} = \frac{1}{2}\langle\sigma\mathrm{v}\rangle t \ . \tag{6.34}$$

Introducing the fuel burn-up fraction

$$f_b = 1 - \frac{n_f}{n} \tag{6.35}$$

and equating t with $\bar{t}$ given by (6.27), (6.35) with the help of (6.34) can be written as follows:

$$f_b = \frac{nr}{nr + \dfrac{8c_s}{\langle\sigma\mathrm{v}\rangle}} \ . \tag{6.36}$$

or with $n = \rho/AM = \rho/2.5M$

$$f_b = \frac{\rho r}{\rho r + \dfrac{20Mc_s}{\langle\sigma\mathrm{v}\rangle}} \ . \tag{6.37}$$

At $kT = 50$ keV $= 8 \times 10^{-8}$ erg ($T = 5.8 \times 10^8$ °K), one has $c_s = 2.5 \times 10^8$ cm/s and $\langle\sigma\mathrm{v}\rangle \simeq 10^{-15}$ cm^3/s, hence

$$f_b = \frac{\rho r}{\rho r + 8.3} \ . \tag{6.38}$$

If the fuel burn up is $f_b \simeq 0.1$ (i.e. 10%), we have $\rho r \sim 1$ g/cm^2. On the other hand if $F = 1$ and hence $\rho r = 0.14$ g/cm^2, one has $f_b \simeq 0.012$, i.e. only a 1.2% fuel burn up.

To obtain a large fuel burn up one has to go to large ρr values. This poses no problem for large (fission triggered) thermonuclear explosive devices, but for small (non-fission triggered) thermonuclear assemblies it compels us to precompress the thermonuclear material to densities larger than solid state densities. For the charged fusion products to dissipate their kinetic energy in the assembly requires that $\rho r \sim 1$ g/cm^2, the same as for a large fuel burn up fraction f_b. This shows that for large values of f_b self-heating of the assembly by the charged fusion products can become important.

To treat this problem we write as before (6.24):

$$E_{in} = \left(\frac{4\pi}{3}r^3\right) 3nkT_0 \tag{6.39}$$

but now instead of (6.25):

$$\frac{dE_{out}}{dt} = \left(\frac{4\pi}{3}r^3\right)\left(\frac{n^2}{4}\right)\langle\sigma\mathrm{v}\rangle\varepsilon_0 \tag{6.40}$$

and therefore

$$\frac{1}{E_{in}}\frac{dE_{out}}{dt} = \frac{n\varepsilon_0}{12kT_0}\langle\sigma\mathrm{v}\rangle\ . \tag{6.41}$$

The heat added to the burning assembly by the charged fusion products is

$$3nk\frac{dT}{dt} = \frac{n^2}{4}\langle\sigma\mathrm{v}\rangle\varepsilon_0 fP \tag{6.42}$$

where f is the fraction of the energy ε_0 going into charged fusion products, and $P = P(n,T)$ given by (6.17). For the DT reaction $f = 0.2$. With the substitution $dt = dr/c_s = (3M/4kT)^{1/2}dr$ one obtains from (6.42):

$$\frac{40\sqrt{3}\,k^{3/2}}{\sqrt{M}\varepsilon_0}\int_{T_0}^{T_1}\frac{\sqrt{T}}{\langle\sigma\mathrm{v}\rangle P\,(n,T)}\,dT = nr\ . \tag{6.43}$$

A second relation is obtained by eliminating dt from (6.41) and (6.42). If F^* is the gain with self-heating one now has

$$F^* = \frac{E_{out}}{E_{in}} = \frac{1}{fT_0}\int_{T_0}^{T_1}\frac{dT}{P\,(n,T)}\ . \tag{6.44}$$

If at $T = T_0$, $F = F_0$ one has from (6.29):

$$F_0 = \frac{1}{32\sqrt{3}}\,\frac{\langle\sigma\mathrm{v}\rangle_0\,\varepsilon_0\sqrt{M}}{(kT_0)^{3/2}}\,nr \tag{6.45}$$

whereby one can write for (6.43)

$$F_0 = \frac{5}{4} \frac{\langle \sigma \mathrm{v} \rangle_0}{T_0^{3/2}} \int_{T_0}^{T_1} \frac{\sqrt{T}}{\langle \sigma \mathrm{v} \rangle \, P(n,T)} \, dT \; . \tag{6.46}$$

From (6.44) and (6.46) one has to eliminate T_1 to obtain a value for F^*. For this (6.44) and (6.46) have to be integrated numerically with the expression for $\langle \sigma \mathrm{v} \rangle$ given by (2.59) and $P(n, T)$ by (6.17) and (6.16).

6.7 Thermonuclear Detonation Waves

Once ignition is achieved the thermonuclear reaction can spread into adjacent material. Because this goes supersonically as in a chemical detonation wave, one may call it a thermonuclear detonation wave. However, there are important differences between a chemical detonation wave in high explosives and thermonuclear detonation waves. In thermonuclear detonation waves the range of the charged fusion products driving the wave is large compared to the thickness of the shock discontinuity, unlike the range of the combustion products in a chemical detonation wave.

Let us consider the case of a plane thermonuclear detonation wave of infinite extension. We first neglect bremsstrahlung losses. This is a good assumption for the DT reaction where the ratio of the thermonuclear energy production into charged fusion products and the bremsstrahlung loss rate has a maximum $g_{max} = 280$. The assumption appears to be not so good for the DD reaction where $g_{max} \simeq 3.1$, but only if the energy released by secondary reactions with the DD reaction products is ignored. Later on we will show how the calculation can be corrected to include bremsstrahlung losses.

With the thermonuclear burn wave propagating supersonically there is a shock discontinuity between the burnt and the unburnt thermonuclear explosive. In a somewhat different notation than the one used in chapter 5.1, the velocity of the shock discontinuity shall be v_0 with the fluid velocity behind the discontinuity equal to v. The atomic number densities in front and behind the shock shall be n_0 and n_1. From (5.10) and (5.12) one has for $\gamma = 5/3$

$$n_1 = 4 n_0 \tag{6.47}$$

$$\mathrm{v} = \frac{3}{4} \mathrm{v}_0 \tag{6.48}$$

The temperature behind the shock, called here T, is given by (5.13)

$$T = \frac{1}{2}\frac{\mathrm{v}^2}{c_\mathrm{v}} \,. \tag{6.49}$$

With c_v given by (3.6b) one has

$$T = \frac{AM}{3\,(1+Z)\,k}\,\mathrm{v}^2 \,. \tag{6.50}$$

The velocity of the shock discontinuity then is

$$\mathrm{v}_0 = \sqrt{\frac{16\,(1+Z)\,kT}{3MA}} \,. \tag{6.51}$$

From (6.50) it follows that

$$\frac{1}{2}\,MA\mathrm{v}^2 = \frac{3}{2}\,(1+Z)\,kT \tag{6.52}$$

which means that the energy deposited in the shock goes into equal amounts of kinetic fluid energy and heat, or that the kinetic energy of the fluid behind the shock is equal to its thermal energy. The total amount of energy needed to drive the shock is the sum of both, or

$$E = 3\,(1+Z)\,n_0kT \,. \tag{6.53}$$

For the shock to be sustained an energy flux

$$\phi_e = E\mathrm{v}_0 = 3\,(1+Z)\,nkT\mathrm{v}_0 = \frac{12}{\sqrt{3}}\,\frac{(1+Z)^{3/2}}{(MA)^{1/2}}\,n_0\,(kT)^{3/2} \tag{6.54}$$

must pass through the shock front.

The thermonuclear energy release rate behind the shock front is given by (2.66):

$$\varepsilon_f = \frac{1}{4}\,n_1^2\,\langle\sigma\mathrm{v}\rangle\varepsilon_0 = 4n_0^2\,\langle\sigma\mathrm{v}\rangle\,\varepsilon_0 \,. \tag{6.55}$$

(For the DD reaction the r.h.s. would be $8n_0^2\,\langle\sigma\mathrm{v}\rangle\,\varepsilon_0$.)

The flux ϕ of charged fusion products released behind the detonation front positioned at $z = 0$ is

$$\phi = \phi_0 e^{-z/\lambda_0} \tag{6.56}$$

where

$$\lambda_0 = a \, \frac{(kT)^{3/2}}{n_1} = a \, \frac{(kT)^{3/2}}{4n_0} \tag{6.57}$$

is the range of the charged fusion products behind the shock front. Of the charged fusion products only those coming from the distance

$$L = \int_{-\infty}^{z} e^{-(z-z')/\lambda_0} \, dz' = \lambda_0 \tag{6.58}$$

behind the shock contribute to the energy flux needed to drive the shock. Because of the six spatial directions and because only the fraction f of the energy is released into charged fusion products, the energy flux driving the shock is

$$\phi_c = \frac{f}{6} \, \varepsilon_f \lambda_0 = \frac{f}{6} \, n_0 \langle \sigma \mathrm{v} \rangle \, \varepsilon_0 a \, (kT)^{3/2} \, . \tag{6.59}$$

To sustain a thermonuclear detonation wave $\phi_c \geq \phi_e$, hence

$$\frac{24\sqrt{3}\,(1+Z)^{3/2}}{(MA)^{1/2} \, af\varepsilon_0 \langle \sigma \mathrm{v} \rangle} \leq 1 \, . \tag{6.60}$$

What is remarkable about this result is that both n_0 and $(kT)^{3/2}$ have dropped out.

In the presence of a strong azimuthal magnetic field set up along a thin cylindrical rod by a current larger than I_c given by (6.23), the radial confinement of the charged fusion products changes the factor $(f/6)$ into $(f/2)$, because in a one-dimensional geometry there are only two spatial directions the magnetically confined charged fusion products can go, unlike the six directions in three dimensions. There then the l.h.s. of (6.60) has to be divided by the factor $(1/2)/(1/6) = 3$. And the condition $\rho r \geq 1$ g/cm^2, has to be replaced by $\rho z \geq (1/3)$ g/cm^2. However, if the current and thus the magnetic field substantially penetrates into the rod, the required ρz value may be even smaller due to the Larmor radius reduction of the charged fusion product range.

6.8 Growing Thermonuclear Detonation Wave

Next we will treat the problem of a thermonuclear detonation wave propagating into the z-direction of a rotational symmetric horn of growing cross sectional area $S(z)$ (see Fig. 6.3). We assume that there are no energy losses by expansion or heat conduction through the sidewalls of the horn. To ignite the detonation wave an energy

$$E = 3\,(1+Z)\,n_0 k T_{ign} V_0 \tag{6.61}$$

must be deposited into the volume $V_0 \approx 2\lambda_0 S_0$, where S_0 is the initial cross sectional area of the horn with the length $2\lambda_0$ required to stop the charged fusion products over this length.

Instead of (6.54) one now has

$$\phi_e = 3\,(1+Z)\,n_0 kT \mathrm{v}_0 S(z) = \frac{12}{\sqrt{3}} \frac{(1+Z)^{3/2}}{(MA)^{1/2}}\, n_0\,(kT)^{3/2}\, S(z) \tag{6.62}$$

and instead of (6.59)

$$\phi_c = \left(\frac{f}{6}\right) \varepsilon_f \int_{z-\lambda_0}^{z} S(z)\, dz = \frac{2}{3} f n_0^2 \langle \sigma \mathrm{v} \rangle \varepsilon_0 \int_{z-\lambda_0}^{z} S(z)\, dz\ . \tag{6.63}$$

The condition for a self-sustained detonation wave is $\phi_c = \phi_e$ and one has

$$S(z) = \frac{f n_0 \langle \sigma \mathrm{v} \rangle \varepsilon_0\,(MA)^{1/2}}{6\sqrt{3}\,(1+Z)^{3/2}\,(kT)^{3/2}} \int_{z-\lambda_0}^{z} S(z)\, dz\ . \tag{6.64}$$

This integral equation has the solution

$$S(z) = S_0 e^{z/z_0} \tag{6.65}$$

where z_0 is determined as a solution of the equation

$$z_0 \left(1 - e^{-\lambda_0/z_0}\right) = 6\sqrt{3}\, \frac{(1+Z)^{3/2}\,(kT)^{3/2}}{f n_0 \langle \sigma \mathrm{v} \rangle \varepsilon_0\,(MA)^{1/2}}\ . \tag{6.66}$$

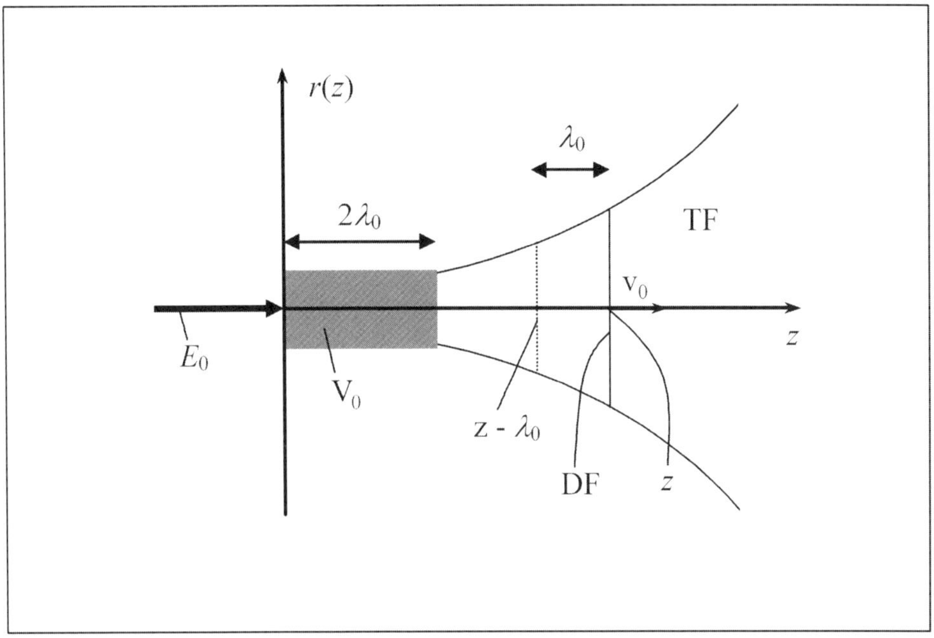

Figure 6.3: The physics of a growing thermonuclear detonation wave in a rotationally symmetric configuration. The trigger energy E_0 is deposited into the thermonuclear fuel occupying the volume V_0.

With the abbreviations

$$\left.\begin{aligned} x &= \frac{\lambda_0}{z_0} \\ c = c\,(T) &= \frac{24\sqrt{3}\,(1+Z)^{3/2}}{fa\varepsilon_0\langle\sigma \mathrm{v}\rangle\,(MA)^{1/2}} \end{aligned}\right\} \tag{6.67}$$

(6.66) is brought into the form

$$1 - e^{-x} = cx \tag{6.68}$$

as before, n_0 and $(kT)^{3/2}$ have dropped out.

The temperature dependence $c = c\,(T)$ results from the temperature dependence of $\langle\sigma \mathrm{v}\rangle$, and a growing wave with $z_0 > 0$ is possible only if $c < 1$. For $c = 1$ the solution of (6.68) is $x = 0$, for a wave of constant cross section. For $c > 1$ one has $x < 0$ and a wave of decreasing cross section. An example of a wave with a decreasing cross section is a convergent spherical detonation wave.

The largest positive root of (6.68) occurs for the smallest value of c, reached for the largest value of $\langle\sigma \mathrm{v}\rangle = \langle\sigma \mathrm{v}\rangle_{max}$, but only if $\langle\sigma \mathrm{v}\rangle$ is larger that the minimum permitted value of $\langle\sigma \mathrm{v}\rangle$ which is

$$\langle\sigma \mathrm{v}\rangle_{min} = \frac{24\sqrt{3}\,(1+Z)^{3/2}}{fa\varepsilon_0\,(MA)^{1/2}} \tag{6.69}$$

obtained by putting $c = 1$ in (6.68). With this definition of $\langle\sigma \mathrm{v}\rangle_{min}$ one can write for c in (6.67)

$$c = \frac{\langle\sigma \mathrm{v}\rangle_{min}}{\langle\sigma \mathrm{v}\rangle} \tag{6.70}$$

with the smallest value

$$c_{min} = \frac{\langle\sigma \mathrm{v}\rangle_{min}}{\langle\sigma \mathrm{v}\rangle_{max}}\ . \tag{6.71}$$

For the DT thermonuclear reaction one has $Z = 1$, $A = 2.5$, $\varepsilon_0 = 17.6$ MeV $= 2.82 \times 10^{-5}$ erg, $f = 0.2$, $a \simeq 2.5 \times 10^{34}$ cm^{-2}erg$^{-3/2}$ and $\langle\sigma \mathrm{v}\rangle_{max}$

$\simeq 10^{-15}$ cm^3/s, reached at $T = 8 \times 10^8$ °K, and finally $\langle\sigma v\rangle_{min} \simeq 4.05 \times 10^{-16}$ cm^3/s, reached at $T \simeq 1.7 \times 10^8$ °K. One obtains $c_{min} = 0.4$, where $x = x_{max}$ $\simeq 2.2$. For a wave propagating in solid DT where $n_0 = 5 \times 10^{22}$ cm^{-3} and $T = 8 \times 10^8$ °K, one finds that $\lambda_0 = 4.5$ cm and $z_0 = \lambda_0/x_{max} \simeq 2$ cm. A thermonuclear detonation wave in DT can rapidly grow, but in thermonuclear microexplosions a growing wave requires compression to high densities to make λ_0 and z_0 much smaller.

In the case of the DD reaction treated next, the r.h.s. of (6.69) must be divided by 2 (because we have here a reaction between identical particles). And we have here to consider the range of three kinds of charged fusion products: protons, tritons and He^3 particles, each of them possessing a different kinetic energy. Averaging the ranges over the kinetic energy of these fusion products, one obtains an average range constant $a = 4.0 \times 10^{34}$ cm^{-2}erg$^{3/2}$. For the maximum value of $\langle\sigma v\rangle$ we must choose a temperature where the bremsstrahlung losses are not too large, otherwise our approximation becomes questionable. At $T = 3.5 \times 10^9$ °K (≈ 300 keV), one has $\langle\sigma v\rangle_{max} \simeq 5 \times 10^{-17}$ cm^3/s. For the value of ε_0 we have to average over the two equally probable branches of the DD reaction (see chapter 2.2) with the result that $\bar{\varepsilon}_0 = 3.63$ MeV $= 5.8 \times 10^{-6}$ erg. We furthermore have here $f = 0.66$, $Z = 1$ and $A = 2$. With these numbers we find $c_{min} = 1.4 > 1$. It therefore seems to follow that a growing thermonuclear detonation burn wave in deuterium is not possible. This however, is only true as long as one neglects the secondary reactions with tritium and helium 3, which are burn products of the DD reaction. Provided $\langle\sigma v\rangle_{DT} \gg \langle\sigma v\rangle_{DD}$ and $\langle\sigma v\rangle_{DHe^3} \gg \langle\sigma v\rangle_{DD}$, the tritium and helium 3 reaction fusion products will rapidly burn with D behind the detonation front, adding the energy of their charged fusion products to ε_0. We thus have to deal with three additional charged fusion products, two α-particles and one proton, one of the α-particles from the DT reaction, the other α-particles and the proton from the DHe^3 reaction. To account for the energy of the secondary charged fusion products we simply set in the expression for c, $f\varepsilon_0 = \varepsilon_t$, where ε_t is the kinetic energy of all charged fusion products. We find that $\varepsilon_t = 13.4$ MeV $= 2.14 \times 10^{-5}$ erg. The approximate expression for c then is

$$c \simeq \frac{24\sqrt{3}}{\bar{a}\varepsilon_t \langle\sigma v\rangle_{DD} (M)^{1/2}} \, . \tag{6.72}$$

At the temperature $T = 3.5 \times 10^9$ °K where $\langle\sigma v\rangle_{DD}$ is largest, $\langle\sigma v\rangle_{DD}$ is much smaller than $\langle\sigma v\rangle_{DT}$ and $\langle\sigma v\rangle_{DHe^3}$. With $\langle\sigma v\rangle_{DD} \simeq 5 \times 10^{-17}$ cm^3/s at this temperature one finds $c = 0.4$ with $x \simeq 2.2$, as for the DT thermonuclear detonation wave. This shows that a rapidly growing detonation wave is possible in deuterium, but in comparison to the DT detonation wave it requires a much higher temperature.

To take into account the bremsstrahlung losses one has to make the substitution:

$$\begin{aligned} f\varepsilon_0 \frac{n^2}{4}\langle\sigma \mathrm{v}\rangle &\longrightarrow f\varepsilon_0 \frac{n^2}{4}\langle\sigma \mathrm{v}\rangle - \gamma\alpha Z^3 n^2 \sqrt{T} \\ &= f\varepsilon_0 \frac{n^2}{4}\langle\sigma \mathrm{v}\rangle \left(1 - \frac{1}{gf}\right) \end{aligned} \tag{6.73}$$

where g is the ratio of the thermonuclear energy production rate to the bremsstrahlung loss rate, defined in (6.10). The substitution (6.73) is equivalent with the introduction of a reduced $\langle\sigma \mathrm{v}\rangle$ value

$$\langle\sigma \mathrm{v}\rangle^* = \langle\sigma \mathrm{v}\rangle \left(1 - \frac{1}{gf}\right) . \tag{6.74}$$

For the DT reaction $gf = 280 \times 0.2 = 56$, and thus $\langle\sigma \mathrm{v}\rangle^* = 0.92\langle\sigma \mathrm{v}\rangle$. For the DD reaction we must take the secondary reactions into account and also the increased bremsstrahlung from the He^3-DD fusion product. As before we set $f\varepsilon_0 \to \varepsilon_t$ and hence $f = 3.8$. For the increased bremsstrahlung loss we estimate that 1/3 of the ions are He^3 ions with $Z = 2$, hence we put $Z^3 \to 2^3/3 = 8/3$. We thus have to put $gf \to g\left(\varepsilon_t/\varepsilon_0\right)/Z^3 = 3.1 \times 3.8/(8/3) \simeq 4.4$, with the result that $\langle\sigma \mathrm{v}\rangle^* = 0.77\langle\sigma \mathrm{v}\rangle$. For the DT reaction one has $c_{min} \to 0.4/0.92 = 0.43$, and for the DD reaction $c_{min} \to 0.4/0.77 = 0.52$. Both are well below $c = 1$, as required for a thermonuclear detonation wave.

6.9 Ignition and Thermonuclear Gain for Spherical Assemblies

The gain of a thermonuclear explosive is defined as the ratio of the energy output E_{out} and the energy input E_{in} for ignition:

$$G = \frac{E_{out}}{E_{in}} \tag{6.75}$$

while the yield Y is defined by

$$Y = E_{out} \,. \tag{6.76}$$

For thermonuclear microexplosions in particular, a large gain with a small yield is desirable. Of special interest is a spherical thermonuclear assembly because it permits its ignition by a convergent shock wave.

We will now derive the scaling laws for the gain of such an assembly by making the simplifying assumption that the temperature rise in a convergent shock wave is given by

$$T = T_0 \left(\frac{R}{r} \right) \tag{6.77}$$

where at the initial radius R the temperature is T_0. The thermal energy input by the convergent shock wave is

$$E_{in} = \int_0^R 3nkT \times 4\pi r^2 \, dr \tag{6.78}$$

and because of (6.77) this is

$$E_{in} = 3nkT_0 \times 2\pi R^3 \tag{6.79}$$

or

$$E_{in} = 6\pi \, (kT) \, (nr) \, R^2 \,. \tag{6.80}$$

The energy output at the other hand is

$$E_{out} = \left(\frac{4\pi}{3} R^3 \right) \times n\varepsilon_0 \tag{6.81}$$

provided the assembly is ignited at the center and consumed by an outgoing radial detonation wave following ignition. For ignition to occur the nr product at the center must be above the critical value for thermonuclear burn, with the ignition temperature reached at the distance r from the center of convergence.

From (6.80) and (6.81) one has

$$G = \frac{2}{9}\frac{n\varepsilon_0 R}{(kT)\,(nr)} \tag{6.82}$$

and from (6.80)

$$R = \sqrt{\frac{E_{in}}{6\pi\,(kT)\,(nr)}}\,. \tag{6.83}$$

Inserting (6.83) into (6.82) one has

$$G = \frac{1}{9}\sqrt{\frac{2}{3\pi}}\,\frac{n\varepsilon_0}{(kT)^{3/2}\,(nr)^{3/2}}\,E_{in}^{1/2}\,. \tag{6.84}$$

Repeating the same calculation with the more accurate temperature dependence of a convergent shock wave, where $T \propto r^{-0.9}$ (see equation 5.29), one finds that

$$G \propto E_{in}^{0.47}\,. \tag{6.85}$$

The same result is obtained by computer calculations.

The disadvantage of the convergent shock wave approach is the large, unnecessary amount of energy needed for the convergent shock wave to reach the ignition temperature in the center of convergence. It has therefore been suggested to "drill" a hole into the center of the sphere, for example with an intense laser beam, with the radius r of the hole large enough to satisfy the nr condition for ignition, while at the same time heating the plasma in the hole to the ignition temperature T. For this so-called fast ignitor concept one has

$$E_{in} = \pi r^2 R \times 3nkT \tag{6.86}$$

with E_{out} as before. One has now

$$G = \frac{4}{3}\frac{R^2\varepsilon_0}{r^2\,(kT)} \tag{6.87}$$

or expressing R by E_{in} with (6.86)

$$G = \frac{4}{27\pi^2} \frac{\varepsilon_0}{(kT)^3 (nr)^2 r^4} E_{in}^2 . \tag{6.88}$$

Unlike the gain formula (6.84), where $G \propto \sqrt{E_{in}}$, the $G \propto E_{in}^2$ dependence is a significant improvement. This is of particular importance for thermonuclear microexplosions where a large gain is desirable.

For large thermonuclear explosive devices one may place a small fissile sphere at the center ("spark plug"), which upon compression by the convergent shock wave becomes critical. Still better is a fissile sphere with an implanted lattice of Li^6D. These concepts are important for compact thermonuclear explosive devices. For thermonuclear microexplosions one may put DT in the center surrounding it with D. Once a reaction is ignited in DT, it can launch a detonation wave into D.

For the DT reaction where $\varepsilon_0 = 2.8 \times 10^{-15}$ erg, $T \simeq 10^8$ °K, $nr = 2.4 \times 10^{23}$ cm^{-2} ($\rho r \simeq 1$ g/cm^2), one finds from (6.84) that $G \simeq 7.2 \times 10^{-30} n\sqrt{E_{in}}$. For $E_{in} = 10^{13}$ erg (1MJ), one has for uncompressed DT $G \simeq 1$, but for 10^3 times compressed DT one has $G \simeq 10^3$. With the fast igniter the gain can be much larger. According to (6.88) its effectiveness depends on the smallness of the hole drilled by the laser, which cannot be made smaller than a laser light wavelength. Assuming that $r \simeq 5 \times 10^{-3}$ cm, (about one order of magnitude larger than the laser wave length), one finds that $G \sim 10^4$.

6.10 Various Methods to Achieve Ignition

The two methods to achieve ignition, (a) by a convergent shock wave and (b) by creating a hot spot or "spark", are just two examples out of a variety of many other possibilities. All these methods can be subdivided into three principal categories and combinations of them:

1. Ignition by ablatively driven implosion.

2. Fast ignition by a powerful "spark".

3. Ignition by hypervelocity impact.

Ignition by ablatively driven implosion requires the deposition of energy on the surface of a sphere. In ablating its surface it launches a convergent shock

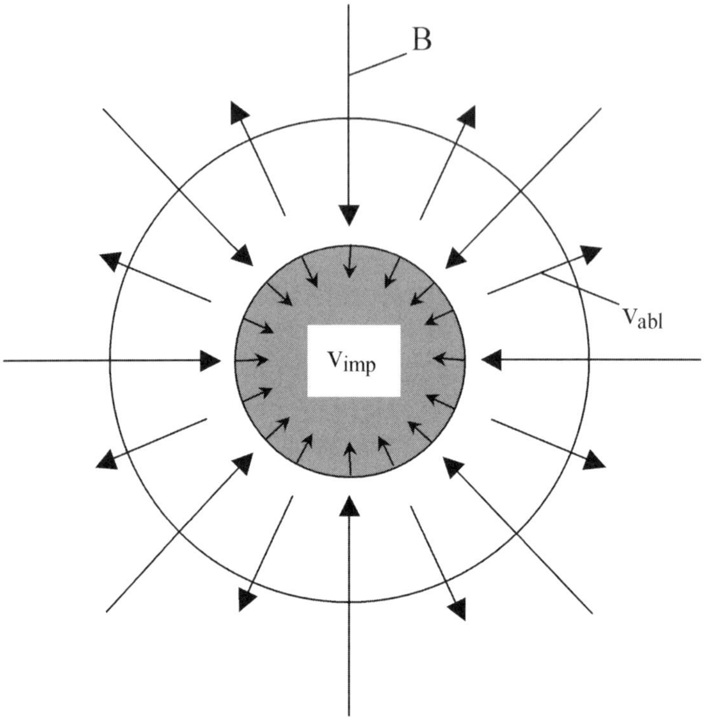

Figure 6.4: Ablation implosion of thermonuclear target bombarded by beams B (either laser or charged particles) from many sides. v_{imp} is the implosion velocity, and v_{abl} is the ablation product velocity.

wave (see Fig. 6.4). The energy can be in the form of photons, electrons, ions or even a stream of fast moving microparticles. They draw their energy from what is called the "drivers", or what may be also called the "ignitor". For large thermonuclear explosive devices the ignitor is an exploding fission bomb, releasing a copious stream of soft x-ray photons. For nonfission ignited thermonuclear microexplosions the ignitor must deliver an energy in excess of 10^6 J, in less than 10^{-8} sec onto a target area smaller than 1 cm^2, with a power in excess of 10^{14} Watt and a power flux density larger than 10^{14} Watt/cm^2. Large thermonuclear explosions can in principle be set off by a fissionless ignitor through staging, beginning with the fissionless ignition of a thermonuclear microexplosion.

In the ablatively driven implosion scheme, the driver energy E_D is deposited in a thin layer of thickness δ on the surface of the sphere of radius R. If this layer is heated to the temperature T_{ab}, the driver energy deposited is

$$E_D = 4\pi R^2 \delta \cdot 3nkT_{ab} \tag{6.89}$$

with the heated surface material ablated with the velocity $u \propto \sqrt{T_{ab}}$. As a result, a convergent shock wave is launched into the sphere, accelerating its material to the velocity $\mathrm{v} \propto \sqrt{T_0}$, where T_0, different from T_{ab}, is the temperature of the shock at the surface of the sphere. From the rocket equation

$$\mathrm{v} = u \ln\left(\frac{m_0}{m}\right) \tag{6.90}$$

where m_0 is the initial and m the final mass of the accelerated material of the sphere, and from the energy equation for the ablated material

$$E = \frac{1}{2}\left(m_0 - m\right) u^2 \tag{6.91}$$

one obtains the fraction η of the energy going into the mass m accelerated to the velocity v

$$\eta = \frac{m\mathrm{v}^2}{2E} = \frac{(\mathrm{v}/u)^2}{(e^{\mathrm{v}/u} - 1)} \simeq \frac{\mathrm{v}}{u} = \sqrt{\frac{T_0}{T_{ab}}}\,, \qquad \mathrm{v} \ll u\,. \tag{6.92}$$

Since E is equal the driver energy E_D, one has to equate ηE_D with E_{in} given by (6.79). One obtains

$$\delta = \frac{1}{2}\sqrt{\frac{T_0}{T_{ab}}}\,R\,. \tag{6.93}$$

For a heated thin surface layer, a large ablation temperature seems preferable, however, the temperature cannot be made too large, otherwise hot electrons are produced in the layer, preheating the interior of the sphere, making it more difficult to compress it to high densities. (This problem though is only important for thermonuclear microexplosions.)

If the "ablator" (i.e. the material to be ablated) is a layer of a high atomic number material with the atomic weight A_{ab}, and if the material of the sphere has an atomic number A, one has

$$\delta = \frac{1}{2}\sqrt{\frac{T_0 A}{T_{ab} A_{ab}}}\, R\,. \tag{6.94}$$

For $A \sim 2$ and $A_{ab} \sim 200$, δ turns out to be ten times smaller than R.

In the fast ignitor concept, an ignition spark is delivered in or near the center of the precompressed assembly. It appears there are several ways this can be done. One proposal is to use very intense laser beams. Normally, if the frequency of a laser beam is less than the electron plasma frequency, the laser radiation cannot penetrate the plasma and is reflected. With $\nu = \nu_p$ where ν is the laser frequency and $\nu_p = \omega_p/2\pi$ the electron plasma frequency ($\omega_p = \sqrt{4\pi n_e e^2/m}$), one can define a critical electron number density n_c above which the laser beam cannot penetrate the plasma

$$n_c = \left(\frac{\pi m}{e^2}\right)\nu^2\,. \tag{6.95}$$

This though is valid only if the radiation pressure of the laser beam does not exceed the pressure $n_c mc^2$ of the heated electron gas, or if the laser beam intensity I_L is smaller than a critical intensity I_c:

$$I_L < I_c \simeq n_c mc^3 = \left(\frac{\pi m^2 c^3}{e^2}\right)\nu^2\,. \tag{6.96}$$

With $\lambda = c/\nu$ one has

$$\left.\begin{aligned} I_c &\simeq 2.8\times 10^{-11}\lambda^{-2}\ [\mathrm{W/cm^2}] \\ n_c &\simeq 10^{13}\lambda^{-2}\ [\mathrm{cm^{-3}}] \end{aligned}\right\} \tag{6.97}$$

If $I_L > I_c$ the laser "drills" a hole into the plasma, reducing its density from $n > n_c$ down to n_c (see Fig. 6.5a). For yellow laser light with $\lambda = 6\times10^{-5}$ cm and $n_c = 3\times10^{21}$ cm^{-3} one has $I_L \sim 3\times10^{19}$ [W/cm^2]. Instead of an intense laser beam one may use for fast ignition an intense relativistic electron beam provided the small stopping length (4.57a) by the two-stream instability can be realized.

The condition for thermonuclear detonation wave ignition in DT is $\rho r \gtrsim 1$ g/cm^2 or $nr \gtrsim 2.5\times10^{23}$ cm^{-2}. Therefore, if the DT is compressed $\sim 10^3$

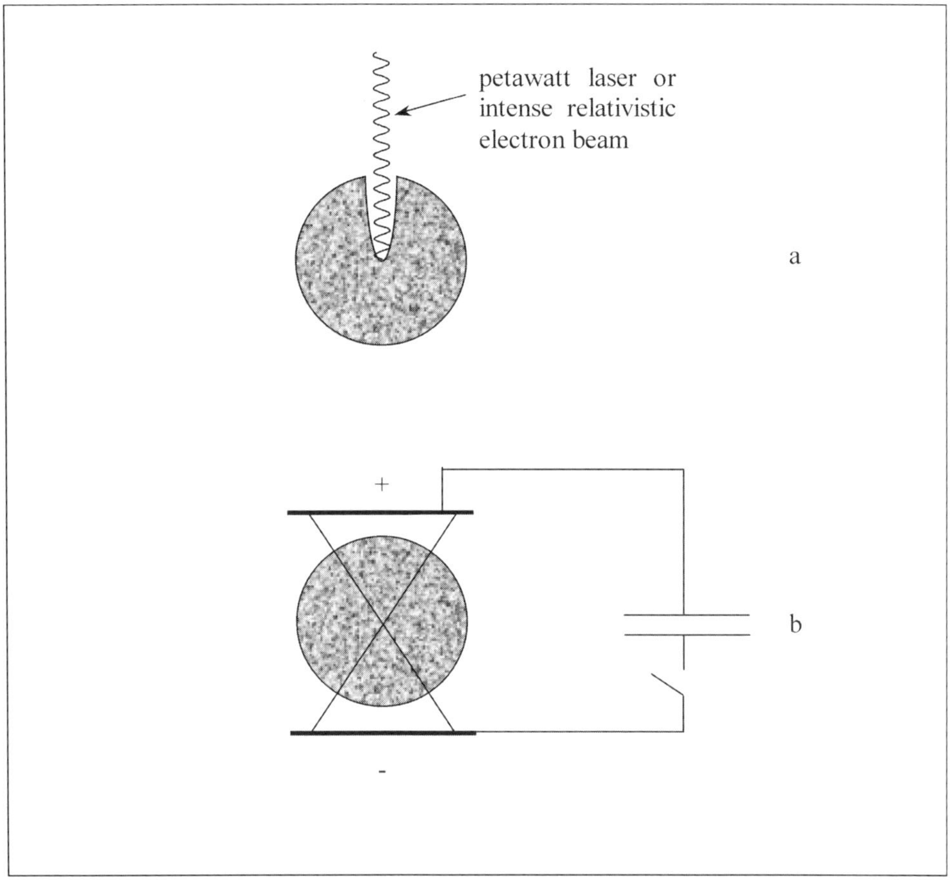

Figure 6.5: Fast ignition: (a) With a petawatt laser or intense relativistic electron beam and a compressed target. (b) With an x-pinch and a magnetized target.

times solid density one has $r \geq 5 \times 10^{-3}$ cm for the radius of the ignition spark. For the laser light intensity $I_L \sim 3 \times 10^{19}$ W/cm^2, the laser power required would thus be $P \geq I_L \pi r^2 \simeq 2 \times 10^{15}$ W. Lasers with this power (petawatt lasers) have recently been developed. High voltage intense relativistic electron beams may be even better.

A power in excess of $\sim 10^{15}$ W can in principle be also reached with intense ion beams, with a beam current and voltage of $\sim 10^7$ ampere and 10^8 volts. And the same power could be reached by fast microparticles with a radius of 5×10^{-3} cm having a velocity of $\gtrsim 10^7$ cm/s.

At lower densities fast ignition would require less power. This is the case for magnetized targets. One possibility might be to combine magnetized targets with the x-pinch, the latter consisting of two crossed thin wires over which a fast capacitor bank is discharged. Placing the cross point inside a magnetized target, a thermonuclear detonation, assisted by the magnetic field, becomes possible. One might even use the x-pinch both for fast ignition and producing the strong magnetic field (see Fig. 6.5b).

Placing a small deuterium-tritium sphere in the center of a sphere made of deuterium only, a detonation wave in the deuterium can be launched from the ignited DT, provided the DT is heated to the DT ignition temperature and $\rho r > 1$ gm/cm^2. This demonstrates the importance of fast ignition as a road to almost pure deuterium burn.

Finally, in the impact fusion concept, ignition in combination with compression is possible through the impact of a fast moving projectile (see Fig. 6.6). To reach a power of $\gtrsim 10^{14}$ Watt/cm^2, the projectile (with a density ~ 10 g/cm^3) would have to move with a velocity of $\gtrsim 10^7$ cm/s. For an ignition energy of $\sim 10^7$ J, the projectile mass would have to be of the order ~ 1 g. At a projectile velocity of $\mathrm{v}_p \sim 10^7$ cm/s, the impact pressure is $p \sim \rho \mathrm{v}_p^2 \sim 10^{15}$ dyn/cm^2. If one isentropically compresses a cold DT target with this pressure, its density would be ~ 100 times solid density. However, if the pressure of $\sim 10^{15}$ dyn/cm^2 is balanced by the DT plasma pressure $p = 2nkT$ at $T = 10^8$ °K, one only has $n = n_0 = 5 \times 10^{22}$ cm^{-3}, equal solid state density, but the projectile can here hold together the hot DT plasma much longer, provided the projectile mass is large enough, or at least as large as the mass of the DT. In this case, corresponding to a "momentum-rich" particle beam, the inertial confinement time is equal to $\tau \sim R/\mathrm{v}_p$, where R is the target radius. Without the momentum-rich projectile $\tau \sim R/\mathrm{v}$, with $\mathrm{v} \sim 10^8$ cm/s the thermal expansion velocity of the DT plasma at $T \sim 10^8$ °K. With $\mathrm{v}/\mathrm{v}_p \simeq 10$, the inertial confinement time is much larger

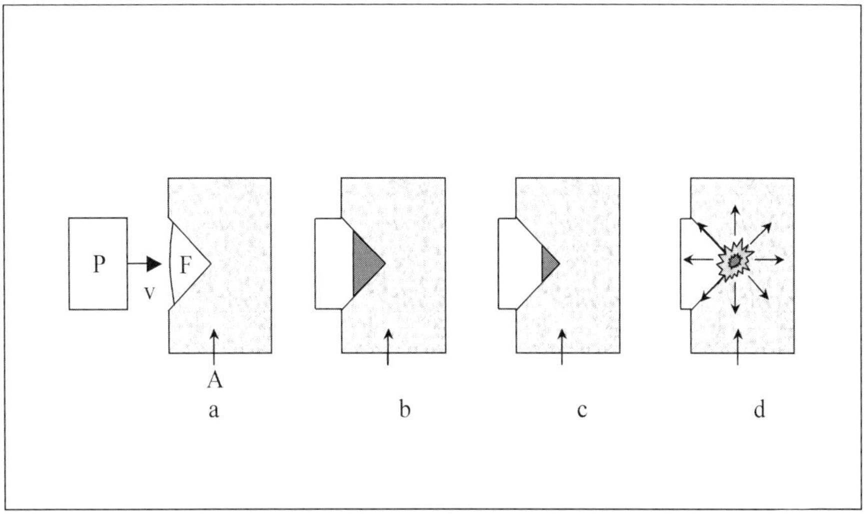

Figure 6.6: Impact fusion concept and sequence of events: Projectile P strikes anvil A holding thermonuclear fuel F in conical cavity. The configuration **a-d**, shows: **(a)** the moment before impact, **(b)** shock heating, **(c)** isentropic compression up to ignition, and **(d)** thermonuclear burn.

by the same factor.

To compute the burn-up fraction for impact fusion we have to substitute in (6.36) v_p for c_s and obtain instead of (6.38)

$$f_b = \frac{\rho r}{\rho r + 0.83} \,. \tag{6.98}$$

For a burn-up fraction $f_b \simeq 0.1$, we need only $\rho r \geq 0.2$, which means that volume ignition is here sufficient.

Upon impact, the projectile preheats the DT to the initial temperature given (5.13):

$$kT_0 = \frac{1}{2} A M \mathrm{v}_p^2 \tag{6.99}$$

where $AM = 4.15 \times 10^{-24}$ g is the DT ion mass. Following the impact, the DT is further heated by isentropic compression, according to

$$\frac{T}{T_0} = \left(\frac{p}{p_0}\right)^{\gamma-1} = \left(\frac{p}{p_0}\right)^{2/3} . \tag{6.100}$$

We now set $p = 2n_0kT$, where n_0 is the solid state density, and $T \simeq 10^8$ °K the DT ignition temperature. With $p_0 = p = 2nkT_0$, we obtain

$$\frac{n}{n_0} = \sqrt{\frac{T_0}{T}} = \frac{\mathrm{v}_p}{\mathrm{v}} \tag{6.101}$$

where $\mathrm{v}_p = (2kT_0/AM)^{1/2}$, $\mathrm{v} = (2kT/AM)^{1/2}$. For $\mathrm{v}_p/\mathrm{v} \simeq 0.1$, and $T \simeq 10^8$ °K, one must have $T_0 \simeq 10^6$ °K, and $n/n_0 = 0.1$. With $\rho r \approx 0.2$ g/cm^2, i.e. $nr \simeq 5 \times 10^{22}$ cm^{-3}, one has for $n = n_0 = 5 \times 10^{22}$ cm^{-3}, $r \simeq 1$ cm. At a 10% burn-up, the yield of this "microexplosion" would be $\simeq 5.6 \times 10^{17}$ erg ~ 10 tons of TNT.

We must check if our assumption of isentropic compression is satisfied taking radiation and heat conduction losses into account. The radiation loss time by bremsstrahlung is

$$\tau_R = \frac{3nkT}{\varepsilon_r} = 2.9 \times 10^{11} \frac{\sqrt{T}}{n} \text{ [s]} \tag{6.102}$$

where ε_r is given by (4.67b). The characteristic compression velocity needed to overcome the bremsstrahlung then is

$$\mathrm{v}_b = \frac{r}{\tau_R} = 3.4 \times 10^{-12} T^{-1/2} (nr) \text{ [cm/s]} . \tag{6.103}$$

For $T = 10^8$ °K and $nr = 5 \times 10^{22}$ cm^{-2}, one finds $\mathrm{v}_b \simeq 1.7 \times 10^7$ cm/s.

With $T \propto n^{2/3}$, τ_R scales as $n^{-2/3}$. And with $r \propto n^{-1/3}$, v_b scales as $n^{1/3}$. If at the completion of the isentropic compression $\mathrm{v}_b \sim \mathrm{v}_p \sim 2 \times 10^7$ cm/s the bremsstrahlung losses can just be overcome. At the beginning of the compression process, where $n \simeq (1/10)n_0 = 5 \times 10^{21}$ cm^{-3}, these losses are half as large.

The heat conduction losses into the projectile are obtained from the equation

$$3nk\frac{\partial T}{\partial t} = \kappa \nabla^2 T \tag{6.104}$$

where κ is given by (4.18). With $\ln \Lambda \simeq 10$ one has for a DT plasma

$$\kappa = 2 \times 10^{-6} T^{5/2} \text{ [erg/s °K cm]} . \tag{6.105}$$

Putting $\nabla^2 T \sim T/r^2$ and $\partial T/\partial t \sim T/\tau_c$ where τ_c is the characteristic heat conduction loss time, one has

$$\tau_c = \frac{3nkr^2}{\kappa} = 2.1 \times 10^{-10} T^{-5/2} r\,(nr) \text{ [s]} \tag{6.106}$$

and for the characteristic compression velocity to overcome the heat conduction losses

$$\mathrm{v}_c = \frac{r}{\tau_c} = \frac{4.8 \times 10^9 T^{5/2}}{nr} \text{ [cm/s]} . \tag{6.107}$$

For $T = 10^8$ °K and $nr = 5 \times 10^{-22}$ cm^{-2} one has $\mathrm{v}_c \geq 10^7$ cm/s.

In summary, to overcome the bremsstrahlung and heat conduction losses, a projectile velocity $\mathrm{v}_p \gtrsim 2 \times 10^7$ cm/s is needed.

In the presence of a strong magnetic field the heat conduction losses can be substantially reduced, but to fully exploit this possibility, the magnetic lines of force must be closed within the plasma as shown in Fig. 6.7, with the magnetic field compressed together with the DT. A velocity of $\sim 2 \times 10^7$ cm/s is difficult to reach, but a velocity of $\sim 10^6$ cm/s can already be reached with chemically driven two-stage light gas guns, raising the question if a lowering of the velocity for $\mathrm{v}_p \sim 2 \times 10^7$ cm/s down to $\mathrm{v}_p \sim 2 \times 10^6$ cm/s is possible with the heat insulating effect of a strong magnetic field. According to (6.103) lowering v_b by a factor 20 from $\mathrm{v}_b \sim 2 \times 10^7$ cm/s down to $\mathrm{v}_b \sim 2 \times 10^6$ cm/s, allows us to reduce nr from $nr = 5 \times 10^{22}$ cm^{-2} to $nr = 2.5 \times 10^{21}$ cm^{-2}, but such a reduction of the nr value would increase v_c from 10^7 cm/s to $\mathrm{v}_c \sim 2 \times 10^8$ cm/s, were it not for the heat insulating effect of a strong magnetic field. In the presence of a strong magnetic field κ in (6.106) has to be replaced by $\kappa_\perp$, given by (4.28b). For a DT plasma one has (with $\ln \Lambda \simeq 10$)

$$\kappa_\perp = 2.4 \times 10^{-16} \frac{n^2}{\sqrt{T} H^2} \text{ [cgs]} \tag{6.108}$$

and for the heat conduction loss time

$$\tau_\perp = 1.72 \sqrt{T}\, \frac{(Hr)^2}{n} \text{ [s]} \tag{6.109}$$

with $nr = 2.5 \times 10^{21}$ cm^{-2} this becomes

$$\tau_\perp = 6.9 \times 10^{-22} \sqrt{T} \, (Hr)^2 \, r \ [\mathrm{s}] \ . \tag{6.110}$$

For the heat conduction losses to be insignificant $\tau_\perp \gg r/\mathrm{v}_p$, or

$$\mathrm{v}_\perp = \frac{r}{\tau_c} = 1.45 \times 10^{21} T^{-1/2} \, (Hr)^{-2} \ll \mathrm{v}_p \ . \tag{6.111}$$

For $T = 10^8$ °K, $\mathrm{v}_p = 10^6$ cm/s, one has $Hr \gg 3.8 \times 10^5$ Gcm.

If the heat conduction losses are determined by the microturbulence Bohm diffusion, $\kappa_\perp$ has to be replaced by κ_B, given by (4.37). There one has

$$\tau_B = \frac{6e}{ck} \frac{r^2 H}{T} = 6.9 \times 10^{-4} \frac{r^2 H}{T} \ [\mathrm{s}] \tag{6.112}$$

and with $\tau_B \gg r/\mathrm{v}_p$ one has

$$1.45 \times 10^3 \, T \, (Hr)^{-1} \ll \mathrm{v}_p \ . \tag{6.113}$$

For $T = 10^8$ °K, $\mathrm{v}_p = 10^6$ cm/s, one there has $Hr \gg 1.45 \times 10^5$ Gcm. It follows that for cm-size cavities the heat conduction losses are insignificant for multi-megagauss fields even with the Bohm diffusion. But to exploit this possibility, the time r/v_p must be short compared with the time the magnetic field can diffuse out of the cavity. If the wall of the cavity is a good electrical conductor, the loss of the magnetic field into the cavity wall is determined by (3.85b). Putting in (3.85b) $\nabla^2 H \sim H/r^2$, and $\partial H/\partial t \sim H/\tau_d$, where τ_d is the diffusion time of the magnetic field into the wall, one has

$$\tau_d = 4\pi\sigma \frac{r^2}{c^2} \ . \tag{6.114}$$

The losses into the wall are insignificant if $\tau_d \gg r/\mathrm{v}_p$ or if

$$\frac{c^2}{4\pi\sigma r} \ll \mathrm{v}_p \ . \tag{6.115}$$

For a good conductor like copper $\sigma \simeq 10^{18}$ s^{-1}, and inequality (6.115) implies that $r \gg 10^{-4}$ cm. Therefore, for megagauss fields and spherical cm-size cavities, the heat conduction and magnetic field diffusion losses can be ignored within the cavity implosion time $r/\mathrm{v}_p \sim 10^{-6}$ s. It is therefore a reasonably good approximation to assume that the temperature of the DT plasma in the cavity rises as it does under the law of isentropic compression:

$$\frac{T}{T_0} = \left(\frac{r_0}{r}\right)^2 \tag{6.116}$$

and the magnetic field under the law of magnetic flux conservation:

$$\frac{H}{H_0} = \left(\frac{r_0}{r}\right)^2 \tag{6.117}$$

where r_0 is the initial cavity radius $r_0 > r$, where $T = T_0$, and $H = H_0$.

Equating the kinetic energy density of the projectile $\frac{1}{2}\rho_p \mathrm{v}_p^2$, where ρ_p is the density of the projectile, with the magnetic energy density $H^2/8\pi$, one obtains the maximum magnetic field which can be generated by magnetic flux compression through the impact of a fast moving projectile:

$$H = \sqrt{4\pi\rho_p}\mathrm{v}_p \, . \tag{6.118}$$

For $\rho_p = 7$ g/cm^3 (iron) and $\mathrm{v}_p = 10^6$ cm/s, one has $H \simeq 3 \times 10^7$ G.

During the implosive compression the magnetic lines of force become closed by field reversal even without a central conducting rod as in Fig. 6.7a. For field reversal to occur $H \sim 2H_0$ or $r \lesssim r_0/\sqrt{2}$.

One other condition which has to be satisfied is that the plasma pressure $2nkT$ does not exceed the magnetic pressure $H^2/8\pi$, which implies that

$$n \leq \frac{H^2}{16\pi kT} \, . \tag{6.119}$$

For the numbers $H \simeq 3 \times 10^7$ G, $kT \simeq 10^{-8}$ erg, one finds that $n \leq 2 \times 10^{21}$ cm^{-3}. Together with $nr = 2.5 \times 10^{21}$ cm^{-2}, it follows that $r \sim 1$ cm.

For a magnetic field $H \simeq 3 \times 10^7$ G and $r \simeq 1$ cm, one computes from (6.110) that $\tau_\perp \approx 10^{-2}$ s. And from (6.112) that $\tau_B \simeq 2 \times 10^{-4}$ s. Both $\tau_\perp$ and τ_B are large compared to the implosion time $\tau_{imp} = r/\mathrm{v}_p \simeq 10^{-6}$ s.

To obtain a value for the needed kinetic projectile energy, we must equate it with the sum of the magnetic and thermal plasma energy of the imploded

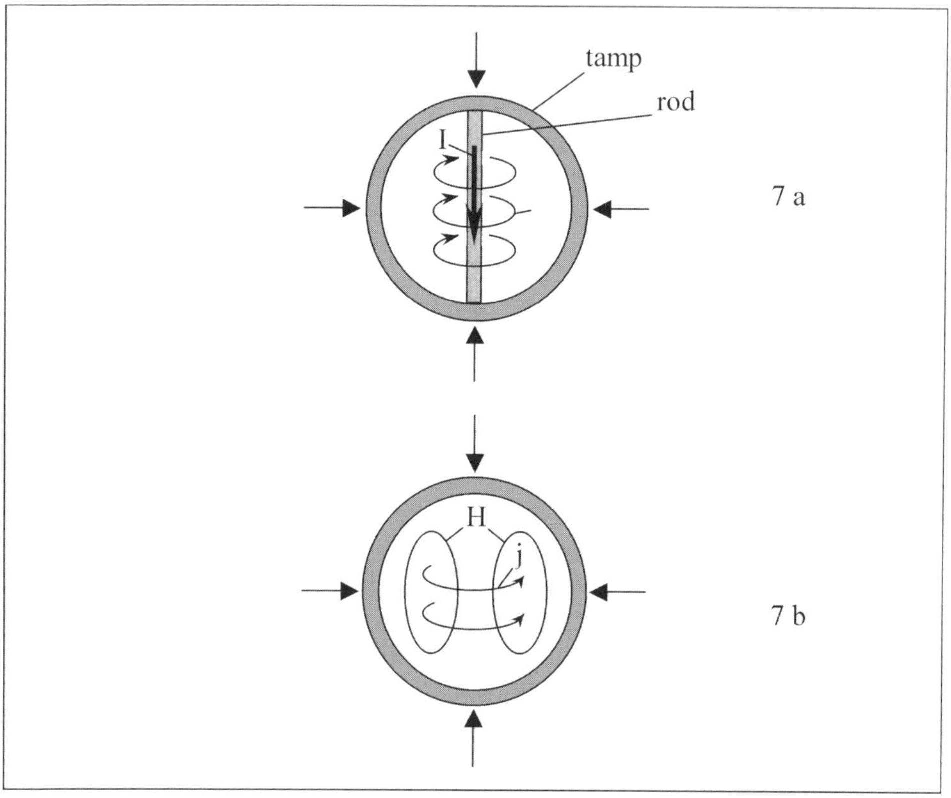

Figure 6.7: Magnetized fusion targets with closed magnetic field lines. **(a)** with internal rod as a conductor and **(b)** without such a conductor.

cavity. (We actually should add the plastic deformation work of the projectile and target cavity.) With the magnetic and thermal plasma energy about equal we have

$$E_{kin} \simeq 2 \times \left(\frac{H^2}{8\pi}\right)\left(\frac{4\pi}{3}r^3\right) = \frac{1}{3}r^3 H^2 \tag{6.120}$$

in our example $E_{kin} \simeq 3 \times 10^{14}$ erg $= 30$ MJ. The total number of particles confined in the cavity is $N = (4\pi/3)r^3 n \simeq 8 \times 10^{21}$. According to (6.21) and table 6.1, the Larmor radius of the DT fusion α-particles at $H \simeq 3 \times 10^7$ G is $r_L \sim 10^{-2}$ cm, small compared with the cavity radius $r \sim 1$ cm. The α-particles are therefore effectively bootstrapped leading to an effective self-heating, justifying the assumption of $a \sim 10\%$ fuel burn-up. With each DT particle pair releasing $\varepsilon_0 = 2.8 \times 10^{-5}$ erg, and with an assumed 10% burn-up one has a yield $Y \simeq 10^{16}$ erg $= 10^3$ MJ, with a gain $G = Y/E_{kin} \sim 30$. This gain though is small in comparison to the gain which can be reached with thermonuclear detonation waves.

Besides a small gain, one other disadvantage of magnetized fusion targets is a consequence of (6.116) and (6.117). It shows that for reasonable ratios r/r_0, one has to begin with a highly preheated (to the temperature T_0) DT plasma, and also a high initial field H_0. For $r_0/r = 10$ one would need $T_0 \sim 10^6$ °K and $H_0 \sim 3 \times 10^5$ G. For $r = 1$ cm, this means $r_0 \simeq 10$ cm, too large for a compact projectile. This suggests projectile velocities in between those for pure impact fusion, $v_p \simeq 2 \times 10^7$ cm/s and $v_p \simeq 10^6$ cm/s. One advantage of magnetized targets though is that they permit the reduction of the ρr value from $\rho r \sim 1$ g/cm^2 down to $\rho r \sim 10^{-2}$ g/cm^2. With the tamping effect of a fast moving projectile the ρr value could have already been reduced to $\rho r \sim 0.1$ g/cm^2, but here a high projectile velocity is required.

In a sense magnetized fusion targets are somewhere in between inertial confinement and magnetic fusion. For the latter $n \lesssim 10^{16}$ cm^{-3} and $r \gtrsim 10^2$ cm, with $nr \sim 10^2$ cm^{-2} or $\rho r \sim 10^{-8}$ g/cm^2.

A much better way to make use of a strong magnetic field is in a magnetic field assisted thermonuclear detonation wave. There, the fast ignitor concept can be directly applied, for example with the x-pinch (Fig. 6.5b), where at the cross point a hot spot can launch a thermonuclear detonation wave if $I > I_c$, with I_c given in table 6.1. A hot spot can, of course, also be generated by a pulsed laser beam, as in the original fast ignitor concept. We return to this important and promising approach in the context of the ultra fast z-pinch.

The presence of a strong magnetic field reduces the velocity for impact fusion, but there is a quite different concept which reduces the required velocity as well. In it a hypervelocity projectile upon impact first converts part of its kinetic energy into heat which is released as a burst of intense blackbody radiation which in turn ablatively implodes and ignites a thermonuclear target.

If a fast moving disk of thickness δ collides with a second disk at rest possessing the same thickness and density, 1/2 of the kinetic energy is converted into heat if the collision is completely inelastic. The kinetic energy per unit volume dissipated into heat is thus $\frac{1}{4}\rho \mathrm{v}^2$. Since in most circumstances the disk remains optically thick, the temperature is obtained by the Stefan-Boltzmann law $aT^4 = \frac{1}{4}\rho \mathrm{v}^2$, $a = 7.67 \times 10^{-15}$ erg/cm^3 °K. One therefore has $T \propto \sqrt{\mathrm{v}}$. The intensity of the blackbody radiation released from the surface of the hot disk is given by (4.72):

$$|j_r| = \frac{\lambda_{opt} c}{3} \nabla \left(aT^4\right) \sim \frac{1}{3} \frac{\lambda_{opt} c}{\delta} aT^4 \tag{6.121}$$

where $\lambda_{opt} = (\kappa\rho)^{-1}$ with κ given by (4.71), hence

$$\left.\begin{aligned} &\lambda_{opt} \propto T^{3.5}/\rho^2 \propto \mathrm{v}^{1.75} \\ &j_r \propto T^{7.5} \propto \mathrm{v}^{3.75} \end{aligned}\right\} \tag{6.122}$$

which shows a strong dependence on the impact velocity.

As an example we take an aluminum disk with $Z^2/A = 6.2$ and $\rho = 2.7$ g/cm^3. There we find

$$\lambda_{opt} = (\kappa\rho)^{-1} = 3 \times 10^{-27}\, T^{3.5} \,. \tag{6.123}$$

For $\mathrm{v} = 50$ km/s we find that $\frac{1}{4}\rho \mathrm{v}^2 = 1.7 \times 10^{13}$ erg/cm^3 $= aT^4$, hence $T = 6.3 \times 10^6$ °K, and $\lambda_{opt} = 1.9 \times 10^{-3}$ cm. Finally we find that $j_r \approx 3 \times 10^{20}/\delta$ erg/cm^{2}s $= 3 \times 10^{13}/\delta$ W/cm^2. Therefore, if the disk has a thickness $\delta \approx 0.3$ cm, $j_r \approx 10^{14}$ W/cm^2, and with an energy of $\frac{1}{2}$ MJ/cm^2 stored in the disk, this is sufficient to ablatively implode a capsule with an implosion velocity of $\sim 10^7$ cm/s.

The foregoing can be further refined with the concept of the dynamic "hohlraum" where intense blackbody radiation inside a cavity (i.e. hohlraum) is isentropically compressed by imploding the hohlraum[1], for the

[1]The name "hohlraum" is German and means cavity. It was introduced by Kirchoff in the theory of blackbody radiation as the radiation in a cavity.

implosion of a DT target placed inside the hohlraum. The feasibility of this concept requires that the implosion velocity is larger than the diffusion velocity v_D of the radiation into the cavity wall. Putting $|j_r| = aT^4 v_D$ one obtains from (6.121)

$$\frac{v_D}{c} = \frac{1}{3}\frac{\lambda_{opt}}{\delta} \tag{6.124}$$

For $T = 10^7$ °K and $r = 18$ g/cm^3 (uranium) one finds that $\lambda_{opt} = 2 \times 10^{-4}$ cm, and for $\delta = 0.1$ cm, that $v_D \simeq 2 \times 10^7$ cm/s. The implosion velocity can be reduced by filling the cavity with a high Z-number gas or foam which reduces the radiation diffusion velocity.

With $\gamma = 4/3$, the specific heat ratio valid for blackbody radiation, one has $TV^{\gamma-1} = TV^{1/3} = TR = const.$ or $T \propto 1/R$ where R is the cavity radius.

The dynamic hohlraum concept is used in the implosion of small DT assemblies by the blackbody radiation of multiple wire implosions, explained in chapter 8.13, where subsequent to its generation the blackbody radiation is further compressed and amplified.

Fig 6.8 displays two examples for a dynamic hohlraum. In both cases the hohlraum is filled with a low density high Z-number gas which upon impact from the imploding cavity (hohlraum) wall is transformed into a high temperature blackbody radiation dominated plasma. In Fig. 6.8a the implosion of the cavity is caused by the ablation of a layer (ablator) bombarded with intense particle or laser beams. In Fig. 6.8b it is caused by the impact of a high velocity projectile.

We remark that this kind of indirect impact fusion is utilized in the so-called mini-nukes explained in chapter 7.14.

6.11 Autocatalytic Fission-Fusion Implosions

As in Fig. 6.7 we now consider a magnetized thermonuclear DT target, except that the tamp is now made from fissile material. And as in the magnetized fusion target the shell is imploded, but now the neutrons released by the thermonuclear reactions cause fission reactions in the fissile shell. If the rate of these reactions is large enough, the fissile shell is heated up to high temperatures exploding it outwards but also inwards. By its inward implosion it increases the thermonuclear reaction rate in the thermonuclear target with more neutrons released, making more fissions in the shell. It is this coupling

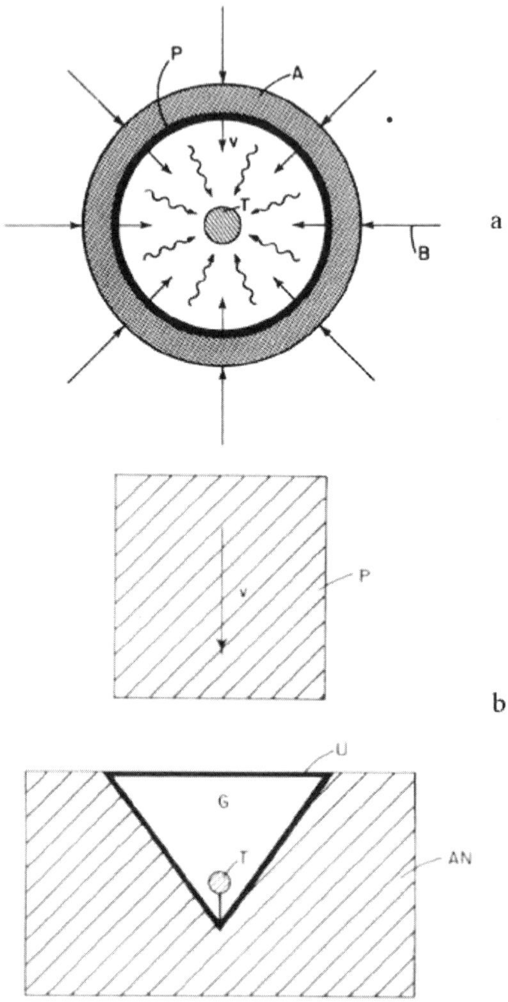

Figure 6.8: Dynamic "hohlraum" target configurations. **(a)** shows the implosion of blackbody radiation by an ablatively driven shell. B are the incoming laser of charged particle beams, A is the ablator, P the pusher and T the thermonuclear target inside a cavity filled with blackbody radiation. **(b)** shows the implosion by hypervelocity impact of blackbody radiation entrapped inside a conical cavity. P is the hypervelocity projectile and G the high atomic weight gas inside the cavity. U is a thin but dense high atomic weight material (for instance uranium) covering the inner surface of the cavity and which is surrounded by the anvil AN. T is the thermonuclear target.

of the fission and fusion process which we call an autocatalytic fission-fusion implosion. This concept shall now be analyzed. For an incompressible fissile shell with an outer and inner radii R and r, mass conservation requires

$$R^3 - r^3 = const. \tag{6.125}$$

By differentiation with regard to time this gives

$$\frac{\dot{r}}{\dot{R}} = \left(\frac{R}{r}\right)^2 \tag{6.126}$$

and hence for the implosion velocity

$$\mathrm{v} = \mathrm{v}_0 \left(\frac{r_0}{r}\right)^2 \tag{6.127}$$

where $r = r_0$ is the inner shell radius at the time $t = 0$, and v_0 the initial implosion velocity imparted on the shell by the high explosive.

For a compressible shell with an equation of state of the form (p pressure, ρ density, γ specific heat ratio)

$$p = A\rho^\gamma, \qquad A = const. \tag{6.128}$$

the implosion velocity rises less rapidly, and is as a function of γ obtained from a gas dynamic similarity solution (see chapter 5.5). One there has

$$\mathrm{v} = \mathrm{v}_0 \left(\frac{r_0}{r}\right)^m \tag{6.129}$$

where $m = m(\gamma)$. For an incompressible shell one has $\gamma \to \infty$ and $m \to 2$. In general $\gamma = \gamma(p)$, with $\gamma \simeq 10$ a typical value for a metallic shell under megabar pressures.

The equation of continuity requires that

$$r^2 \rho \mathrm{v} = r_0^2 \rho_0 \mathrm{v}_0 \tag{6.130}$$

where ρ_0 is the initial density at $p = 0$. One thus has

$$\frac{\rho}{\rho_0} = \left(\frac{\mathrm{v}_0}{\mathrm{v}}\right) \left(\frac{r_0}{r}\right)^2 \tag{6.131}$$

or because of (6.130)

$$\frac{\rho}{\rho_0} = \left(\frac{r_0}{r}\right)^{2-m} . \tag{6.132}$$

For an incompressible shell where $m = 2$ one has, as expected, $\rho = \rho_0$. For $\gamma = 10$ one has $m \approx 1$ hence $\rho/\rho_0 \approx r_0/r$.

The DT reaction releases neutrons at the rate

$$S = \left(\frac{n^2}{4}\right)\langle\sigma \mathrm{v}\rangle\left(\frac{4\pi}{3}r^3\right) \tag{6.133}$$

where n is the DT atomic number density of the magnetized plasma, and $\langle\sigma \mathrm{v}\rangle$ the nuclear reaction cross section particle velocity product averaged over a Maxwell velocity distribution. With $N = (4\pi/3)r^3 n$ the total number of nuclei in the DT plasma, one has

$$S = \left(\frac{N}{4}\right)\langle\sigma \mathrm{v}\rangle n . \tag{6.134}$$

The implosion starts from the radius $r = r_0$, reaching ignition at $r = r_1$ where $n = n_1$. Then, for $r < r_1$ one has $n = n_1(r_1/r)^3$ and hence

$$S = S_1\left(\frac{r_1}{r}\right)^3 \tag{6.135}$$

where $S_1 = (N/4)\langle\sigma \mathrm{v}\rangle n_1$.

The number of fission reactions made by the fusion neutrons in passing through the fissile shell of thickness δ is

$$f = S n_f \sigma_f \delta \tag{6.136}$$

where n_f is the atomic number density of the fissile shell and σ_f the fission cross section. With ε_f the energy released per fission reaction one has for the total rate of the fission energy in the shell:

$$E_f = f\varepsilon_f = S\delta n_f \sigma_f \varepsilon_f . \tag{6.137}$$

Because of (6.132) one has

$$\frac{n_f}{n_{f1}} = \left(\frac{r_1}{r}\right)^{2-m} \tag{6.138}$$

where $r_1 < r_0$ is the inner shell radius below which the number of fission reactions becomes important. Because of $\rho\delta^3 = \rho_1\delta_1^3$ one has

$$\frac{\delta}{\delta_1} = \left(\frac{\rho_1}{\rho}\right)^{1/3} = \left(\frac{r}{r_1}\right)^{(2-m)/3} \tag{6.139}$$

and for the rate of the fission energy released in the shell

$$E_f = S_1 \left(\frac{r_1}{r}\right)^3 \delta_1 \left(\frac{r}{r_1}\right)^{(2-m)/3} n_{f1} \left(\frac{r_1}{r}\right)^{2-m} = E_{f1} \left(\frac{r_1}{r}\right)^{\alpha} \tag{6.140}$$

where

$$E_{f1} = S_1 \delta_1 n_{f1} \sigma_f \varepsilon_f \,, \qquad \alpha = \frac{(13-2m)}{3}$$

with S_1, δ_1, n_{f1}, ρ_1 the respective values for $r = r_1$.

For $r < r_1$ the implosion velocity is increased by the fission reactions in the shell resulting in its heating and expansion. To take this effect into account we put

$$\mathrm{v} = \mathrm{v}_1(t) \left(\frac{r_1}{r}\right)^{m} \tag{6.141}$$

where $\mathrm{v}_1(t)$ is a function of time. Assuming that the energy released by the fission reactions goes in equal parts into heat and kinetic energy of the shell one may put

$$\frac{M}{2}\frac{d\mathrm{v}_1^2}{dt} = \frac{E_f}{2}\,. \tag{6.142}$$

Since $d\mathrm{v}_1^2/dt = \mathrm{v}_1\, d\mathrm{v}_1^2/dr = \frac{2}{3} d\mathrm{v}_1^3/dr$ one has

$$\frac{d\mathrm{v}_1^3}{dr} = \frac{3}{2}\frac{E_{f1}}{M}\left(\frac{r_1}{r}\right)^{\alpha} \tag{6.143}$$

which by integration gives

$$\mathrm{v}_1^3(t) - \mathrm{v}_1^3(r_1) = \frac{\frac{3}{2}E_{f1} r_1}{M(\alpha-1)}\left[\left(\frac{r_1}{r}\right)^{\alpha-1} - 1\right] \tag{6.144}$$

with the asymptotic solution

$$\mathrm{v}_1 = \mathrm{v}_1(0)\left(\frac{r_1}{r}\right)^{(\alpha-1)/3}, \qquad r \ll r_1 \tag{6.145}$$

where

$$\mathrm{v}_1(0) = \left[\frac{\frac{3}{2}\left(E_{f1} r_1\right)}{(\alpha-1)\,M}\right]^{1/3}. \tag{6.146}$$

Inserting (6.145) into (6.141) one obtains

$$\mathrm{v} = \mathrm{v}_1\left(r_1\right)\left(\frac{r_1}{r}\right)^{\beta}, \qquad r \ll r_1 \tag{6.147}$$

where

$$\beta = \frac{(10+7m)}{9}. \tag{6.148}$$

One then has

$$\left.\begin{aligned} \frac{n_f}{n_{f1}} &= \left(\frac{r_1}{r}\right)^{2-\beta} \\ \frac{\delta}{\delta_1} &= \left(\frac{r}{r_1}\right)^{(2-\beta)/3} \end{aligned}\right\} \tag{6.149}$$

If, for example, $m = 1$ (corresponding to $\gamma \approx 10$) one finds $\beta \approx 2m = 2$, as if the shell would be incompressible. There one has

$$\left.\begin{aligned} \mathrm{v} &= \mathrm{v}_1\left(\frac{r_1}{r}\right)^{2} \\ n_f &\approx n_{f1} \\ \delta &\approx \delta_1 \end{aligned}\right\} \tag{6.150}$$

In a useful approximation one may match the asymptotic solution (6.145) for $r \ll r_1$ where the fission reactions are taken into account, with the solution (6.129) valid for $r \gg r_1$ above which the fission reactions are small. From this matching condition one can determine a value of $\langle\sigma \mathrm{v}\rangle$ above which fusion neutrons become important. With the following choice of parameters: $n_1 = 5 \times 10^{20}$ cm^{-3}, $N_1 = 10^{21}$, one obtains $S_1 = 1.25 \times 10^{41}$ $\langle \sigma \mathrm{v} \rangle$ s^{-1} and $r_1 = 0.78$ cm. Further assuming that $\delta_1 \approx 1$ cm, $n_{f1} \approx 10^{23}$ cm^{-3}, $\sigma_f = 2 \times 10^{-24}$ cm^2, $\varepsilon_f = 3 \times 10^{-4}$ erg, one finds $E_{f1} = 7.5 \times 10^{36}$ $\langle \sigma \mathrm{v} \rangle$ erg/s. Assuming that $M = 18$ g (1 cm^3 of U235) and $\alpha = 11/3$ (corresponding to $\gamma \approx 10$), we obtain from (6.146) that $\mathrm{v}_1(0) = 5.55 \times 10^{11}$ $(\langle \sigma \mathrm{v} \rangle)^{1/3}$ cm/s.

Let us assume that the uranium shell is imploded with the initial velocity of ~ 3 km/s, and that its implosion velocity goes as $1/r$. If it implodes from $r_0 \sim 3$ cm to $r_1 \sim 1$ cm, it would reach an implosion velocity $\mathrm{v}_1 \sim 10$ km/s. To match this velocity with $\mathrm{v}_1(0)$ would require that $\langle \sigma \mathrm{v} \rangle \approx 6 \times 10^{-18}$ cm^3/s, which is reached at a plasma temperature of $T \sim 4 \times 10^7$ °K. This is below the ignition temperature for the DT reaction, but large enough to generate a sufficient number of neutrons to make a substantial number of fission reactions.

With the temperature rising in proportion to r^{-2}, the implosion from an initial radius of $r_0 \approx 3$ cm to a radius of $r_1 \approx 1$ cm, would raise the temperature from 10^6 °K to 4×10^7 °K, high enough to get the autocatalytic reaction started. At a temperature of $T = 4 \times 10^7$ °K, and a DT plasma particle density of $n_1 = 5 \times 10^{20}$ cm^{-3}, the plasma pressure is $p = 2n_1kT$ $= 5 \times 10^{12}$ dyn/cm^2. This pressure is less than the pressure of the imploding fissile shell $p \sim \rho \mathrm{v}^2$ at a density $\rho \geq 20$ g/cm^3 and $\mathrm{v} \sim 10^6$ cm/s.

With the rise of the implosion velocity the DT plasma is further compressed and heated until the ignition temperature is reached where $\langle \sigma \mathrm{v} \rangle$ $\sim 10^{-15}$ cm^3/s. The burn time τ for the DT plasma is there of the order

$$\tau \sim (n \langle \sigma \mathrm{v} \rangle)^{-1} \ . \tag{6.151}$$

For $n \sim 10^{22}$ cm^{-3} it is $\tau \sim 10^{-7}$ s, of the same order of magnitude as the disassembly time for the fissile shell moving with a velocity of $\sim 10^7$ cm/s. A large burn-up can therefore be expected.

The burn of $N = 10^{21}$ DT nuclei, sets free an energy $N\varepsilon \sim 10^{16}$ erg $= 1$ GJ, where $\varepsilon \sim 10^{-5}$ erg is the fusion energy per nucleon. The amount of fission energy set free is $Nn_f\sigma_f\varepsilon_f\delta$ which for $n_f \sim 10^{23}$ cm^{-3}, $\sigma_f \sim 10^{-24}$ cm^2, $\varepsilon_f \sim 10^{-4}$ erg, $\delta \sim 1$ cm, is also equal to ~ 1 GJ. Making δ larger (using more fissile material), the fission energy output could be further increased, with the fusion energy output remaining the same. A fission chain reaction in the shell, even if the shell is subcritical, would also increase the fission energy output. With a subcritical neutron multiplication factor $k < 1$, the fission energy output would be increased by $1 + k + k^2 + \cdots = 1/(1+k)$. As long as a fission chain reaction can be ignored one may use U^{238} or even B^{10} for the shell material. Neutron induced fission of B^{10} into Li^7 and He^4 releases an energy of 3.0 MeV $= 4.8 \times 10^{-6}$ erg. By comparison, the fission of U^{238} releases an energy of 180 MeV $= 2.9 \times 10^{-4}$ erg. With the fission cross sections of U^{238} and B^{10} of the same order of magnitude, it follows that about the same amount of fission energy would be released in a B^{10} shell of a mass comparable to the mass of a U^{238} shell. The use of B^{10} is of special interest since it does not lead to the kind of undesirable fission products released by U^{238}.

6.12 Bibliography for Chapter 6

R. F. Post, Reviews of Modern Physics **28**, 338 (1956).

S. Glasstone and R. H. Loveberg, Controlled Thermonuclear Reactions, Robert E. Krieger Publishing Company, Huntington, New York, (1975).

M. O. Hagler and M. Kristiansen, An Introduction to Controlled Thermonuclear Fusion, Lexington Books, D. C. Heath and Company, Lexington Massachusetts, (1977).

T. J. Dolan, Fusion Research, Pergamon Press, New York, (1982).

F. Winterberg in "Physics of High Energy Density", International School of Physics "Enrico Fermi", edited by P. Caldirola and H. Knoepfel, Academic Press, New York, (1971).

M. Tabak, J. Hammer, M. E. Glinsky, W. L. Kruer, S. C. Wilks, J. Woodworth, E. M. Campbell, and M. D. Perry, Physics of Plasmas **1**, 1626 (1994).

J. G. Linhart, et al., Nucl. Fusion Supplement **Pt2**, 733 (1962).

F. Winterberg, Atomkernenergie/Kerntechnik **44**, 145 (1984).

Chapter 7

Ignition by Fission Explosives

7.1 Temperature and Radiation Flux of a Fission Explosion

The energy released in a fission explosion goes in part into particle energy and in part into radiation. The part which goes into particle energy per unit volume is

$$\varepsilon_p = \frac{f}{2} nkT \tag{7.1}$$

where f is the number of degrees of freedom for the particles. If the energy only goes into translational kinetic energy, as in a fully ionized plasma, one has $f = 3$, but if internal degrees of freedom are excited as well, f can be much larger. This is especially true for the plasma of a fission explosion with uranium (or plutonium), where the temperature is far below the temperature for complete ionization given by (3.3), which for $Z = 92$ is $T_i \simeq 10^{10}$ °K. If $R_0 \gg (\rho\kappa)^{-1}$, where R_0 is the critical radius given by (2.42), $\rho = 18$ g/cm^3 the density of the uranium metal, and κ the opacity coefficient (4.71), the fission bomb plasma is opaque. There then, a large portion of the energy goes into blackbody radiation with the energy density

$$\varepsilon_r = aT^4 \tag{7.2}$$

($a = 7.67\times10^{-15}$ erg/cm^3°K^4). If $\varepsilon_r > \varepsilon_p$ the energy density of the blackbody radiation will be predominant, which happens if

$$T \gtrsim 10^7\, f^{1/3}\ [°\mathrm{K}]\,. \tag{7.3}$$

In a uranium (or plutonium) plasma, with a temperature of many million-degree Kelvin, f can become quite large. Let us assume that $f \simeq 100$, whereby $T \gtrsim 4.6 \times 10^7$ °K. The inequality (7.3) is valid for uncompressed uranium, otherwise it must be multiplied on the r.h.s. by $(n/n_s)^{1/3}$, where $n > n_s = 4.5 \times 10^{22}$ cm^{-3} (n_s atomic number density in uncompressed uranium). According to published reports, the yield of the first fission bomb, (an uncompressed, slightly above critical mass of uranium) was equivalent to 20 kilotons of TNT or equal to 8×10^{20} erg. The critical radius, computed with (2.42), gives for metallic uranium $R_0 = 7.5$ cm. To be 5% above critical, one has $R \simeq 7.9$ cm. At this radius the number of uranium atoms is $N = (4\pi/3)\, R^3 n_s = 9.2 \times 10^{25}$. The energy per fission is $\varepsilon_f = 180$ MeV $= 2.9 \times 10^{-4}$ erg, and if all nuclei N are fissioned, the energy output would be 2.7×10^{22} erg. With a yield of 8×10^{20} erg, the burn-up rate is 3%. For such a burn-up rate the energy density in the fission explosion is 3.9×10^{17} erg/cm^3. Equating this with $\varepsilon_r = aT^4$, one obtains $T \simeq 8.7 \times 10^7$ °K, satisfying (7.3) for the predominance of blackbody radiation. From (4.71) one obtains ($\rho = 18$ g/cm^3, $g/t \simeq 1$) $\rho\kappa = 2.2$ cm^{-1} or $\lambda_{opt} = (\rho\kappa)^{-1} = 0.45$ cm. With $R = 7.9$ cm, $\lambda_{opt}/R \ll 1$, as required for blackbody radiation, is well satisfied.

For $T = 8.7 \times 10^7$ °K, the blackbody radiation flux at the surface of the sphere is equal to $\phi_r = \sigma T^4 = (ac/4)\, T^4 = 3.2 \times 10^{27}$ erg/cm^2. This radiation is emitted in passing through a surface layer of thickness $\lambda_{opt} = 0.45$ cm, with the radiation flux through this layer, according to (4.72):

$$j_r = -\frac{\lambda_{opt} c}{3} \nabla \left(aT^4\right) \approx \frac{ac}{3}\, T^4 \approx \sigma T^4 \,. \tag{7.4}$$

In the transparent surface layer of thickness λ_{opt}, the uranium plasma ions are decoupled from the photons, retaining their thermal velocity

$$\mathrm{v}_{th} = \left(\frac{3kT}{M}\right)^{1/2} = 9.6 \times 10^6 \text{ cm/s} \tag{7.5}$$

where $M = 3.9 \times 10^{-22}$ g is the mass of an uranium atom.

The time the uranium sphere is inertially held together is of the order $t \approx 2R/\mathrm{v}_{th} \simeq 1.7 \times 10^{-6}$ s. According to equations (2.37-2.45) the neutron avalanche goes as

$$n = n_0 e^{\lambda t} \,. \tag{7.6}$$

For $\Delta R/R = 0.05$, we had found that $\lambda = 10\lambda_0 = 3 \times 10^7 \text{ s}^{-1}$, hence $\lambda t \simeq 51$. For a 3% burn-up rate $n = 0.03 \times 4.5 \times 10^{22} = 1.35 \times 10^{21}$ cm^{-3}. We thus have $n_0 = 1.35 \times 10^{21}\ e^{-51} \simeq 3 \times 10^{-2}$ cm^{-3}, or for the volume $(4\pi/3)\, R^3 = 2 \times 10^3$ cm^3, about 60 neutrons needed to start the chain reaction. The neutron flux in the sphere is $\phi = n\mathrm{v}_0 = 1.35 \times 10^{21} \times 2 \times 10^9 = 2.7 \times 10^{30}$ cm^{-2}s^{-1}.

With the kinetic neutron energy equal to 2 MeV $\simeq 3.2 \times 10^{-6}$ erg, we summarize all three energy flux quantities (1) for radiation, (2) for neutron radiation, (3) for uranium kinetic energy

$$\left.\begin{aligned} \phi_r &= \sigma T^4 = 3.2 \times 10^{27} \text{ erg/cm}^2\text{s} \\ \phi_n &= n\mathrm{v}_0 \times 2 \text{ MeV} = 8.6 \times 10^{24} \text{ erg/cm}^2\text{s} \\ \phi_k &= \frac{1}{2}\rho \mathrm{v}_{th}^3 = 7.9 \times 10^{22} \text{ erg/cm}^2\text{s} \end{aligned}\right\} \quad (7.7)$$

Also of interest are the radiation pressure p_r, the kinetic neutron pressure p_n and the uranium plasma pressure p

$$\left.\begin{aligned} p_r &= \frac{1}{3} a T^4 = 1.3 \times 10^{17} \text{ dyn/cm}^2 \\ p_n &= \frac{\phi_n}{\mathrm{v}_0} = 4.0 \times 10^{15} \text{ dyn/cm}^2 \\ p &= \frac{\phi_k}{\mathrm{v}_{th}} = 8.2 \times 10^{15} \text{ dyn/cm}^2 \end{aligned}\right\} \quad (7.8)$$

The critical radius and with it the critical mass can be substantially reduced (a) by a neutron reflector and (b) by compression of the fissile material. A good neutron reflector is gold, but also beryllium, the latter because of its n, $2n$ reaction, is a neutron multiplier. One can, of course, also use natural uranium where the fast fission process enhances the neutron economy as in the n, $2n$ reaction of a beryllium reflector.

7.2 The Ignition Problem

Superficially, it seems that in order to achieve ignition one would simply have to place the thermonuclear explosive in direct contact with an exploding fission bomb (Fig. 7.1). In this hypothetical configuration, some of the thermonuclear material would serve as a fuse to ignite a much larger thermonuclear explosion. One drawback of this approach is that it could only work (if at all) with DT as the thermonuclear explosive, since only in DT is the temperature of the fission explosion $T \simeq 9 \times 10^7$ °K higher than the ignition temperature $T \simeq 5 \times 10^7$ °K of this reaction, but not high enough for the ignition of the DD reaction where $T_1 \simeq 3 \times 10^8$ °K.

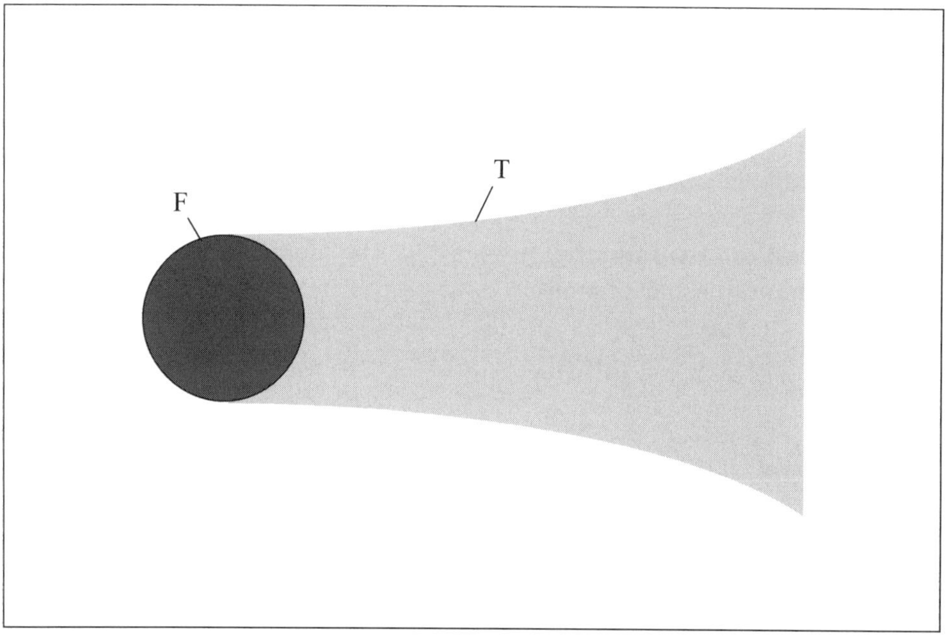

Figure 7.1: General schematic arrangement of fissile and thermonuclear explosives in which a fission bomb F will ignite a thermonuclear explosive T.

A better chance for ignition is obtained in the reaction

$$n + Li^6 \longrightarrow He^4 + T + 4.8 \text{ MeV} \tag{7.9}$$

through the intense neutron flux of $\sim 3 \times 10^{30}$ cm^{-2}s^{-1} of the fission explosion. This reaction has a neutron cross section $\sigma \simeq 3 \times 10^{-25}$ cm^2. If Li^6D is used as the explosive, with $n_L \simeq 2 \times 10^{22}$ cm^{-3} lithium atoms, the reaction rate is $n_L \sigma \phi \simeq 2 \times 10^{28}$ cm^{-3}s^{-1}. The reaction takes place over the slowing down length of the fast fission neutrons in Li^6D. This length should be of the same order of magnitude as the slowing down length in D_2O which is ≈ 10 cm. In addition to the 4.8 MeV reaction energy of (7.9), the 2 MeV kinetic neutron energy has to be added resulting in a total of 6.8 MeV $\simeq 10^{-5}$ erg. The energy released per unit volume in the time $t = 1.7 \times 10^{-6}$ s, the bomb explosion time, is then

$$\begin{aligned} \varepsilon &= n\sigma\phi \times 10^{-5} \text{ erg} \times 1.7 \times 10^{-6} \text{ s} \\ &= 3.4 \times 10^{17} \text{ erg/cm}^3 \end{aligned} \tag{7.10}$$

about the same as the energy density of the fission explosion. Accordingly, the temperature, computed from $\varepsilon = aT^4$, is $T \simeq 8.6 \times 10^7$ °K, about the same as the temperature of the fission explosion. The tritium formed in the reaction (7.9) can then react with the deuterium in the Li^6D salt.

A more serious problem with the idea to place the thermonuclear explosive side by the exploding fission bomb is the large radiation pressure from the exploding fission bomb, by which the thermonuclear explosive is blown aside. The radiation pressure p_r of the exploding fission bomb was estimated to be $p_r \simeq 1.3 \times 10^{17}$ dyn/cm^2. Suppose that the thermonuclear "fuse", the part of the horn made up with the thermonuclear explosive and in contact with the fission bomb, (as shown in Fig. 7.1) has a diameter $d \sim 2R \sim 15$ cm. The radiation pressure will exert a force $F = p_r d^2$ onto the fuse of mass $M \sim \rho_F d^3$, where $\rho_F \simeq 0.4$ g/cm^3 is the density of Li^6D. The radiation pressure induced acceleration of the fuse is $b = p_r/\rho d$, and the time needed to displace the fuse by the distance d is

$$\tau_d = d\sqrt{\frac{2\rho_F}{p_r}} \simeq 2R\sqrt{\frac{2\rho_F}{p_r}} \tag{7.11}$$

for $R \simeq 15$ cm one has $\tau_d \simeq 2.7 \times 10^{-7}$ s, or about one order of magnitude shorter than the explosion time $t \simeq 1.7 \times 10^{-6}$ s of the fission bomb. Accordingly, the energy density (7.10) is actually $3.4 \times 10^{17} \times (2.7 \times 10^{-7}/1.7 \times 10^{-6})$ $= 5.4 \times 10^{16}$ erg/cm^3, and the temperature $T \simeq 6 \times 10^7$ °K, just above the ignition temperature of the DT reaction, but too low for the Li^6D reaction.

The problem of igniting a thermonuclear explosion with a fission bomb "match" has been aptly compared to the problem of igniting, with a burning match, a cigarette in a wind storm. Long before the cigarette can catch fire, the "match" will be blown out. Another way to look at this problem is the impossibility to ignite a piece of wood with a comparable piece of gun cotton placed side by side with the wood. The temperature of the burning gun cotton of several 10^3 °K is certainly high enough to ignite the wood, but it does not last long enough.

7.3 The Thermonuclear Booster Concept

Rather than placing the thermonuclear material outside and in direct contact with the exploding fission bomb, one may place it inside as shown in Fig. 7.2. There the thermonuclear material is compressed, further raising its temperature. This is important for thermonuclear fuels with an ignition temperature above $\sim 10^8$ °K, such as Li^6D. The compression not only raises the temperature, but also the reaction rate, the latter going with the square of the density. If a neutron producing thermonuclear fuel is used, the fusion neutrons can accelerate the fission process as was explained in Chapter 2.5. DT as a fuel is not very useful for military applications, because it must be kept liquid requiring very low temperatures. Li^6D is a salt at room temperature but its reaction does not produce a net surplus in neutrons. Such a surplus is released in a mixture of Li^6D with Li^6T.

A drawback of the configuration shown in Fig. 7.2 is that it leads to a larger outer radius in order to be able to become critical, because it has to give space for the thermonuclear material placed in its center. A better way is to put the $Li^6D + Li^6T$ in between shells of U^{235} (Pu), or as a lattice inside the fissile material as shown in Fig. 7.3.

7.4 The Polyhedron Configuration

The temperature of one fission bomb explosion falls short of what is required to ignite the Li^6D reaction, but also the DD reaction, the latter having an ignition temperature $T_1 \gtrsim 3 \times 10^8$ °K. This though is not the case if several bomb explosions are set off simultaneously, with the bombs equidistantly placed on a virtual spherical surface. In the limit of an infinite number of

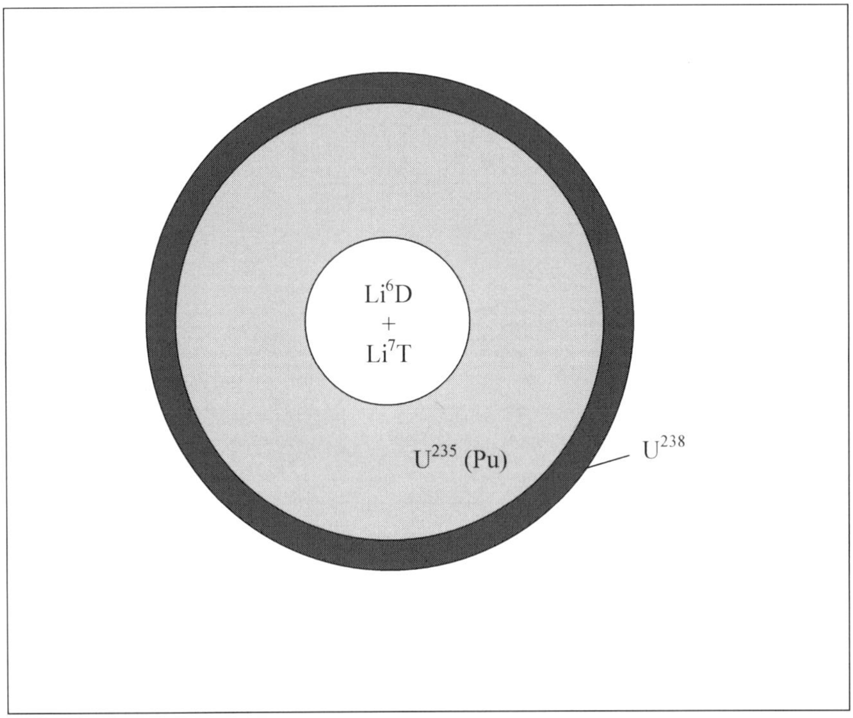

Figure 7.2: Fusion boosted fission bomb.

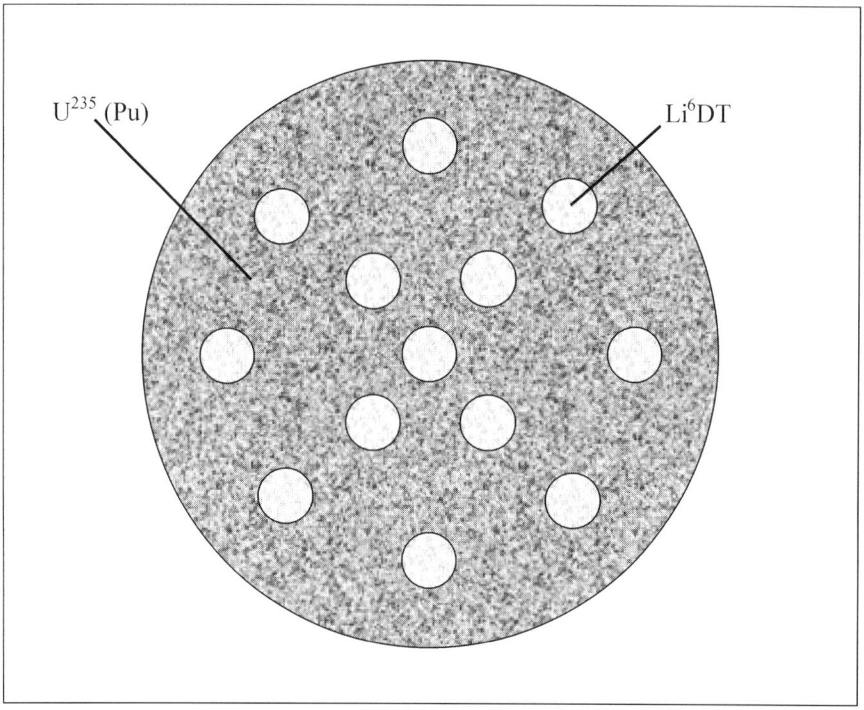

Figure 7.3: The booster concept: Li^6DT pellets inside U^{235}.

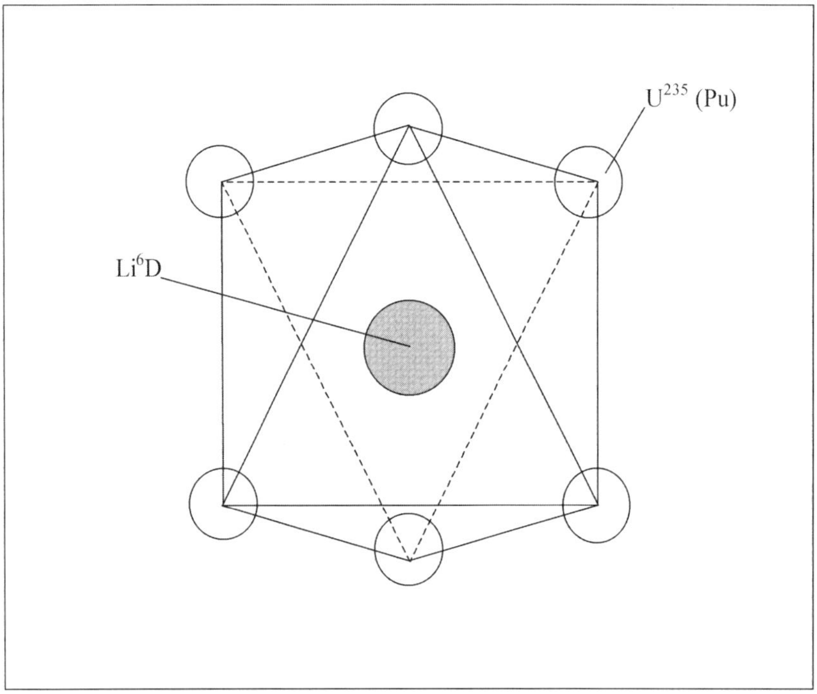

Figure 7.4: Polyhedron configuration with 6 fission bombs.

fission bombs placed on the surface, they would launch a spherical convergent shock wave with a temperature rising in proportion to $r^{-0.9}$, with r the distance measured from the center of the sphere (eq. 5.29). The minimum number of bombs needed to produce a quasispherical shock wave is obviously four, placed on the four corners of a tetrahedron, but a minimum of 6, placed on the six surfaces of a cube, is probably needed (Fig. 7.4).

With the temperature of the exploding fission bombs equal to $T_0 \simeq 8 \times 10^7$ °K, the temperature rises tenfold if $R_0/r_1 \sim 10$, where R_0 is the radius of the sphere on which the fission bombs are placed, and r_1 the radius of the spherical fusion explosive. For a large fusion yield $r_1 \gg R$, where R is the critical radius of one fission bomb. This means R_0 must be at least a few meters, leading to a very large device.

An arbitrarily large yield with a smaller radius R_0 would require that the fusion explosive of radius r_1 acts as a fuse to ignite a much larger amount

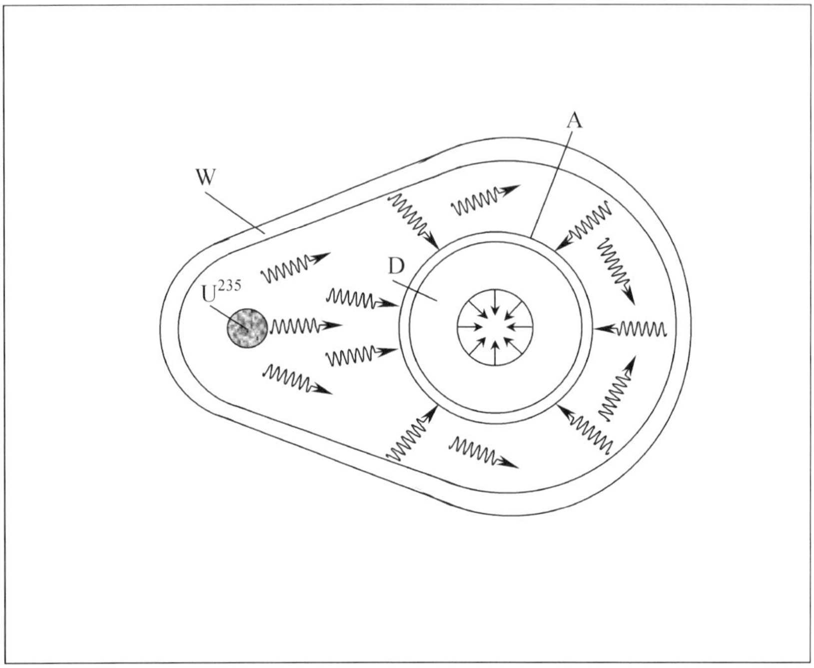

Figure 7.5: Teller-Ulam hohlraum configuration.

of fusion fuel in a horn attached to the fuse. However, for the same reasons stated in Chapter 7.3 this is not likely possible. The extent to which large yields with just one fission bomb can be reached will be explained next.

7.5 The Teller-Ulam Configuration

Very large (but not arbitrarily large) yields with just one fission-bomb-trigger are realized in the Teller-Ulam configuration (so named after its inventors). The principle of this configuration is explained in Fig. 7.5. A fission bomb (U^{235}, Pu) is placed in one "focus" of a pear-shaped cavity (called a hohlraum), with the thermonuclear explosive D (deuterium, Li^6D) placed in the other focus. According to (7.7) the exploding fission bomb is the source of blackbody radiation, neutron radiation and an expanding hot plasma. With the lion's share emitted as blackbody radiation and with most of the thermal energy in the fission bomb plasma stored in blackbody radiation, the time dependence of this energy in the bomb plasma is ruled by the equation

$$\frac{4\pi}{3} R^3 \frac{d}{dt} \left(aT^4\right) = -4\pi R^2 \sigma T^4 \tag{7.12}$$

or with $\sigma = ac/4$

$$\frac{dT^4}{dt} = -\frac{1}{\tau_E} T^4 , \qquad \tau_E = \frac{4R}{3c} \tag{7.13}$$

where τ_E is the energy loss time of the bomb plasma. For the example $R = 7.9$ cm one has $\tau_E \simeq 3.5 \times 10^{-10}$ s.

If the thickness d of the wall material is larger than $(\kappa\rho)^{-1}$, where κ is given by (4.71) and ρ the density of the wall material, the blackbody radiation is confined by the wall for a time of the order

$$\tau_c = \sqrt{\frac{2d}{b}} . \tag{7.14}$$

There $b = p_c/\rho d$ is the acceleration of the wall material by the radiation pressure p_c in the cavity, and where it is assumed that τ_c is determined by a wall displacement equal to the wall thickness d. Hence

$$\tau_c \simeq d\sqrt{\frac{2\rho}{p_c}} . \tag{7.15}$$

With the cavity volume V the temperature of the wall-confined blackbody radiation obeys the adiabatic law $TV^{\gamma-1} = const.$ For blackbody radiation one has $\gamma = 4/3$ and hence $TV^{1/3} = const.$ If R_c is the mean cavity radius, R and T the radius and temperature of the exploding fission bomb, the temperature of the radiation confined in the cavity is

$$T_c = T\left(\frac{R}{R_c}\right) \tag{7.16}$$

and the radiation pressure

$$p_c = \frac{a}{3}\left(\frac{R}{R_c}\right)^4 T^4 . \tag{7.17}$$

As an example we take the previously determined $R = 7.9$ cm, $T = 8.7 \times 10^7$ °K, valid for a uranium fission bomb, and furthermore $R/R_c = 10$. We obtain $T_c = 8.7 \times 10^6$ °K and $p_c = 1.3 \times 10^{13}$ dyn/cm^2. At this temperature and a density $\rho \simeq 10$ g/cm^3 of the wall material one obtains from (4.71) $(\rho\kappa)^{-1} \simeq 5 \times 10^{-4}$ cm. With $d \gg 10^{-4}$ cm, one has from (7.15) $\tau_c \sim 10^{-6} d$ [s]. By comparison, the time needed to fill the cavity with radiation is of the order $\tau_R \sim R_c/c \sim 3 \times 10^{-9}$ [s], and the time for the bomb plasma to reach and destroy the cavity wall is $\tau_p \sim R_c/\mathrm{v}_{th} \sim 10^{-5}$ [s]. For a cavity wall with a thickness $d > 0.1$ cm, there is thus plenty of time for the confined radiation to heat the ablator surrounding the spherical assembly of a thermonuclear explosive and to launch a convergent spherical shock wave into the explosive, igniting the explosive at the center of convergence, followed by a divergent detonation burn wave moving supersonically in the radial direction from the center of convergence.

7.6 Ignition and Burn in the Teller-Ulam Configuration

To estimate the gain and yield in the Teller-Ulam configuration we go to Chapter 6.9.

According to published reports, the 1952 Mike test was a deuterium explosion with a yield of $\sim 10^7$ tons of TNT equivalent, or 4×10^{23} erg. The total energy released by the DD reaction, including the reactions with the secondary T and He^3 products of the DD reaction, was computed in Chapter 6.3, to be $\varepsilon_t = 13.4$ MeV $= 2.14 \times 10^{-5}$ erg. The fusion energy density in liquid deuterium with $n = 5 \times 10^{22}$ cm^{-3}, thus is $n\varepsilon_t = 1.1 \times 10^{18}$ erg/cm^3. For the reported output of 10^7 tons of TNT $= 4 \times 10^{23}$ erg, a volume of liquid deuterium equal to 3.74×10^5 cm^3 is therefore needed, or a liquid deuterium sphere with a radius $R = 94$ cm.

We compare these reports with the simple model developed in Chapter 6.9. For this we have to set in the gain formula (6.84) $\varepsilon_0 = \varepsilon_t = 2.14 \times 10^{-5}$ erg, $n = 5 \times 10^{22}$ cm^{-3}, $T = 3 \times 10^8$ °K, $nr \simeq 3 \times 10^{22}$ cm^{-2}. We obtain

$$G = 4 \times 10^{-8} \sqrt{E_{in}} \,. \tag{7.18}$$

To obtain a gain of 10^3, $E_{in} = 6 \times 10^{20} \simeq 15$ kt (TNT) is required. If the fission trigger was a Hiroshima type bomb, with an output of ~ 20 kt

$\simeq 8 \times 10^{20}$ erg, 75% of the energy released would have gone into igniting the fusion explosive, with 15% or 5 kt dissipated into the cavity wall.

With the temperature in the cavity of the order $T_0 \sim 10^7$ °K, rising in the convergent shock wave as $(R/r)^{0.9}$, the temperature at the radius $r = 1$ cm would be $T_1 = 6 \times 10^8$ °K, above the ignition temperature of the DD reaction. As we had pointed out in Chapter 5.3, the density in a convergent spherical shock wave, following its reflection at the center of the convergence at $r = 0$, is increased by the factor $2 \times (\gamma + 1) / (\gamma - 1) = 8$. Because of it, the temperature rises to $T = 8^{\gamma-1} T_1 = 4T_1 = 2.4 \times 10^9$ °K, or roughly to the temperature where $\langle \sigma \mathrm{v} \rangle$ (for the DD reaction) has its maximum.

The range of the charged fusion products of the DD reaction, with a density 32 times solid density (incoming wave 4 times, reflected wave 4 times, convergence 2 times), is $\lambda \simeq 1$ cm. Therefore, all the conditions for the ignition of a thermonuclear detonation wave launched from the center are met.

7.7 Nuclear "Spark Plugs"

To facilitate the ignition in the center of convergence in the Teller-Ulam configuration one may put there a DT fuse or "spark plug". Instead of DT one can use a mixture in equal amounts of Li^6D and Li^6T, which is a solid salt at room temperature. In doing so, one can lower the ignition temperature by at least one half order of magnitude, and with it the outer radius of the thermonuclear explosive by about the same factor, from say 100 cm down to 50 cm, with a reduction in the yield by one order of magnitude, from 10^7 tons to $\sim 10^6$ tons. And one can, of course, use a smaller fission bomb to act as a trigger.

A different kind of nuclear spark plug is a piece of fissile material compressed to high densities in or near the center of convergence of the convergent shock wave. The critical radius of a fission explosive scales as $1/n$ where n is the number density of fissile nuclei (see (2.42) where N stands for n). In the center of a convergent shock wave the pressure rises as $r^{-0.9}$, but due to the reflection of the wave at $r = 0$ and convergence the density actually rises by the factor 8, and the pressure thus by the factor $8^{5/3}$ =32. Therefore, if at $R \simeq 10^2$ cm the pressure was $\sim 10^{13}$ dyn/cm^2 (the radiation pressure for $T \sim 10^7$ °K), the pressure would rise $r \simeq 1$ cm to $10^{13} \times 32 \times 10^{1.8} \simeq 10^{16}$ dyn/cm^2. According to (3.171) this implies that $nZ \sim 4 \times 10^{25}$ cm^{-3}, hence for $Z = 92$, that $n \sim 4 \times 10^{23}$ cm^{-3} or $n = 10n_0$, where $n_0 = 4 \times 10^{22}$ cm^{-3} is the density of metallic uranium. The critical radius thus goes down from 7.9 cm to less than 1 cm. With the 10 fold density increase, the energy density in the fission spark plug is increased by the same amount to $\sim 4 \times 10^{18}$ erg/cm^3, and the temperature by $10^{1/4}$, from 8.7×10^7 °K to 1.55×10^8 °K.

With several fission spark plugs very large thermonuclear yields (like the Russian 100 MT device) are then possible. The spark plugs (for example six) are placed near the center of the spherical thermonuclear explosive in a polyhedral configuration, as in the six atomic bomb polyhedron configuration described in Chapter 7.4 (and shown in Fig. 7.4), except that the spark plugs are $\sim$ 10 times smaller (see Fig. 7.6). The incoming convergent shock wave compresses the spark plugs making them critical whereby they explode. The incoming shock wave is thus greatly amplified, reducing the temperature to launch a convergent shock wave from the spherical surface of the explosive to reach ignition at the center. With a reduced surface temperature the radius of the spherical explosive can be increased and with it the yield, without requiring a larger fission explosion.

This concept somehow resembles the fast ignitor explained in Chapter 6.9. In (spherical) convergent shock wave ignited assemblies, the gain is proportional to the square root of the input energy, making it difficult to reach high gains. This however, is possible with the fast ignitor, where the laser pulse provides the ignition spark. The same happens in the Teller-Ulam configuration with fission spark plugs.

7.8 Fission-Fusion-Fission Bombs

Large nuclear yields are possible if one makes the walls of the cavity confining the radiation from natural (U^{238}) uranium. There, the many neutrons released in the fusion explosion can make fission reactions in the cavity wall. This concept though, has fallen somehow into "discredit", because of the large radioactive fallout, which is also a problem for nonmilitary applications, such as canal excavation etc.

If the radiation confining wall has a thickness d, the probability for a neutron passing through the wall to make a fission reaction is

$$P = n_f \sigma_f d \tag{7.19}$$

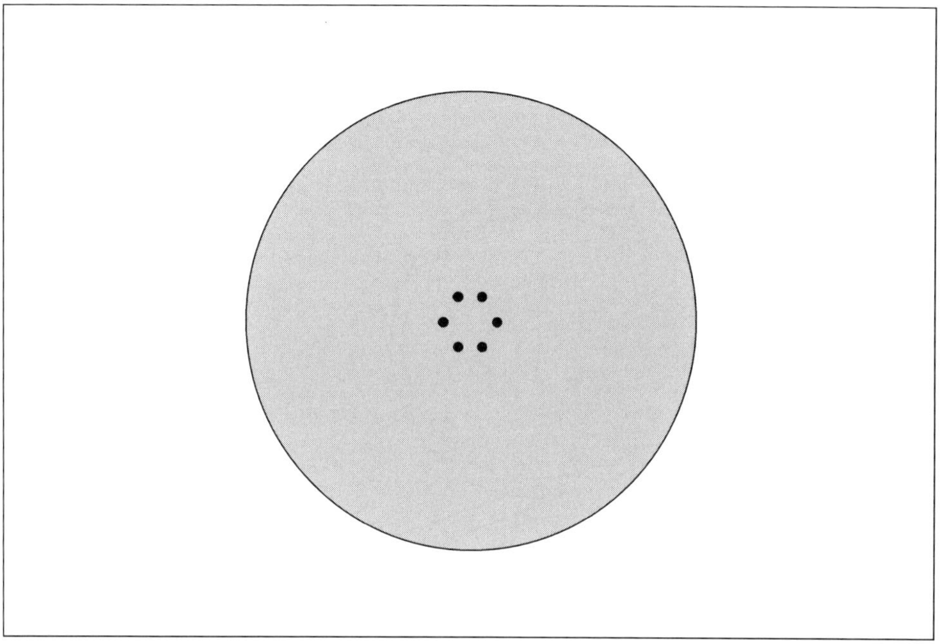

Figure 7.6: Six fission spark plugs to amplify a convergent shock wave near the center of convergence.

where $n_f = 4 \times 10^{22}$ cm^{-3} is the atomic number density of uranium, $\sigma_f = 2.3 \times 10^{-24}$ cm^2, the fast fission cross section in natural uranium. If N neutrons are released in the thermonuclear explosion, the fission energy released is

$$E = PN\varepsilon_f \tag{7.20}$$

where $\varepsilon_f = 2.9 \times 10^{-4}$ erg is the energy released per fission.

The number of neutrons is of the same order of magnitude as the number of nuclei in the fusion explosion. With $n = 5 \times 10^{22}$ cm^3, one has $N \simeq 2 \times 10^{28}$. Assuming $d \simeq 1$ cm one then has $E \simeq 5 \times 10^{23}$ erg $\sim 10^7$ tons of TNT.

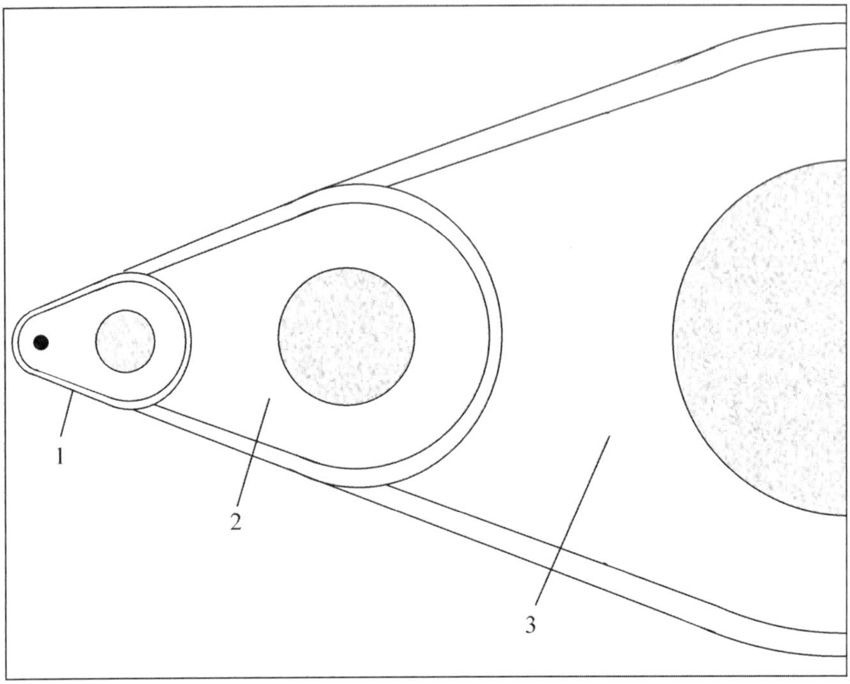

Figure 7.7: Staged thermonuclear explosion.

7.9 Staging of Thermonuclear Explosions

For the deflection of comets or asteroids which threaten the earth, and for the future prospect of lunar, asteroidal or planetary mining, very large thermonuclear explosions are required.

With the gain of a thermonuclear explosion (according to equation 6.84) going in proportion to $\sqrt{E_{in}}$, where E_{in} is provided by a fission explosion, and with the limited yield of a fission explosion, very large yields are not possible with the Teller-Ulam configuration.

There are two ways which promise arbitrarily large gains. The first is by a staged Teller-Ulam configuration. As shown in Fig. 7.7, a fission bomb, which is the zeroth stage, ignites the first stage thermonuclear explosive in cavity 1, with the energy for ignition delivered by the zeroth stage. By blasting a hole through the wall of cavity 1, the radiation from the first thermonuclear stage enters cavity 2, where it ignites the larger second stage thermonuclear explosive. The burn of the second stage will likewise blast a hole through the wall of cavity 2, with the radiation from the second stage entering cavity 3, igniting the still larger third stage, and so on.

For the energy E_1 set free in the first stage, we rewrite the gain formula (6.87) as follows

$$E_1 = cE_0^{3/2} = c\left(\eta E_f\right)^{3/2} \tag{7.21}$$

$$c \simeq 4 \times 10^{-8}\ \left[\mathrm{erg}^{-1/2}\right] \tag{7.22}$$

where E_f is the energy released by the fission bomb trigger, with the amount $E_0 = \eta E_f$ driving the convergent shock wave for ignition, and the amount $(1-\eta)\,E_f$ mainly dissipated into the cavity wall.

For the second stage one has

$$E_2 = c\left(\eta E_1\right)^{3/2} \tag{7.23}$$

and for the nth stage

$$E_n = c\left(\eta E_{n-1}\right)^{\alpha}\ , \qquad \alpha = \frac{3}{2}\ . \tag{7.24}$$

It is then easy to show by induction that

$$E_n = \left(\frac{1}{\eta}\right)\left(c\eta\right)^{\beta(n)}\left(\eta E_f\right)^{\alpha^n} \tag{7.25}$$

where

$$\beta(n) = \frac{\alpha^n - 1}{\alpha - 1}\ . \tag{7.26}$$

The total amount of energy released is obtained by summing up over all the n stages and adding the energy E_f of the zeroth stage fission bomb trigger:

$$E_{tot} = \sum_{n=1}^{n} E_n + E_f = \left(\frac{1}{\eta}\right)\sum_{n=0}^{n}\left(c\eta\right)^{\beta(n)}\left(\eta E_f\right)^{\alpha^n}\ . \tag{7.27}$$

For large thermonuclear explosive devices it is convenient to express the energy in tons t of TNT. There then $c = 8\left[t^{-1/2}\right]$. Let us take the example $E_f = 2 \times 10^4[\mathrm{t}]$ and $\eta = 1/2$. We have

$$\begin{aligned} E_1 &= c\,(\eta E_f)^{3/2} = 8 \times 10^6\ [\mathrm{t}] \\ E_2 &= c\,(\eta E_1)^{3/2} = 7.2 \times 10^{10}\ [\mathrm{t}] \\ E_3 &= c\,(\eta E_2)^{3/2} = 5.6 \times 10^{16}\ [\mathrm{t}] \end{aligned}$$

This example demonstrates that two stages are sufficient with the third stage quite extravagant. But it also demonstrates the frightening potential that if a way can be found to ignite with a chemical high explosive a thermonuclear microexplosion, a large thermonuclear explosion could be set off without a fission trigger.

To explore this possibility in some detail we turn to Chapter 6.9. There from the gain formula for the DT reaction and a 10^3-fold compressed DT, one has for $E_{in} = 3 \times 10^{13}$ erg (3 MJ), $E_{out} = 8 \times 10^{16}$ erg $= 2$ [t]. Replacing in (7.21) ηE_f with ηE_{out} one has ($\eta = \frac{1}{2}$, $c = 8\ [\mathrm{t}^{-1/2}]$)

$$\begin{aligned} E_1 &= c\,(\eta E_{out})^{3/2} = 8\ [\mathrm{t}] \\ E_2 &= c\,(\eta E_1)^{3/2} = 64\ [\mathrm{t}] \\ E_3 &= c\,(\eta E_2)^{3/2} = 1.4 \times 10^3\ [\mathrm{t}] \\ E_4 &= c\,(\eta E_3)^{3/2} = 1.5 \times 10^5\ [\mathrm{t}] \\ E_5 &= c\,(\eta E_4)^{3/2} = 1.6 \times 10^8\ [\mathrm{t}] \end{aligned}$$

One sees that with 5 stages one can, at least in principle, make a $\sim$100 megaton explosion.

7.10 Autocatalytic Thermonuclear Detonation

In staged thermonuclear explosions the yield rises rapidly from stage to stage. This raises the question if staging cannot be replaced by a continuous process with any desired yield in between those yields reached by staging. This appears possible with the concept of the autocatalytic thermonuclear detonation wave.

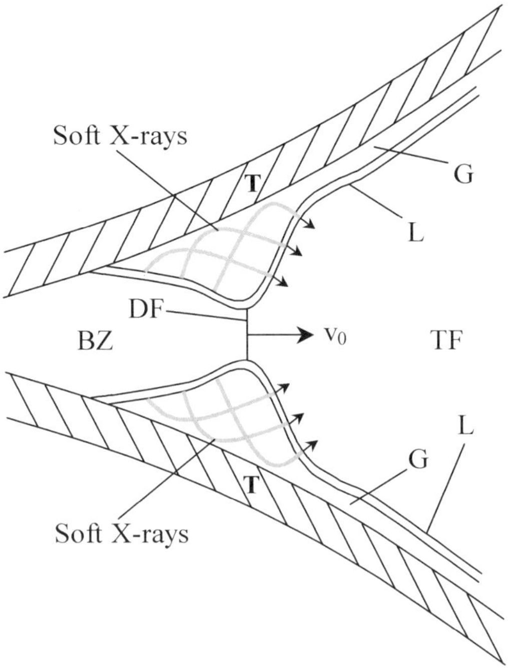

Figure 7.8: Autocatalytic thermonuclear detonation using a soft x-ray precursor from the burn zone BZ to precompress the thermonuclear fuel TF ahead of the detonation front DF. The soft x-rays travel through the gap G between the tamp T and the liner L.

In the Teller-Ulam configuration the thermonuclear fuel is compressed 32-fold in the center of the convergent shock wave. In the concept of the autocatalytic thermonuclear detonation, soft x-rays generated through the burn of the thermonuclear plasma behind the detonation front, compress the still unburned thermonuclear fuel ahead of the front. The increase in the fuel density, both in the Teller-Ulam configuration and the autocatalytic thermonuclear detonation wave, is of crucial importance, with the reaction rate proportional to the square of the density.

The concept of the autocatalytic thermonuclear detonation wave is explained in Fig. 7.8, showing a detail near the position of the detonation wave front. Ahead of the front is still unburned fuel TF, and behind it the thermonuclear burn zone BZ. The fuel is enclosed in a liner L, followed by a gap G between the liner and the tamp T. Both the liner and tamp consist of high Z material.

From the burning plasma behind the detonation front, energy flows into all spatial directions. Part is by bremsstrahlung and part by electronic heat conduction. Roughly half of the energy flows into the liner, a quarter into the still unburned fuel ahead of the wave and one quarter into the opposite direction.

The flux of the bremsstrahlung which goes into the liner is $\varepsilon_r r/2$, where ε_r is given by (4.67b), and where r is the radial coordinate of the cylindrical coordinate system with the rotational symmetric assembly made up of the fuel, the liner and the tamp centered on the z-axis. For a burn temperature $T \simeq 2 \times 10^8$ °K one has

$$\varepsilon_r r/2 \simeq 10^{-23} n^2 r \text{ [erg/cm}^2\text{ s]} \tag{7.28}$$

or with $nr \simeq 2.5 \times 10^{23}$ cm^{-2} (corresponding to $\rho r \simeq 1$ g/cm^2)

$$\varepsilon_r r/2 \simeq 3n \text{ [erg/cm}^2\text{ s]} \,. \tag{7.29}$$

The energy flux by electronic heat conduction is

$$j = -\kappa \frac{dT}{dr} \sim \kappa \frac{T}{r} \tag{7.30}$$

where κ given by (4.18) is $\kappa \simeq 2 \times 10^{-6} T^{5/2}$ [erg/s °K cm]. For $T = 2 \times 10^8$ °K this is

$$j \sim \frac{2 \times 10^{23}}{r} \text{ [erg/cm}^2\text{ s]} \,. \tag{7.31}$$

Comparing (7.29) with (7.31) one sees that bremsstrahlung is predominant for $nr \gtrsim 6 \times 10^{22}$ cm^{-2}. Because $nr \geq 2.5 \times 10^{23}$ cm^{-2} is required for thermonuclear burn, it is mainly bremsstrahlung which heats the liner. The energy absorbed by the liner is reemitted as blackbody radiation into the gap separating the liner from the tamp (provided the thickness of the liner is larger that the optical path length $(\kappa\rho)^{-1}$). The temperature T_ℓ of the liner is then obtained by equating $\varepsilon_r r/2$ with σT_ℓ^4 ($\sigma = 5.75 \times 10^{-5}$ erg/cm^{2}s°K). For the example $\varepsilon_r r/2 \simeq 3n$ [erg/cm^2 s], one has $T_\ell \sim 10^7$ °K. As for the radiation confining wall of the Teller-Ulam configuration one has $(\kappa\rho)^{-1} \sim 10^{-4}$ cm, small compared to a millimeter thick liner.

One half of the soft x-ray blackbody radiation emitted from the hot liner will flow in the same direction as the detonation wave, the other half into the opposite direction. Because the radiation flows with a velocity $\simeq c/3 = 10^{10}$ cm/s it outruns the detonation propagating with a velocity of $\simeq 10^9$ cm/s. With the detonation front positioned at $z' = z$, one can then compute the intensity ϕ of the radiation flux for $z' > z$ by the equation

$$\frac{d\phi}{dz'} = -\frac{1-R}{\delta}\,\phi \tag{7.32}$$

where R is the reflection coefficient of the inner tamp surface and δ the width of the gap. If $\delta = const.$, (7.32) has the solution

$$\phi\left(z'\right) = \phi\left(z\right) e^{-(z'-z)/\ell} \tag{7.33}$$

where $\ell = \delta/\left(1-R\right)$. For perfect (100%) reflection one has $R = 1$ and $\ell = \infty$. In this limit there is no attenuation of the radiation flowing through the gap. Assuming that $R = 0.75$, one has $\ell = 4\delta$, approximately reproducing the pattern shown in Fig. 7.8.

With half of the radiant energy flowing in the direction of the burn wave, one has

$$\phi\left(z\right) \simeq \frac{\varepsilon_r r}{4} = 1.5n \;\left[\mathrm{erg/cm^2\;s}\right]. \tag{7.34}$$

Assuming a ~ 30 fold compression, comparable to the compression in the Teller-Ulam configuration one has $\phi\left(z\right) \sim 2\times10^{24}$ erg/cm^2 s $= 2\times10^{17}$ Watt/cm^2, powerful enough to lead to a substantial precompression of the fuel in front of the detonation wave.

The autocatalytic thermonuclear detonation wave therefore has the potential for arbitrarily large yields.

If $c \ll 1$, with c defined by (6.67), a growing autocatalytic thermonuclear wave is possible as shown in Fig. 7.8. Not only is this possible with DT, D and Li^6D thermonuclear explosives, but, of course, also for a thermonuclear detonation wave of constant cross section. With such a burn wave one can contemplate the "pencil-bomb" shown in Fig. 7.9. It may have a military application for multiple warheads carried by one missile, but it appears also very practical for nuclear mining, by fitting into a narrow bore hole.

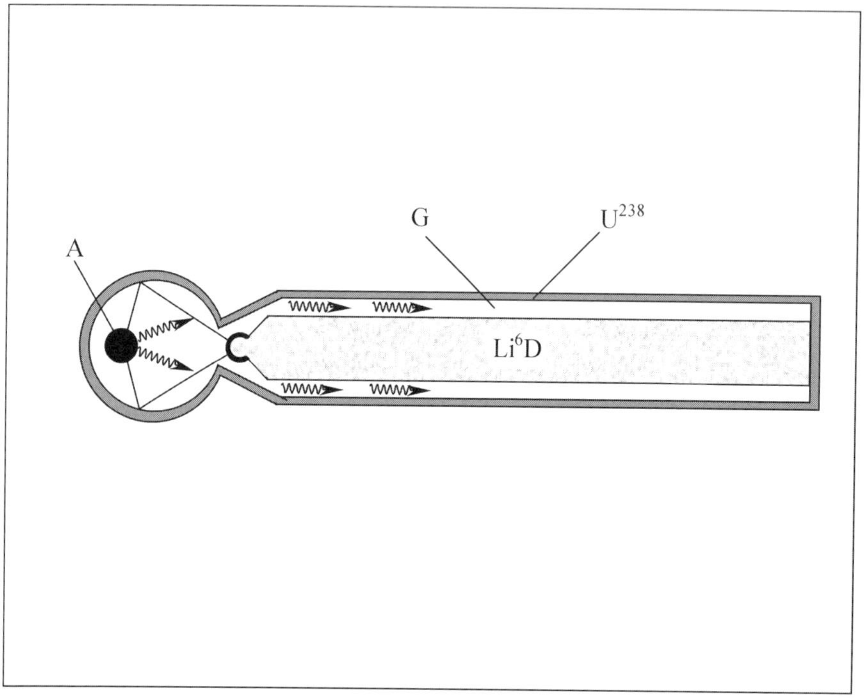

Figure 7.9: H-bomb using the autocatalytic principle, where the atom bomb A sends soft x-rays through the gap G between the U^{238} liner and the Li^6D thermonuclear fuel.

7.11 Magnetized Thermonuclear Explosive Devices

In an x-ray tube an electron beam penetrating a solid target releases x-rays by bremsstrahlung. The opposite, the making of an electron current by x-rays penetrating a target must then also be possible. The intense soft x-rays emitted from an exploding fission bomb can for this reason generate large electron currents in a solid target. In penetrating the target, the electron current becomes the source of a magnetic field, extending into the target up to the optical path length, resp. range of the x-rays. The range is short for high Z material targets, but can be quite large for low Z thermonuclear explosives. If the magnetic field is strong enough, or the electron current larger than I_c given by (6.23), the size of the thermonuclear explosive device can be reduced, replacing the stopping length λ_0 (4.25) with twice the ion Larmor radius.

Expressed in spherical polar coordinates, the differential and total photon-electron scattering cross sections of photons with frequency ν is:

$$d\sigma_e = \left(\frac{r_0^2}{137^4}\right)\left(\frac{mc^2}{h\nu}\right)^{7/2} 4\sqrt{2}\,\frac{\sin^2\theta\cos^2\phi}{(1-\beta\cos\theta)^4}\,d\Omega \tag{7.35}$$

$$\sigma_e = \left(\frac{8\pi r_0^2}{3\,(137^4)}\right) 4\sqrt{2}\left(\frac{mc^2}{h\nu}\right)^{7/2} \tag{7.36}$$

where $\beta = \mathrm{v}/c$ and $r_0 = e^2/mc^2$ the classical electron radius.

The velocity of the scattered electrons is

$$\mathrm{v} = \sqrt{\frac{2h\nu}{m}} = \sqrt{\frac{2kT}{m}} \tag{7.37}$$

the electrons are not scattered isotropically and the fraction $\overline{\cos\theta}$, where

$$\overline{\cos\theta} = \frac{1}{\sigma_e}\int \cos\theta\, d\sigma_e \tag{7.38}$$

contributes to a photoelectric current, with the photon flux vector along the z-axis. For $\beta \ll 1$ one has

$$\overline{\cos\theta} = \frac{2}{5}\beta\,. \tag{7.39}$$

With the photon flux ϕ and the macroscopic photon-electron scattering cross section $\sum_e = n\sigma_e$, the photocurrent density vector is

$$j = e\sum_e \phi\lambda\left(\frac{2}{5}\beta\right) \tag{7.40}$$

where $\lambda = 1/n\sigma_d$ is the electron mean free path with σ_d given by (4.9), hence

$$j = \frac{2\sigma_e}{5\sigma_d}\,e\beta\phi\,. \tag{7.41}$$

The photon flux $\phi = n_p\bar{c}$ (n_p photon number density and $\bar{c}$ the averaged forward photon velocity) is computed by setting

$$\sigma T^4 = n_p\bar{c}h\nu = n_p\bar{c}kT \,. \tag{7.42}$$

With $\sigma = ac/4$ one has

$$\phi = \frac{ac}{4k}T^3 \tag{7.43}$$

and finally

$$\beta = \sqrt{\frac{2kT}{mc^2}} \,. \tag{7.44}$$

Inserting the expressions for β, (σ_e/σ_d) and ϕ into (7.41) one has

$$\begin{aligned} j &\simeq 50T^2 \text{ [esu]} \\ &\simeq 2\times 10^{-8}T^2 \text{ [A/cm}^2\text{]} \end{aligned} \tag{7.45}$$

In Chapter 7.1 we had computed the temperature of an exploding fission bomb to be $T \simeq 8.7\times 10^7$ °K. This would lead to a photoelectric current density of $\sim 1.5\times 10^{10}$ [A/cm^2].

In the configuration shown in Fig. 7.10, the spherical fission explosive and the cylindrical fusion explosive are placed in a conducting cylinder which serves as the return current conductor for the photoelectric current in the fusion explosive. If the cylinder has a length $\sim$ 10 times larger than the diameter of the fission explosive, the radiation expands $\sim$ 10-fold reducing the temperature by the factor $10^{1/3}$ from 8.7×10^7 °K to 4×10^7 °K. At this temperature $j \simeq 3\times 10^7$ [A/cm^2], and for a cross sectional area of the cylindrical thermonuclear explosive equal to $\sim 10^2$ cm^2, the current would be $\sim 3\times 10^9$ [A], with a magnetic field of $\sim 10^8$ [G].

Part of the soft x-ray flux from the fission explosion can compress the cylinder, but since the current is so large, there is actually no need to compress the fusion explosive. The compressive effect of the x-rays still helps to hold the cylindrical explosive together more than would otherwise be possible.

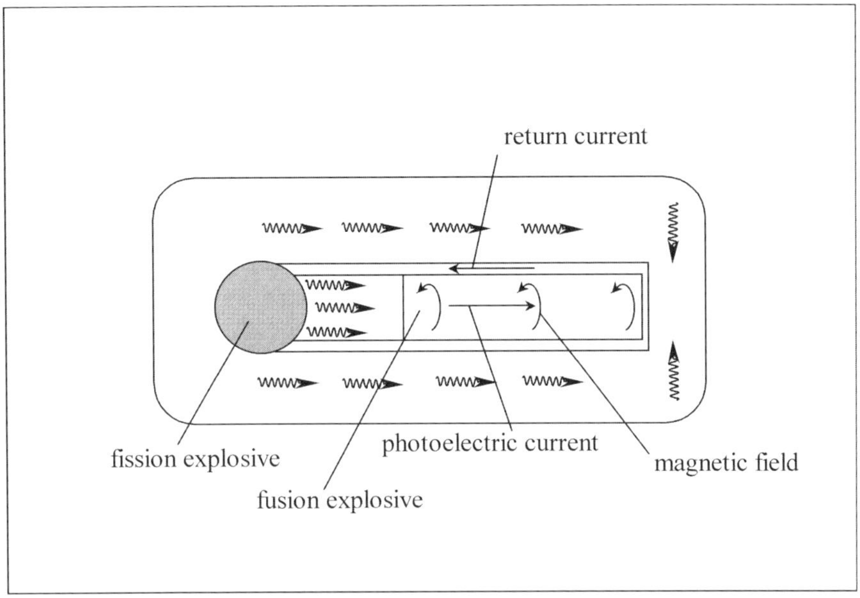

Figure 7.10: Magnetized thermonuclear explosive device.

7.12 Miniaturized Thermonuclear Explosive Devices

The concept of a magnetized thermonuclear explosive device is just one out of a large number of concepts for compact thermonuclear explosive devices. These concepts are of principal significance in the quest for non-fission-ignited thermonuclear microexplosions treated in the next part. One concept is the implosion of a shell at its inner side coated with a layer of frozen DT. As we had seen in Chapter 5.5 the shell implosion leads to high velocities. With a sufficiently high velocity the DT inside the shell can be compressed to $\sim 10^3$ times solid density, after which it is ignited at its center by a convergent shock wave. The same can be done in the Teller-Ulam configuration, first by imploding a shell of Li^6D, followed by the compression to high densities and ignition by a convergent shock.

The implosive compression of a fissile shell to high densities is also important for small fusion explosives, with the critical mass reduced by the factor $(\rho_0/\rho)^2$, where ρ_0 is the uncompressed and ρ the compressed density. Quite obviously for a compact thermonuclear explosive device a small fission trigger is important.

One particular concept is to fill the "hohlraum" of the Teller-Ulam configuration with foam consisting in part of high Z material. The density of the foam must be sufficiently low to keep the cavity optically transparent with regard to the $\sim 10^7$ °K soft x-rays stored in it. For the opacity of the foam we may use (4.71). With the temperature of $T \simeq 8.7 \times 10^6$ °K as in previous examples, we compute with (4.71) the optical path length in dependence of the foam density ρ [g/cm^3]:

$$\lambda_{opt} \simeq \frac{3 \times 10^{-2}}{\rho^2} [\text{cm}] \, . \tag{7.46}$$

Next we compare an empty volume V_0 storing blackbody radiation, with a volume V_1 filled with the foam and storing the same amount of energy E. For the volume V_0 we have

$$E = V_0 a T^4 \tag{7.47}$$

and for the volume V_1

$$E = V_1 \left[\frac{3}{2} (Z_i + 1) \, nkT + \frac{f}{2} nkT + aT^4 \right] . \tag{7.48}$$

The first term in the square bracket of (7.48) is the kinetic energy of the Z_i-times ionized plasma, the second term the energy stored in internal degrees of freedom of the Z_i-times ionized atoms, and the third term the energy stored as blackbody radiation.

The kinetic degrees of freedom for a 5-times ionized plasma are: $(3/2)(5+1) = 9$, but for a high Z-atom the internal degrees of freedom excited at this temperature are likely to be much larger. This implies that a large amount of energy can be stored in these internal degrees of freedom. However, before this potential can be assessed it must be made certain that the plasma pressure $(Z_i + 1) \, nkT$ does not exceed the blackbody radiation pressure $(a/3) \, T^4$, or that

$$nkT \leq \frac{(a/3) \, T^4}{(Z_i + 1)} \, . \tag{7.49}$$

From (7.47-7.48) it then follows that

$$\frac{V_1}{V_0} \geq \left[\frac{3}{2} + \frac{f}{6\left(Z_i + 1\right)}\right]^{-1} . \tag{7.50}$$

Let us assume there are $\sim 10^3$ internal degrees of freedom (corresponding to the many spectral lines of heavy elements), which would lead to $V_1/V_0 \geq 3.6 \times 10^{-2}$, or to a $\sim$ 3-fold reduction in the hohlraum radius. In the Teller-Ulam configuration a meter-size hohlraum radius was assumed. The reduced hohlraum radius of $\sim$ 30 cm cannot be larger than λ_{opt} (7.46). From this condition it follows that $\rho \lesssim 3 \times 10^{-2}$ g/cm^3. This number must be compared with the number n from (7.49)

$$n \leq \frac{a}{3k\left(Z_i + 1\right)} T^3 \tag{7.51}$$

which for $Z_i = 5$, $T = 8.7 \times 10^6$ °K one has $n \leq 2 \times 10^{21}$ cm^{-3}. The density is $\rho = nAM_\mu$, where A is the atomic number and M_μ the proton mass. One finds for $A \simeq 50$ that $\rho \lesssim 0.1$ g/cm^3. With a properly chosen foam material this condition can be met. We have to remember that the formula (4.71) is not very accurate and κ may actually be smaller for a certain combination of high-Z materials. From (7.47-7.49) one has for the maximum energy density ε in the foam-filled cavity

$$\varepsilon = gaT^4 \tag{7.52}$$

where

$$g = \left(\frac{3}{2} + \frac{f}{6\left(Z_i + 1\right)}\right) . \tag{7.53}$$

In a departure from (4.72) one then has for the radiation flux

$$j_r = -g\frac{\lambda_{opt} c}{3} \nabla \left(aT^4\right) . \tag{7.54}$$

A further reduction in the size of the thermonuclear explosive device is possible by combining a foam-filled cavity with a pre-compression of the explosive.

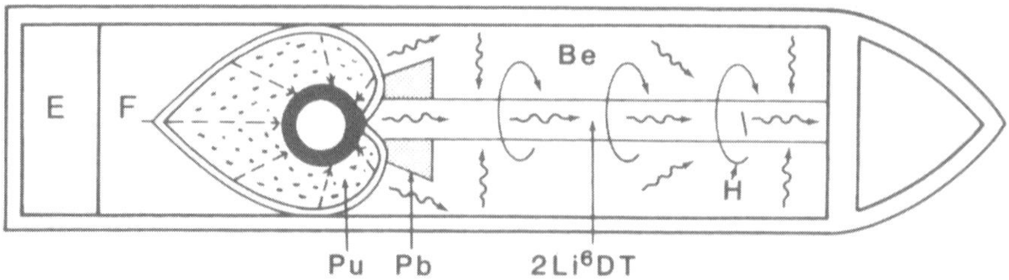

Figure 7.11: Neutron radiation enhanced fusion explosive (neutron bomb).

Finally, we would like to make a few comments about the so-called neutron bomb. This is (most likely) a small LiDT fusion bomb, surrounded with a layer of beryllium to increase the neutron output by n, $2n$ nuclear reactions. Because the cross section of the n, $2n$ reaction increases with lower neutron energies, one may use beryllium hydride, because hydrogen is a very good neutron moderator.

According to (3.3) the complete ionization temperature for beryllium is $T_i \simeq 6 \times 10^6$ °K. With a blackbody radiation temperature of $T_i \simeq 3 \times 10^7$ °K from the fission explosive, T is well above T_i, so that the beryllium is fully ionized. One can therefore use (4.70) to compute σ_{opt}, and one finds $\sigma_{opt} \simeq 10^{-47} n$. And for solid beryllium with $n \simeq 5 \times 10^{22}$ cm^{-3}, one has $\lambda_{opt} \simeq 1/n\sigma_{opt} \simeq 40$ cm. In a volume $\lambda_{opt}^3 = (40)^3$ cm^3, the energy $\lambda_{opt}^3 a T^4 \simeq 6 \times 10^{20}$ erg can be stored, as estimated in Chapter 7.6 sufficiently large for the ignition by a convergent shock wave.

The large optical path length in solid beryllium at a temperature above 10^7 °K suggests the kind of configuration shown Fig. 7.11, as a candidate for a neutron radiation enhanced fusion bomb.

7.13 Thermonuclear Explosion Driven X-Ray Lasers

Optical lasers can be pumped by intense light sources. Replacing the light sources by soft x-ray flashes from nuclear explosions (fission or fusion), one can pump x-ray lasers. Fusion explosions, in particular the neutron enhanced version (neutron bomb), offer the possibility to pump x-ray lasers with neu-

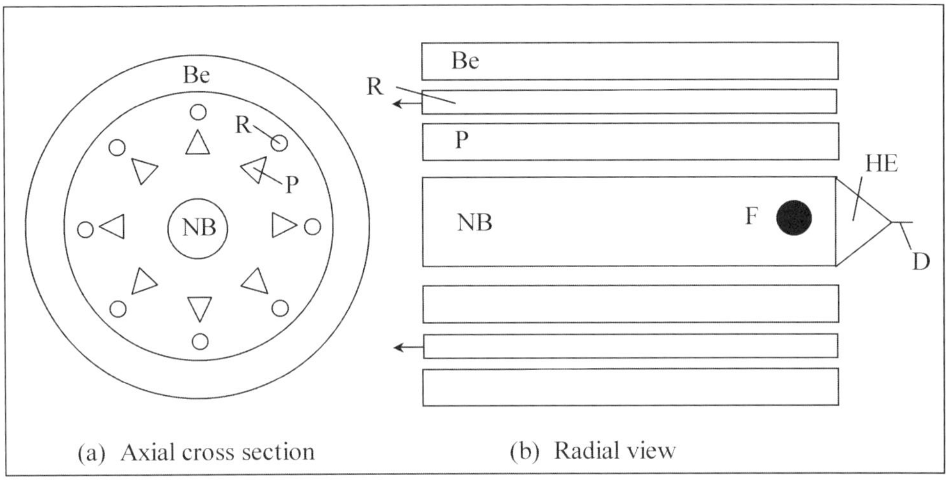

Figure 7.12: Nuclear X-ray laser pumped by a neutron bomb. The cylindrical neutron bomb NB is placed within a cylindrical neutron reflector Be made of beryllium 9. The detonator D sets off a high explosive HE, which in turn explodes the fission trigger F for the neutron bomb. The prisms P surrounding the neutron bomb prevent the laser rods R from being vaporized prematurely. The neutrons from the bomb penetrate the laser rods, which produce laser beams of intense X-rays.

trons. There are several reasons why pumping with neutrons should be better than pumping with x-rays. First, x-ray lasers require high Z material into which neutrons can penetrate much better than x-rays. Second, pumping the laser with neutrons implies pumping it with charged nuclear reaction products from neutron induced nuclear reactions. Third, as it is with electron beam pumping compared to optical pumping, laser pumping with charged nuclear reaction products should be much more efficient compared to pumping with x-rays.

Neutron pumped x-ray lasers can be realized by mixing the lasing material with substances which upon absorption of neutrons release charged nuclear reaction products. Two examples for such substances are Uranium 235 and Boron 11. In the first case the charged nuclear reaction products are fission products and in the second case they are α-particles.

One way a neutron bomb pumped x-ray laser might work is shown in Fig. 7.12. There a pencil-like neutron bomb, designed by the principle previously explained in Fig. 7.9, is positioned in the center of a beryllium 9 neutron reflector. Surrounding the cylindrical neutron bomb are several prisms which prevent the several laser rods from being prematurely vaporized by the x-ray flash of the nuclear explosion.

7.14 Mini Fission-Fusion Explosive Devices

What has come under the name "mini-nukes" are mini-fission-fusion explosive devices where the thermonuclear fusion reaction aides a fission chain reaction and vice versa, reducing the critical mass of the fission chain reaction, the latter providing heat to enhance the thermonuclear fusion reaction. Because of the large number of energetic fusion reaction neutrons and the low ignition temperature it works best with the DT thermonuclear reaction. As it had been shown in Chapter 2.6, astonishingly small critical masses are here possible.

As a typical example we take a "mini-nuke" with a critical mass of 10 g. Assuming a 10% fuel burn up it would release an energy of $\sim$ 100 GJ, equivalent to more than 20 tons of TNT.

A cross section through such a mininuke is shown in Fig. 7.13. At its center is the core made from fissile material with a radius $r_0 \simeq 0.5$ cm, surrounded by a shell with a radius $r_1 \simeq 1$ cm, serving as pusher and confining DT gas under a pressure of 200 atm, with a DT particle number density 5×10^{21} cm^{-3}, equal to 1/10 solid DT particle number density. At the outer surface of the pusher is the ablator, which ideally is a layer of beryllium. Surrounding the pusher-ablator shell is an aluminum radiator shell with a radius $r_2 \simeq 1.5$ cm, in turn surrounded by a larger aluminum shell with a radius $r_3 \simeq 15$ cm on its outer side covered with a several cm thick layer of a high explosive. In between r_1 and r_2, and in between r_2 and r_3, is vacuum.

Following the simultaneous ignition of the high explosive at its outer surface, the aluminum shell of radius $r_3 \simeq 15$ cm is with an initial velocity of ~ 5 km/s imploded onto the inner aluminum shell of radius $r_1 \simeq 1$ cm with an impact velocity of ~ 50 km/s. With the inner aluminum shell becoming the source of intense blackbody radiation, the pusher with a radius $r_1 \simeq 1$ cm is accelerated towards the fissile core with the radius $r_0 \simeq 0.5$ cm, reaching a velocity of ~ 200 km/s, shock-heating and adiabatically compressing the DT to a temperature of a few 10^7 °K and to a pressure of a few 10^{13} dyn/cm^2, resulting in the onset of a fission-fusion chain reaction.

Higher yields or still smaller mini-nukes are possible if the pusher is made from natural uranium or fissile material like plutonium, because there the autocatalytic fission-fusion process described in Chapter 6.11 increases the implosion velocity of the pusher.

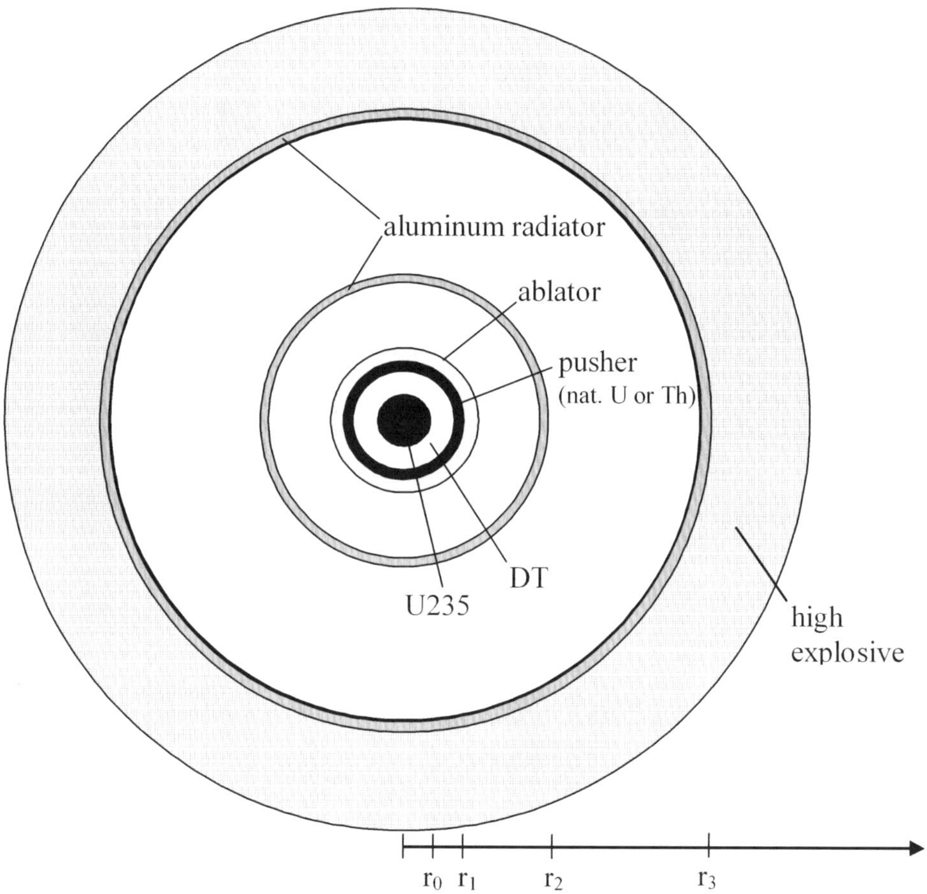

Figure 7.13: Mini-nuke cross section.

7.15 Bibliography for Chapter 7

F. Winterberg, The Physical Principles of Thermonuclear Explosive Devices, Fusion Energy Foundation, New York (1981).

F. Winterberg, Vom griechischen Feuer zur Wasserstoffbombe, E. S. Mittler & Sohn (1992)

F. Winterberg, J. of Plasma Physics **16**, 81 (1976).

F. Winterberg, Atomkernenergie, **39**, 181 (1981); **39**, 265 (1981); **41**, 291 (1982), **43**, 268 (1983); **44**, 145 (1984).

F. Winterberg, Nature, **241**, 449 (1973).

F. Winterberg, Fusion, August 1981, p. 54ff.

Chapter 8

Non-Fission Ignition

8.1 Energy Storage for Non-Fission Ignition

A fission explosive has an energy density of the order $\sim 10^{19}$ erg/cm^3, about 8 orders of magnitude larger than the energy density of chemical explosives, but unlike chemical explosives it is subject to the "tyranny" of the critical mass" (F. Dyson). For that reason, the fusion explosion ignited by a fission explosion should be at least as large as the fission explosion, if the ignition of the fusion explosion does not become extravagant. Chemical energy, second only to nuclear energy, does not live under the tyranny of a critical mass, but if used directly it is not powerful enough to ignite a fusion reaction. The minimum energy for ignition (several 10^6 J) is really not that large, but it has to be delivered in less than 10^{-8} sec, onto an area less than 1 cm^2, with a power in excess of 10^{14} Watt/cm^2. This is the principal problem of non-fission thermonuclear microexplosion ignition.

The problem is illustrated in table 8.1, which displays the properties of the primary energy storage systems. Besides the energy density, the second most important property is the velocity at which the energy can be drawn from the energy storage system. With the exception of nuclear (and anti-matter) energy storage devices, none qualifies for the required energy flux of $\gtrsim 10^{14}$ Watt/cm^2 Therefore some means must be used to compress (i.e. cumulate) the energy in space and time, regardless of its primary origin.

Explosively released chemical energy can be cumulated by many orders of magnitude with convergent shock waves and imploding shells. This has opened the door to speculations that a thermonuclear reaction might perhaps

	e erg/cm^3	v cm/sec	$\phi = $ ev Watt/cm^2	technical
kinetic	10^{10}	10^5	10^8	flywheel
electric	10^6	3×10^{10}	3×10^9	capacitor
magnetic	$\leq 4 \times 10^9$	3×10^{10}	$< 10^{13}$	coil
chemical	10^{11}	10^6	10^{10}	high explosive
nuclear fission*	10^{19}	10^9	10^{21}	fission bomb
nuclear fusion*	10^{18}	10^9	10^{20}	fusion bomb
m_0c^2	$> 10^{21}$	3×10^{10}	3×10^{24}	antimatter

Table 8.1: Primary energy storage systems (*in erg/g nuclear fusion $\sim 10\times$ nuclear fission).

be ignited in this way. However, this seems unlikely, other than with the magnetic booster concept described later. Finally, there is the hypothetical possibility of a chemical explosive which releases the energy directly into laser radiation. If such an explosive should exist it would open the frightening possibility of a compact chemically triggered large thermonuclear explosive device. For this the most likely candidates are explosives made from noble gas compounds.

Next in line are magnetic energy storage systems, inductively storing energy in magnetic field coils. There the principal problem is the rapid opening of a switch to interrupt large electric currents. The density of magnetically stored energy is $e = \mu H^2/8\pi$, where μ is the magnetic permeability. For vacuum (air) one has $\mu = 1$. The upper limit for the magnetic field H is derived from the tensile strength σ_{max} of the coil material, whereby $H_{max} \leq \sqrt{4\pi\sigma_{max}}$. For the typical value $\sigma_{max} \simeq 10^{10}$ dyn/cm^2, one has $H_{max} \simeq 3 \times 10^5$ G. The magnetic permeability of ferromagnetic materials can be quite large but ferromagnets cannot be used to increase the energy density for magnetic fields in excess of the saturation field strength for these materials. Their saturation field strength can be quite large, for gadolinium 6×10^4 G, but they are without exception below H_{max} deduced from the tensile strength limit. A large value of μ implies that below the saturation field strength, where the benefit of a large μ is important, the magnetic energy is stored in magnetic dipoles of the ferromagnetic material, making it difficult to withdraw this energy in the short time needed, thus favoring "air coils".

As stated above the principal problem of magnetic energy storage is the opening of a switch in an inductive circuit, where arcing occurs at the location the current is interrupted. If the opening can be done fast enough, one may place the "load" (i.e. thermonuclear microexplosion assembly) at the position the current is interrupted.

Coming down in the list of table 8.1 to electric energy storage in capacitors, the energy density is there not limited by the tensile strength, but rather by the dielectric breakdown of the material inside the capacitor, with a breakdown field strength of the order $\sim 10^5$ Volt/cm. The energy is there withdrawn by closing (not opening as for magnetic energy storage) a switch. This can be done with a spark gap switch in less than 10^{-8} sec. The density of electrically stored energy is $e = \varepsilon E^2/8\pi$, where ε is the dielectric permeability. For ferroelectric substances ε can be quite large (for barium titanate $\varepsilon \simeq 5000$), but there the breakdown field strength is much lower, and the maximum value of e is therefore not much larger. However, it seems quite conceivable that ferroelectric substances can be stabilized (by giving them the structure of layered single crystals). If this should turn out to be feasible, it would open another frightening prospect for the ignition of large thermonuclear explosions by compact capacitors.

The electric energy stored in capacitors is proportional to ε, but the velocity by which this energy can be withdrawn is equal to $c / \sqrt{\varepsilon}$. The power flux is therefore increased by the factor $\sqrt{\varepsilon}$ only, for barium titanate from 3×10^9 Watt/cm^2 to 2×10^{11} Watt/cm^2. Water where $\varepsilon = 81$ would increase the power flux by the factor $\sqrt{81} = 9$, from 3×10^9 Watt/cm^2 (by about one order of magnitude) to 2×10^{10} Watt/cm^2. This explains the popularity of water filled capacitors for electric pulse power.

A power of $\sim 3 \times 10^{10}$ Watt/cm^2 still falls short by about four orders of magnitude from the required 10^{14} Watt/cm^2, needed for thermonuclear ignition. These four orders of magnitude can be bridged by the idea of magnetic insulation (see also chapter 8.5). In high vacuum electric breakdown is caused by electron field emission. But field emitted electrons can be prevented from crossing the cathode-anode gap between conductors of different polarity by placing a strong magnetic in between and parallel to the conductor surfaces, provided $H \geq E$ (more correctly $H \gtrsim 0.7E$), with H and E in electrostatic cgs units. Under these conditions the field-emitted electrons execute a drift motion perpendicular to both H and E (see equation (3.23)). For $H = H_{max} \simeq 3 \times 10^5$ G, the maximum power flux with magnetic insulation is $\phi = (H_{max}^2/4\pi)\, c \simeq 3 \times 10^{20}$ erg/cm^{2}s $= 3 \times 10^{13}$ Watt/cm^2.

Much higher power fluxes are possible by magnetic self-insulation, which occurs in a coaxial transmission line, where the azimuthal magnetic field set up by a current passing through the line insulates the line against radial breakdown. There the magnetic and electric stresses compensate each other, whereby H can become much larger than $H_{max} = \sqrt{4\pi\sigma_{max}}$. The same thing happens in a relativistic charged particle beam where the repulsive electric beam field is compensated by the equally strong attractive magnetic beam field.

The maximum kinetic energy density which can be stored in a flywheel is determined by the inequality $\frac{1}{2}\rho v^2 \leq \sigma_{max} \simeq 10^{10}$ dyn/cm^2. It is smaller than the chemical energy density, but larger than the electric energy density, and comparable to the magnetic energy density. However, because of the low velocity by which it can be removed, its energy flux is smallest. Nevertheless because of its simplicity and compactness, energy drawn from it can be used to magnetize coils or charge capacitors.

As it is with electric energy storage, where the limitations can be overcome by magnetic insulation, so can the kinetic energy density be substantially increased in projectiles (macroparticles) accelerated to high velocities.

I would like to conclude this chapter with a general comment: Thermonuclear ignition depends on large pulse power. There is plenty of pulse power in a fission explosion, leaving a wide gap between nuclear and non-nuclear pulse power. A breakthrough in closing or narrowing this gap could change the entire picture, making obsolete much of the present research efforts towards fissionless thermonuclear ignition.

8.2 Electric Pulse Power

Be it for laser excitation, intense particle beam acceleration, or ultra short z-pinches, all of them considered for thermonuclear microexplosion ignition, electric pulse power plays a central role. It is here where the whole idea of fissionless thermonuclear microexplosions stands or falls.

One can identify four basic electric pulse power concepts. In a very superficial way they are displayed in Figure 8.1. They are:

1. Capacitive Marx generators.

2. Levitated magnetically insulated Gigavolt capacitors.

3. Inductive Marx generators.

4. Homopolar flywheel generators.

1. Capacitive Marx generator

There one has a bank of say n capacitors, each with a capacitance C, and in parallel with a total capacitance equal to

$$C_p = nC\,. \tag{8.1}$$

The charging of the bank to the voltage V is done by a high voltage transformer and rectifier. After completion of the charging, the n capacitors are switched in series by closing the spark gap switches, with the resistors preventing the capacitors from short-circuiting. By being switched in series their voltages add up from V to nV, with their total capacitance now

$$C_s = \frac{C}{n}\,. \tag{8.2}$$

While the discharge time for the capacitors in parallel is

$$\tau_p = \frac{\pi}{c}\sqrt{LC_p} = \frac{\pi}{c}\sqrt{LnC} \tag{8.3}$$

it is, after they are switched in series,

$$\tau_s = \frac{\pi}{c}\sqrt{LC_s} = \frac{\pi}{c}\sqrt{L\frac{C}{n}} = \frac{\tau_p}{n} \tag{8.4}$$

where L is the self-inductance of the bank. The spark gap switches close in $\sim 10^{-9}$ sec. By going to higher voltages, from V to nV, the discharge time goes down by $1/n$ while the power goes up by n. If, for example, 100 capacitors are charged in parallel to 100 kV, their voltage following the closing of the spark gap switches is 10^7 V.

2. Levitated magnetically insulated Gigavolt capacitor

Electric breakdown in ultrahigh vacuum occurs in the following order, (1) at the triple point conductor-insulator-vacuum, and (2) by electron field emission from the conductor. Breakdown from the triple point can be avoided

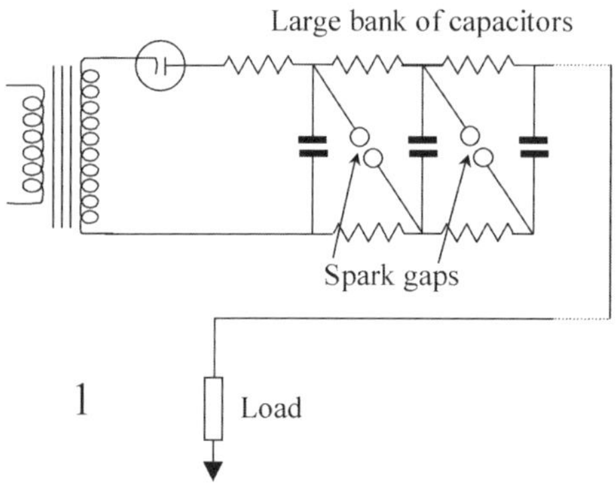

1. In the Marx circuit, capacitors charged in parallel are switched in series by spark gaps. This multiplies the charging voltage by the number of capacitors.

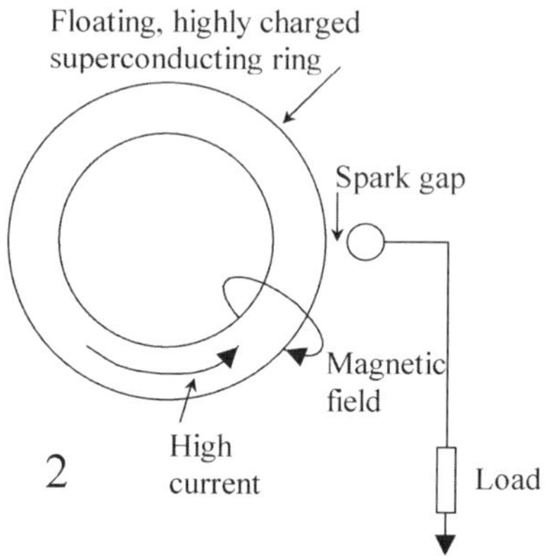

2. A magnetically levitated, superconducting ring, isolated in a high vacuum and carrying a strong electric current, might be charged up to gigavolt potentials before breakdown.

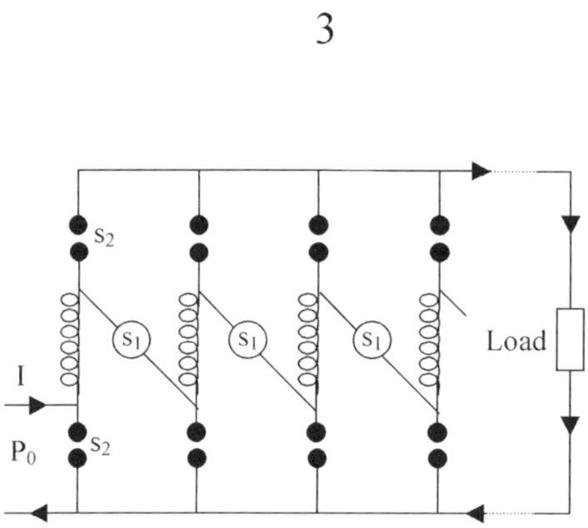

3. In the inductive Marx generator (Xram), magnetic field coils magnetized in series are switched in parallel. This multiplies the charging current by the number of coils.

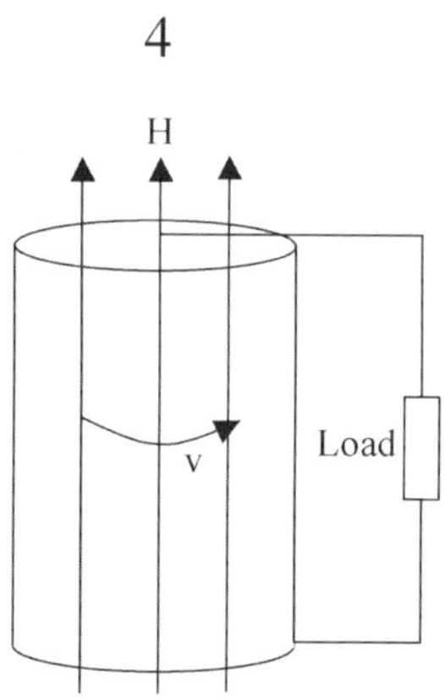

4. Homopolar flywheel generator.

Figure 8.1: The four basic electric pulse power concepts.

by magnetically levitating the conductor, and breakdown by electron field emission with magnetic insulation.

This idea can be realized by a levitated superconducting ring, with the toroidal ring currents in the superconductor supplying the insulating magnetic field. If, for example, the ring currents generate a magnetic field of $H \simeq 3 \times 10^4$ G, this makes it possible to charge the ring to a surface voltage of the same magnitude, that is to 3×10^4 esu $\simeq 10^7$ V/cm. If the ring has meter-size dimensions, it can then be charged up to $\sim 10^9$ V $\simeq 3 \times 10^6$ esu. In electrostatic units the capacitance is of the same order as the linear dimension of the ring, that is $C \sim 10^2$ cm. The electrostatic energy stored with the ring is of the order $\sim CV^2 \sim 10^2 \times (3 \times 10^6)^2 \sim 10^{15}$ erg $= 100$ MJ. With $L \sim C$ the discharge time goes as $\sim \sqrt{LC}/c \sim C/c \sim 3 \times 10^{-9}$ s, with a discharge power $\sim 3 \times 10^{15}$ Watt, delivered to the load by closing a spark gap switch. The charging of the ring to the high voltage can be done by a charged particle beam, for a negatively charged ring by a GeV electron beam, and for a positively charged ring by a GeV ion beam. To prevent deflection of the charging beam by the magnetic field of the ring, the beam must be projected along the major axis, with a charge collector in the middle of the ring. One can also charge the ring with a stream of fast moving charged pellets.

3. Inductive Marx generator

Reading the name of Marx in reverse order, this is sometimes called a Xram. There n coils switched in series are each magnetized by the same current I flowing through the coils. By opening and closing switches, whereby the bank is switched from "in series" into "in parallel", n currents add up to the total current nI. The problem as with any inductive energy storage system is the opening of the switches, but the Xram alleviates this problem, because the added up currents nI can be switched off by n switches, each of which has to interrupt only the current I.

A possible elegant solution for the switch-opening problem will be later given in the context of fast z-pinches. The advantage of the Xram over the Marx is the much higher magnetic energy density in comparison with the electrostatic energy density. The advantage of the Marx over the Xram is the short time in which the spark gaps close.

4. Homopolar flywheel generator

This is a rapidly rotating metallic cylinder placed in a magnetic field parallel to the axis of rotation. With the velocity v of the rotating cylinder perpendicular to the magnetic field H, there is a radial electric field

$$E = \frac{\mathrm{v}}{c} H \, . \tag{8.5}$$

For a velocity $\mathrm{v} = 3 \times 10^4$ cm/s (at the tensile strength limit $\frac{1}{2}\rho \mathrm{v}^2 \simeq \sigma_{max} \simeq 10^{10}$dyn/cm^2, and a magnetic field $H = 2 \times 10^4$ G (typical for powerful electromagnets) one has $E = 10^{-2}$ esu $= 3$ V/cm. For a meter size cylinder one then has a voltage of ~ 300 V between the axis and periphery of the cylinder. For this reason the homopolar generator can be viewed as a capacitor with a dielectric constant ε obtained by equating the kinetic energy density

$$e_k = \frac{\rho \mathrm{v}^2}{2} \tag{8.6}$$

with the electric energy density

$$e_e = \frac{\varepsilon E^2}{8\pi} = \varepsilon \left(\frac{\mathrm{v}}{c}\right)^2 \frac{H^2}{8\pi} \, . \tag{8.7}$$

The result is that

$$\varepsilon = \frac{4\pi \rho c^2}{H^2} \, . \tag{8.8}$$

As an example, we take $\rho = 7$ g/cm^3 (steel) and as before $H = 2 \times 10^4$ G. We find $\varepsilon = 2 \times 10^{14}$. With this large effective dielectric constant, the discharge time is of the order $R\sqrt{\varepsilon}/c \sim 5 \times 10^{-2}$ s where $R \sim 10^2$ cm is the radius of the cylinder. The kinetic energy stored in the cylinder is for a cylinder of equal height and diameter of the order $E_k \sim \rho R^3 \mathrm{v}^2 \sim 10^{16}$ J, delivered in 5×10^{-2} s with a power in excess of $\sim 10^{10}$ Watt.

8.3 Intense Electron and Ion Beams

Intense electron and ion beams are important as a possible means for thermonuclear ignition. However, for the intensities required, the space charge of the beams must be neutralized by projecting them through a tenuous background plasma, otherwise the beams will be radially dispersed. And for beams above a critical current (Alfvén current), the beam current must as well be neutralized by an induced return current in the same background plasma, otherwise the beams are stopped by their self-magnetic field.

We first analyze the situation for intense electron beams and show thereafter how these results must be modified for intense ion beams.

The electric field of the electron beam is determined by Poisson's equation

$$\text{div}\,\mathbf{E} = 4\pi n_b e \tag{8.9}$$

where n_b is the electron number density of the beam. Solving (8.9) in cylindrical coordinates one obtains for the radial electric beam field

$$\left.\begin{aligned} E_r &= 2\pi n_b e r\,, && r < r_b \\ &= 2\pi n_b \frac{r_b^2}{r}\,, && r > r_b \end{aligned}\right\} \tag{8.10}$$

Space charge neutralization requires that $Zn \geq n_b$, where n is the ion number density of a Z times ionized background plasma. The beam current for a fully space charge neutralized beam then is

$$I = \pi r_b^2 n_b e \beta \tag{8.11}$$

where $\beta = \mathrm{v}/c$, with v the velocity of the beam electrons.

The space charge neutralized beam with $E = 0$ has an azimuthal magnetic self-field $H = H_\phi$, which at the beam radius r_b is

$$H = \frac{2I}{r_b c}\,. \tag{8.12}$$

Because of this beam field, electrons are turned around by the magnetic $(e/c)\,(\mathbf{v} \times \mathbf{H})$ force if

$$r_L \leq \frac{r_b}{2} \tag{8.13}$$

where

$$r_L = \frac{m_\perp \mathrm{v} c}{eH} = \frac{\gamma m \mathrm{v} c}{eH} = \frac{\beta \gamma m c^2}{eH} \tag{8.14}$$

is the electron Larmor radius with $m_\perp = \gamma m$ the "transverse" relativistic electron mass and $\gamma \equiv (1-\beta^2)^{-1/2}$. From (8.12-8.14) one obtains

$$I \leq I_A \tag{8.15}$$

where

$$I_A = \frac{\beta\gamma mc^3}{e} = 17\,000\,\beta\gamma\,[\mathrm{A}] \tag{8.16}$$

is the so-called Alfvén current.

Currents in excess of I_A are possible by induced return currents in the background plasma. In order to understand this, suppose that the background plasma temperature is high enough to put in equation (3.85) $\sigma = \infty$, whereby (3.85b) becomes $\partial\mathbf{H}/\partial t = 0$. Therefore, if initially $\mathbf{H} = 0$, it must remain so after a beam is projected into the plasma. This happens because a return current is induced in the background plasma, but, as before, this requires for the background plasma that $nZ \geq n_b$.

If the beam is heating the plasma, and if initially $\sigma \neq 0$, with $\partial\mathbf{H}/\partial t \geq 0$, the magnetic field of the beam can be partially trapped in the plasma, but it cannot become larger than the field produced by the Alfvén current.

If the beam is propagating with incomplete space charge neutralization, through a tenuous, or partially ionized background plasma with $nZ < n_b$, currents in excess of the Alfvén current are possible by a balance of the electric and magnetic forces acting on a beam electron. With incomplete space charge neutralization (henceforth putting $n_b = n$), the radial electric beam field is

$$E = (1-f)\,2\pi ner\,, \quad r < r_b \tag{8.17}$$

where f is the degree of space charge neutralization. The magnetic field is

$$H = \frac{2I}{r_b^2 c}\,r\,, \quad r < r_b\,. \tag{8.18}$$

With $I = \pi r_b^2 ne\beta c$ one has

$$H = 2\pi ne\beta r \tag{8.19}$$

hence

$$E = (1 - f)\,\frac{H}{\beta}\,. \tag{8.20}$$

The radial force acting on a beam electron then is

$$\begin{aligned} F &= e\,(E - \beta H) \\ &= eH\left[\frac{1-f}{\beta} - \beta\right] = \frac{2eI}{r_b^2 \beta c}\left[\frac{1}{\gamma^2} - f\right] \end{aligned} \tag{8.21}$$

The force vanishes if

$$f = \frac{1}{\gamma^2}\,. \tag{8.22}$$

Under this condition arbitrarily large currents, and with it arbitrarily large self-fields are possible, but only for relativistic beams. According to (8.20), in the limit $\gamma \to \infty \quad (\beta = 1)$, one has $E = H$. If f is slightly larger than $1/\gamma^2$, the beam collapses by its self-magnetic field, with the magnetic field becoming larger than the repulsive electric field. Therefore, if a beam is projected into a plasma where $f > 1/\gamma^2$, the beam collapses into a narrow filament. This self-focusing of the beam is important for certain thermonuclear ignition concepts.

For $f > 1/\gamma^2$ one has for the equation of motion of an electron

$$m_\perp \ddot{r} = F \tag{8.23}$$

or

$$\ddot{r} + \omega^2 r = 0 \tag{8.24}$$

where

$$\omega^2 = \omega_\beta^2\left[1 - \frac{1-f}{\beta^2}\right] \tag{8.25}$$

and where

$$\omega_\beta^2 = 2\pi n e^2 \frac{\beta^2}{\gamma m} = \frac{2\beta c^2}{\gamma r_b^2}\,\frac{I}{I_A^0}\,, \qquad I_A^0 = \frac{mc^3}{e} \tag{8.26}$$

ω_β is called the betatron frequency.

For $\mathrm{v}_\perp^2 = (\omega r)^2 = \omega^2 \langle r^2 \rangle = \frac{1}{2}\omega^2 r_b^2$ one has

$$\frac{\langle \mathrm{v}_\perp^2 \rangle}{\mathrm{v}^2} = \frac{1}{\beta\gamma} \frac{I}{I_A^0} \left[1 - \frac{1-f}{\beta^2} \right] . \tag{8.27}$$

For $f \simeq 1$, $\beta \simeq 1$, this becomes

$$\frac{\langle \mathrm{v}_\perp^2 \rangle}{c^2} \simeq \frac{I}{\gamma I_A^0} \tag{8.28}$$

or

$$kT_\perp = \frac{1}{2}\gamma m \langle \mathrm{v}_\perp^2 \rangle = \frac{1}{2} mc^2 \frac{I}{I_A^0} . \tag{8.29}$$

With the kinetic beam energy density $e_k = \frac{1}{2}\gamma mc^2$, this is

$$\frac{kT_\perp}{e_k} = \frac{I}{\gamma I_A^0} . \tag{8.30}$$

Applying these results to ion beams one has to make the substitution $m \to AM_H$, $e \to Ze$, where M_H is the proton mass. One has

$$I_A^i = \left(\frac{AM_H c^3}{Ze} \right) \beta\gamma = 3.12 \times 10^7 \left(\frac{A}{Z} \right) \beta\gamma \ \ [\mathrm{A}] \tag{8.31}$$

If in addition to the azimuthal self-field H_ϕ of the beam, the beam is guided by an externally applied axial field H_z, (8.21) is replaced by

$$F = \frac{eH_\phi}{\beta} \left(\frac{1}{\gamma^2} - f \right) - e\frac{\mathrm{v}_\phi}{c} H_z . \tag{8.32}$$

In the limit $\beta \to 1$, $\mathrm{v}_\phi / c \to 1$ this is

$$F = e \left[\left(\frac{1}{\gamma^2} - f \right) H_\phi - H_z \right] . \tag{8.33}$$

$F = 0$ requires that

$$f = \frac{1}{\gamma^2} - \frac{H_z}{H_\phi} . \tag{8.34}$$

Therefore, $F = 0$ is here possible even if $f = 0$, provided $\gamma = \sqrt{H_\phi / H_z}$. This, in part magnetically self-insulated beam, can propagate with high intensities through a vacuum. Because of its large v_ϕ velocity component the beam is here rotating.

Of interest is also the beam power in vacuum where $f = 0$. For the maximum beam power one has to set $F = 0$. With $F = f = 0$ (8.32) becomes

$$\frac{1}{\beta\gamma^2} H_\phi = \frac{\mathrm{v}_\phi}{c} H_z \tag{8.35}$$

and with $H_\phi = 2I/rc$, one has

$$I = \left(\frac{\beta\gamma^2}{2}\right) (\mathrm{v}_\phi r H_z) \ . \tag{8.36}$$

We are especially interested in the maximum power of a heavy ion beam with ions of mass $M = AM_H$ and charge Ze. There one has

$$I = \pi r^2 n e Z \beta c \tag{8.37}$$

and for the beam power

$$P = \pi r^2 n A M_H c^2 (\gamma - 1) \beta c \, . \tag{8.38}$$

From (8.36-8.38) one then has

$$P = \left(\frac{M_H c^2}{2e}\right) \left(\frac{A}{Z}\right) (\mathrm{v}_\phi r H_z) \beta \gamma^2 (\gamma - 1) \ . \tag{8.39a}$$

With the beam emittance defined by $\varepsilon = 2r\mathrm{v}_\phi/\beta c$ this becomes

$$P = \left(\frac{M_H c^3}{4e}\right) \left(\frac{A}{z}\right) (\varepsilon H_z) (\beta\gamma)^2 (\gamma - 1) \ . \tag{8.39b}$$

It shows that for the same beam power a small emittance with better beam focusing properties requires a large beam voltage.

Finally, we give a physical meaning of the Alfvén current. For the transverse kinetic beam energy density $\varepsilon_k = \frac{1}{2} (\gamma m) \beta^2 c^2 n \, , \;\; n = I / (\pi r_b^2 e \beta c)$ one has

$$\varepsilon_k = \frac{\beta \gamma m c I}{2\pi r_b^2 e} \tag{8.40}$$

and for the magnetic beam energy density $\varepsilon_H = H^2/8\pi$, at the surface of the beam where $H = 2I/r_b c$,

$$\varepsilon_H = \frac{I^2}{2\pi r_b^2 c^2} \ . \tag{8.41}$$

Therefore $\varepsilon_H > \varepsilon_k$ if

$$I > \frac{\beta\gamma mc^3}{e} = I_A \ . \tag{8.42}$$

This result means that for $I \gg I_A$ it is more appropriate to think of the beam as an electromagnetic pulse with the beam acting as a wave guide for the pulse.

8.4 Child-Langmuir Law

In our analysis of intense electron and ion beams we have so far assumed that E had no axial and H no radial component. A radial magnetic field component occurs in a magnetic mirror. There the $\mathrm{v}_z \times H_r$ Lorentz force component leads to a rotation of the beam. But here we are interested in the effect of an axial electric field. It leads to an axial acceleration of the beam changing the space charge distribution in that direction. The physics of this kind of space charge flow is of importance in high voltage diodes, but also in linear particle accelerators, both considered for a variety of thermonuclear microexplosion ignition concepts. As we will see later, it also may have important applications in dense pinch fusion concepts, where the physics of nonrelativistic ion space charge flow is important.

The nonrelativistic space charge flow into the z-direction is ruled by three equations: first by

$$j = ne\mathrm{v} \tag{8.43a}$$

for the electric current density; second by

$$\frac{1}{2}m\mathrm{v}^2 = eV \tag{8.43b}$$

for the kinetic energy of a particle with mass m and charge e accelerated by the potential V; and third by

$$\frac{d^2V}{dz^2} = 4\pi n e \,, \tag{8.43c}$$

Poisson's equation for n charges e per unit volume. From these equations one obtains

$$\frac{d^2V}{dz^2} = \frac{4\pi j}{\mathrm{v}} = \frac{4\pi}{\sqrt{2e/m}}\frac{j}{\sqrt{V}} \,. \tag{8.44}$$

For (8.44) one can also write

$$\frac{1}{2}\frac{d}{dV}\left(\frac{dV}{dz}\right)^2 = \frac{4\pi}{\sqrt{2e/m}}\frac{j}{\sqrt{V}} \,. \tag{8.45}$$

With the boundary condition $dV/dz\,|_0 = 0$, it can be integrated with the result

$$\frac{dV}{dz} = \sqrt{8\pi j}\left(\frac{2m}{e}V\right)^{1/4} \,. \tag{8.46}$$

Integrating (8.46) from $z = 0$ to $z = d$, with $V(0) = 0\,,\ V(d) = V$, where d is the diode gap, or more generally the distance over which the electric field $E = -dV/dz$ is sustained, and solving for j one obtains the Child-Langmuir law

$$j = \frac{\sqrt{2e/m}}{9\pi d^2}V^{3/2} \,. \tag{8.47}$$

Putting $j = I/\pi r^2$, where I is the current of a beam with radius r one finds that

$$I = \frac{\sqrt{2}}{9}\left(\frac{e}{m}\right)^{1/2}\left(\frac{r}{d}\right)^2 V^{3/2} \,. \tag{8.48}$$

8.5 Magnetic Insulation

As we had shown in chapter 3.3, a charged particle placed in a crossed electric and magnetic field executes a cycloidal drift motion perpendicular to both fields provided $H > E$ (in electrostatic units), with the radius of the cycloidal motion equal to the Larmor radius. Therefore, if in a diode a magnetic field is applied in a direction perpendicular to the electric field of the diode, a charged particle released from one side of the diode cannot reach the other side if the diode gap is larger than the Larmor radius. One then says the diode gap is "magnetically insulated" [1] with regard to the particle species which cannot cross the gap.

Both electrons and ions, released from the cathode resp. anode of a high voltage diode, are accelerated by the electric diode field to the same energy, whereby the ratio of their Larmor radius is

$$\frac{r_L^e}{r_L^i} = \sqrt{\frac{m}{M}} Z \tag{8.49}$$

with m/M the electron-ion mass ratio and Z the ion charge number. Therefore, by making $r_L^e < d$, where d is the diode gap, but at the same time making $r_L^i > d$, the cathode is magnetically insulated with regard to the electrons. With the large M/m mass ratio, this condition can be easily met.

In approaching 10^7 V/cm, electron field emission from the cathode sets in, with the critical field for field ion emission more like 10^8 V/cm. To magnetically insulate the cathode against an electric field of $\sim 10^7$ V/cm $\simeq 3 \times 10^4$ esu, requires a magnetic field of the same order, that is $H \gtrsim 3 \times 10^4$ G.

Because ions are not as easily released from the anode as electrons are from the cathode, the anode surface must be heated just before the moment the high voltage is applied to the diode if an ion beam shall be drawn from the anode. The heating of the anode creates a plasma layer from which the ions are easily released. The heating can be done, for example, by a pulsed laser beam (see Fig. 8.2).

For the magnetic insulation of the levitated ring shown in Fig. 8.1, it is sufficient to make $r_L^e < d$, where d is there the distance of the ring surface

[1] The idea of magnetic insulation was first proposed by the author back in 1967, but at that time few experts believed it would work (see article by J. Blewett in Nature 249, 863, (1974)).

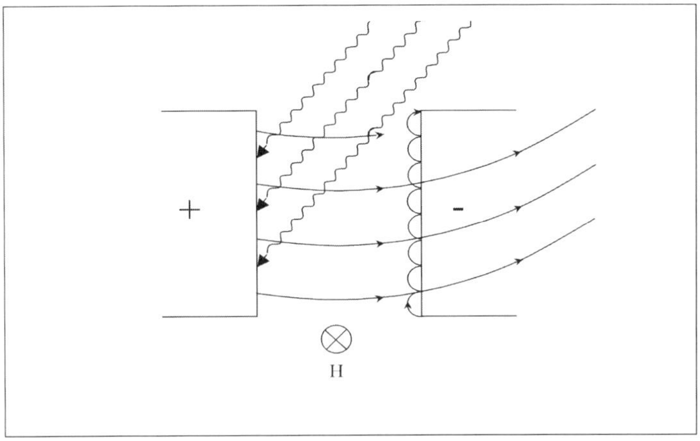

Figure 8.2: Magnetically insulated diode showing electron and ion trajectories.

from the surrounding wall, as long as $E \ll E_{max} \simeq 10^8$ V/cm $\simeq 3 \times 10^5$ esu, (equivalent to $E_{max} = \sqrt{4\pi\sigma_{max}}$ where $\sigma_{max} \simeq 10^{10}$ dyn/cm^2 is a typical value for the tensile strength), where ion field emission sets in.

We finally come to the phenomenon of magnetic self-insulation. It occurs in electric discharges where the magnetic field of the discharge current insulates against electric breakdown. It is of importance for the magnetic self-insulation of a high voltage diode, and of special importance in a coaxial (or co-planar) transmission line.

In electrostatic cgs units the impedance of a coaxial transmission line is

$$Z = \frac{2}{c} \ln\left(\frac{b}{a}\right) \tag{8.50}$$

where a and b are the inner and outer radius of the line. If the transmission line has the length l and if it holds the charge Q, the radial electric field between the inner and outer conductor is $E = 2Q/lr$, hence the voltage between the conductors,

$$V = \int_a^b E\,dr = \frac{2Q}{l} \ln\left(\frac{b}{a}\right) = c\,\frac{QZ}{l}\,. \tag{8.51}$$

With the electromagnetic pulse traveling down the line with the velocity of light c, the discharge time of the line is $\tau = l/c$, hence the current in the line

$$I = \frac{Q}{\tau} = \frac{Qc}{l} = \frac{V}{Z} \,. \tag{8.52}$$

The azimuthal magnetic field in the line therefore is

$$H = \frac{2I}{rc} = \frac{2Q}{rl} = E \tag{8.53}$$

but this is just the condition for magnetic insulation.

At the input position of the line where Z is somewhat larger, magnetic self-insulation is imperfect resulting in losses. These losses though can be reduced with an externally applied magnetic field, by placing the beginning of the line inside a magnetic solenoid.

In a tapered coaxial magnetically self-insulated transmission line (see Fig. 8.3), where $b/a = const.$, the line impedance is the same for each segment, preventing partial reflection of an electromagnetic pulse moving down the line. With $b/a = const.$, both E and H increase as $1/a$, with the power flux density (Poynting vector) increasing as $1/a^2$. As we had mentioned in chapter 8.1, because the repulsive electric forces balance the attractive magnetic forces, very high power flux densities are here possible, reaching those required for thermonuclear microexplosion ignition.

8.6 Ignition with Intense Particle Beams

Because particles can be accelerated to high velocities, intense particle beams can have a high-energy flux density, even though their energy density may be comparatively small. In listing all particle beams, one can include lasers as beams of zero rest mass photons at one end, and fast moving projectiles at the other end. The idea is to use these beams (called drivers), to ignite a small piece of thermonuclear explosive (called a target), by directing the beams onto the target. Apart from a direct bombardment of the target by the beams, one can also use an indirect drive, by placing the target inside a small cavity, with holes in the cavity wall to let the beams pass through. This then resembles the Teller-Ulam configuration, with the beam energy deposited into the cavity and transformed into black body radiation, replacing the energy released into a cavity by a fission bomb.

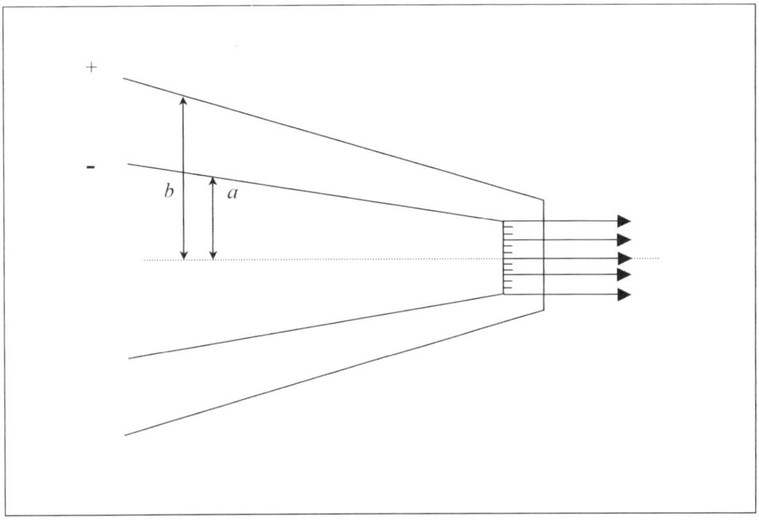

Figure 8.3: Tapered coaxial magnetically self-insulated transmission line.

The direct drive is expected to be more efficient, but it makes a spherical target implosion for ignition more difficult. The indirect drive is less efficient, but (as in the Teller-Ulam configuration) gives a uniform soft x-ray bombardment of the target. As it was with the polyhedron configuration (chapter 7.4), where at least 6 fission bombs are needed to approximate a spherical implosion, there should be at least 6 beams for the direct drive, but principally not more than one beam for the indirect drive.

And as with the miniaturized (fission triggered) compact thermonuclear explosive devices described in chapter 7.12, one can fill the cavity with suitable foam. Because the foam is there transformed into a hot plasma, somewhat resembling the hot gas from the burnt gunpowder in a cannon, these targets are sometimes called cannon-ball targets.

Finally, for intense relativistic electron or ion beams, the beam magnetic field can be used to confine the charged fusion products. There less target compression is required. The four different target and beam bombardment concepts are displayed in Fig. 8.4, together with a list of their problems. The principal problem of beam induced thermonuclear microexplosion ignition is, of course, a beam energy of several megajoule with a power flux of more than 100 terawatt/cm^2. For the commercial release of nuclear fusion energy by thermonuclear microexplosions, there is an additional, equally important problem, dubbed the "stand-off" problem. This occurs for high yield microexplosions with a yield of the order 10^9 J (1 ton TNT $= 4 \times 10^9$ J), requiring meter size cavities to confine the microexplosion. But whatever the yield may be, a particle beam can be a viable driver only if it can be transported over meter size distances from the cavity wall onto the less than cm size target.

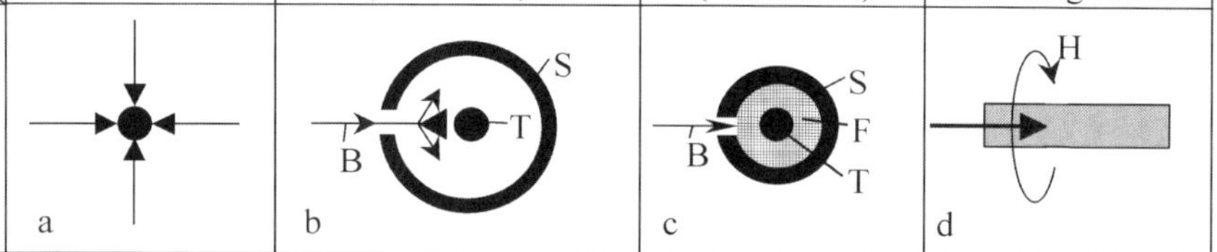

Drive / Driver	direct	indirect (Hohlraum)	indirect (cannonball)	intense relativistic beam magnetized
	problems	problems	problems	problems
laser	Brillouin backscattering	low efficiency	low efficiency	
relativistic electrons	focusing beam stopping	beam stopping	beam stopping	
light ions	focusing (standoff)	focusing (standoff)	focusing (standoff)	
heavy ions	large accelerator	beam stopping	beam stopping	ultra-high voltage power
projectile	large accelerator			

Figure 8.4: Targets and some beam drivers.

In Fig. 8.5 the working principles of the different driver concepts, ranging from lasers to high velocity projectiles, are displayed. Laser drivers, at the top of Fig. 8.5, store their energy in metastable atomic states, from where the energy is withdrawn by a photon avalanche. Relativistic electron beam drivers store their energy in capacitor banks, from where the energy is discharged by the voltage multiplying Marx circuit, with the high voltage terminal of the circuit connected to a relativistic electron beam producing field emission diode. For light ion beams it is (with changed electric polarity of the diode) the same as for relativistic electron beams, except that the diode is there magnetically insulated through the application of a strong

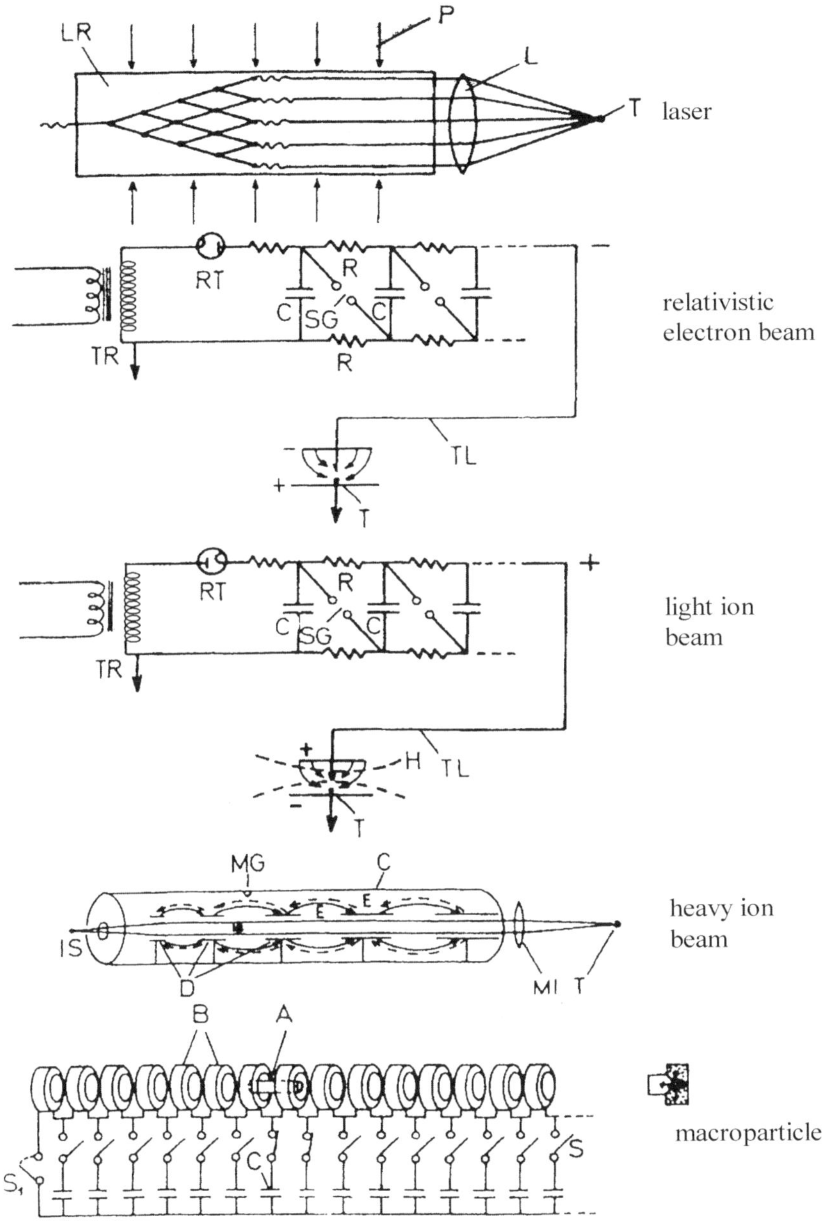

Figure 8.5: Particle beam drivers.

magnetic field perpendicular to the electric field. Next are multi-GeV heavy ion beams, accelerated in conventional linear or circular particle accelerators. Finally there is the idea of the traveling magnetic wave dipole accelerator, accelerating ferromagnetic or superconducting macroparticles to velocities up to 200 km/s, where the targets are of the "hammer and anvil" impact fusion type.

8.7 Laser Drivers

The atomic number density of a solid is $n \sim 5 \times 10^{22}$ cm^{-3}. Therefore, with one excited $\sim$ eV upper laser level per atom, the stored laser energy density could be as large as $\varepsilon \sim 10^{11}$ erg/cm^3. This means a volume of a few liters would be sufficient to store an energy of $\sim 10^{14}$ erg $= 10^7$ J, needed for thermonuclear ignition. At the energy density of $\varepsilon = 10^{11}$ erg/cm^3, the laser radiation power flux would be $\Phi = \varepsilon c = 3 \times 10^{21}$ erg/cm^{2}s $= 3 \times 10^{14}$ W/cm^2. In reality these high population inversion densities cannot be reached and much larger laser volumes are needed.

In glass lasers, the glass is doped with atoms which can be excited into the upper laser level. There the number of atoms which can occupy the upper laser level is equal to the number density of doped atoms in the glass, typically 10^3 times smaller than the number density of glass atoms. Accordingly, the energy density is there 10^3 times smaller and the same is true for the power flux. But even there, the power flux can become so large ($\Phi \gtrsim 10^{10}$ W/cm^2) that the glass is damaged in discharging the laser.

Because of its coherence, a laser beam can be focused down onto a small area, as it is required for thermonuclear microexplosion ignition. This advantage though is offset by the poor efficiency of lasers suitable for thermonuclear microexplosion ignition, where the laser frequency should be just below the plasma frequency of the target to be ablatively imploded. For a solid DT target one has $\nu_p = \omega_p/2\pi = 2 \times 10^{15}$ s^{-1}, and $\lambda = 1.5 \times 10^{-5}$ cm, which is in the ultraviolet. Lasers with high efficiency like the CO_2 and HF chemical laser are in the infrared with a frequency far below ν_p. As can be seen from (3.141), if the laser frequency is well below the plasma frequency, the laser radiation penetrates the plasma by the distance c/ω_p, about equal to a wavelength. In this case then, a thin plasma layer is initially heated to a high temperature. From the equation of motion of an electron in the electric field of the laser radiation

$$m\frac{dv}{dt} = eE = eE_0e^{-i\omega t} \tag{8.54}$$

it follows that the electron undergoes an oscillatory motion perpendicular to the direction of the incident laser beam, with an expectation value of the velocity

$$\langle|\mathbf{v}|\rangle = \frac{eE}{m\omega} \tag{8.55}$$

or expressed in terms of the laser radiation intensity $\Phi = (E^2/4\pi)\,c$:

$$\langle|\mathbf{v}|\rangle = \frac{e}{m\omega}\sqrt{\frac{4\pi\Phi}{c}}\ . \tag{8.56}$$

For $\Phi = 10^{14}$ W/cm^2 $= 10^{21}$ erg/cm^{2}s and $\lambda = 10^{-3}$ cm ($\nu = 3 \times 10^{13}$ s^{-1}, CO_2 laser) one has $\langle|\mathbf{v}|\rangle \simeq 10^{10}$ cm/s $= 30$ keV electrons. The range of 30 keV electrons in solid hydrogen is $\sim 10^{-2}$ cm, large enough to penetrate deep into the DT microexplosion target. With an electron velocity of $\sim 10^{10}$ cm/s it can outrun the compression wave, preheating the DT target, thereby preventing its isentropic compression to the high densities needed. For a laser frequency $\nu = \nu_p = 2 \times 10^{15}$s^{-1} and the same value of Φ, the electron velocity is $\langle|\mathbf{v}|\rangle \simeq 1.5 \times 10^8$ cm $= 6$ eV electrons. However, with $\nu_{laser} \approx \nu_p$, there is a resonance as can be seen from (3.132). Hot electrons produced by this resonance can acquire a much larger velocity than the "quivering" electron velocity (8.56). From (3.141) one can see that if the laser radiation frequency is close to ν_p, the radiation can penetrate deeper into the target than just one wavelength. It is this "soft" laser radiation impact which is used for laser fusion.

Initially, the laser radiation heats and ionizes a surface layer of the target. In blowing off the surface it becomes a plasma corona, surrounding the target. It is in this corona where most of the laser energy is absorbed, with the absorption coefficient given by (3.151). Inserting in (3.151) the expression for the electrical conductivity (4.15b), one has ($\ln\Lambda \simeq 10$):

$$\kappa = \frac{2.4 \times 10^{-4} n^2 Z^2}{T^{3/2}\nu^2\left[1 - (\nu_p/\nu)^2\right]^{1/2}}\ , \qquad \nu > \nu_p\ . \tag{8.57}$$

As an example, we take $n = 5 \times 10^{21}$ cm^{-3} where $\nu_p = 6.3 \times 10^{14}$ s^{-1}, and $\nu = 5 \times 10^{15}$ s^{-1}, $T = 10^7$ °K, $Z = 1$. We obtain $\kappa = 77$cm^{-1}. The absorption therefore takes place in a layer of thickness $\kappa^{-1} = 1.3 \times 10^{-2}$ cm, which is larger than the wavelength $\lambda = 6 \times 10^{-6}$ cm.

Next we compute the heat conduction time from the corona into the target. With (3.6b) and (5.4) we obtain (for $Z = 1$) from (3.105)

$$3nk\frac{\partial T}{\partial t} = \kappa \nabla^2 T \tag{8.58}$$

where κ is here given by (4.18). From (8.58) follows the heat conduction time (ln$\Lambda \simeq 10$):

$$\tau = \frac{3nkx^2}{\kappa} = 2.1 \times 10^{-10} n x^2 T^{-5/2} \text{ [s]} \tag{8.59}$$

For $n = 5 \times 10^{21}$cm^{-3}, $T = 10^7$ °K and $x = 1.3 \times 10^{-2}$ cm, one finds $\tau \sim 5 \times 10^{-10}$ s. The velocity of the heat conduction "wave" is $\mathrm{v}_c \sim x/\tau \sim 3 \times 10^7$ cm/s. If the target implosion velocity is faster, heat conduction cannot preheat the target.

To compute the ablation pressure, we start from the rocket equations for the thrust T and power dE/dt (v ablation velocity):

$$\left.\begin{aligned} T &= \mathrm{v}\,\frac{dm}{dt} \\ \frac{dE}{dt} &= \frac{\mathrm{v}^2}{2}\,\frac{dm}{dt} \end{aligned}\right\} \tag{8.60}$$

With the pressure p equal to the thrust per unit area, and likewise the laser radiation intensity Φ equal the power per unit area, we obtain from (8.60):

$$p = 2\frac{\Phi}{\mathrm{v}} . \tag{8.61}$$

This expression would be correct if all the ablated material leaves the surface in a perpendicular direction. In reality, the material is isotropically ejected into the solid angle 2π. Emitted per unit area of the surface, a jet going into the θ-direction has the cross section $\cos\theta$, and only the fraction $\cos\theta$ of its recoil momentum is transmitted to the surface. We thus have by integrating over all directions

$$p = \frac{2\Phi}{\mathrm{v}} \int_0^{\pi/2} \left(\cos^2\theta\right) \sin\theta \, d\theta = \frac{2}{3}\frac{\Phi}{\mathrm{v}} . \tag{8.62}$$

Furthermore, with $v = \sqrt{\gamma p/\rho} = \sqrt{5/3}\,\sqrt{p/\rho}$ we obtain

$$p = (4/15)^{1/3}\,\Phi^{2/3}\rho^{1/3}\,. \tag{8.63}$$

As an example, we take $\Phi = 10^{16}$ W/cm^2 = 10^{23} erg/cm^{2}s (corresponding to a 10^{14} Watt laser focused onto an area $\sim 10^{-2}$ cm^2), and $\rho \simeq 1$ g/cm^3 (corresponding to ~ 10 times compressed solid hydrogen). We find $p \simeq 10^{15}$ dyn/cm^2. Higher pressures can be reached in a convergent shock wave, where the pressure goes as $r^{-0.9}$, and still higher pressures with imploding shells.

The laser plasma interaction is much more complex than our simple model calculations suggest. If the laser radiation is reflected from the plasma like by a mirror, the incident and reflected wave form a standing wave. This standing wave generates a one dimensional density lattice in the plasma, with a lattice constant equal to $\lambda/2$, leading to increased reflection (Brillouin backscattering), and a self-amplification of the reflection. There a large fraction of the laser energy is lost for compression and heating. The cure for this problem is to make the laser light less coherent. For this though a price has to be paid, because such a less coherent laser light cannot be focused down so easily. This problem is less significant for the indirect drive. But even in the indirect drive, due to the smallness of the target positioned in the hohlraum, CO_2 lasers, for example, are unsuitable, since they still produce energetic electrons by interaction with the hohlraum wall.

8.8 Relativistic Electron Beam Drivers

For relativistic electrons the classical stopping range can be approximated by the expression (4.57b)

$$\lambda_e \simeq \frac{1}{\rho}[0.543E_0 - 0.16]\ [\text{cm}] \tag{8.64}$$

where E_0 is the electron energy in MeV and ρ the target density. For solid DT one has $\rho \simeq 0.2$ g/cm^3, with a range of MeV electrons on the order of several cm, much too large for the direct drive. For the heaviest metals with $\rho \simeq 20$ g/cm^3 and 1 MeV electrons $\lambda_e \simeq 1.5 \times 10^{-2}$ cm. This range is short enough to implode a thin shell of dense material. However, the intense x-rays produced by the electrons hitting the shell preheats a target inside the shell, preventing its isentropic compression to high densities. Preheating prior to implosion though is desired if a magnetized target is placed inside the imploding shell. There an intense electron beam could at the same time magnetize the target. Even though intense relativistic electron beams can be produced with relative ease, not much attention has been given to this possibility.

Much shorter stopping distances are possible with the two stream instability, where the range is given by (4.57a). Thermonuclear microexplosion ignition requires a power of $\sim 10^{15}$ Watt, realized with a 10^8 Ampere, 10^7 Volt intense relativistic electron beam. If the 10^8 Ampere beam is focused onto an area of $\sim 10^{-2}$ cm^2, the number density in the electron beam is $n_b \simeq 2 \times 10^{18}$ cm^{-3}. For solid DT one has $n = 5 \times 10^{22}$ cm^{-3} and $\omega_p \simeq 1.26 \times 10^{16}$ s^{-1}, and for 10^7 eV electrons $\gamma \simeq 20$. The range then is

$$\lambda_D = 1.4 \left(\frac{c}{\omega_p}\right) \left(\frac{n}{n_b}\right)^{1/3} \gamma \simeq 2 \times 10^{-3} \text{ [cm]} \tag{8.65}$$

short enough to stop the beam in a small DT target. The problem though is that for short lengths the two-stream instability saturates and becomes ineffective, but this is less likely to happen in a convergent electrostatic wave, launched by many beams hitting the target from all directions. Calculations show that there the electric field strength for saturation rises in proportion to $1/r^2$, where r is the distance from the center of convergence.

Still another possibility is to use an incompletely space charged neutralized intense relativistic electron beam, as described in chapter 8.3. There, the beam energy is essentially electromagnetic. If such a beam hits a target, it radially collapses onto the target by magnetic forces. Apart from compressing the target, a magnetic field assisted thermonuclear burn is there then possible if the beam current is above I_c given by (6.23). There some target heating does not matter, because the target does not have to be compressed to the high densities needed in the absence of a magnetic field.

Intense relativistic electron beams are most easily produced by electron field emission with the beam extracted from a diode connected to the high voltage terminal of a Marx generator (see Fig. 8.5). According to (8.4), the discharge time of a Marx generator is reduced by the factor $1/n$, however, in reality the inductance L of the bank is so large that the discharge time is not less than 10^{-6} sec, still much too long. Shorter discharge times are possible if the high voltage terminal of the Marx generator is connected to a low inductance high voltage capacitor. A water capacitor with a dielectric constant $\varepsilon = 81$ has been very successful. It can hold a voltage of $\sim 10^7$ Volt just long enough to be charged by the Marx generator. Through a fast switch this capacitor is then discharged onto a magnetically insulated transmission line.

In practical units the impedance of the line is (see eq. (8.50)):

$$Z = 60 \ln\left(\frac{b}{a}\right) [\Omega] \tag{8.66}$$

and if $b = a + d$, where $d \ll a$ is the distance of separation between the inner and outer conductor, one has

$$Z \simeq 60 \left(\frac{d}{a}\right) [\Omega] . \tag{8.67}$$

Typically a current $I \sim 10^6$ [A], with a voltage V $\sim 10^7$ [V], having a power $P = IV = 10^{13}$ [W] runs through one line, requiring $Z = 10\ \Omega$, or $d/a = 1/6$. Therefore, if $d = 3$ cm, then $a = 18$ cm. Increasing the power is possible with more than one transmission line in parallel, where for N transmission lines the overall impedance is $Z^* = Z/N$. Therefore, to reach a power of 10^{15} [W] as is required for thermonuclear ignition one needs 100 lines, which can be arranged in a spherically symmetric way around the thermonuclear target.

At its end each transmission line is connected to a field emission diode, with the cathode surface opposite to an anode window, the latter made from a metallic foil, for example titanium. If the electric field at the cathode surface reaches $\sim 10^7$V/cm, intense electron field emission sets in, with the electrons accelerated toward the anode window. In passing through the anode window, the electrons enter a space-charge and current neutralizing background gas or plasma, whereby the net beam current can be well above the Alfvén limit. The background plasma can initially be a neutral gas, ionized by the beam. Because the beam can have a residual unneutralized current less than the Alfvén current, the beam can be guided through metallic tubes, being repelled by image currents in the conducting wall.

To improve the field emission from the cathode, one may cover its surface with a "brush" of needles. Each needle has at its tip a voltage of the order $E \simeq V/r$, where r is the radius of the semispherical needle tip and V the voltage applied to the needle.

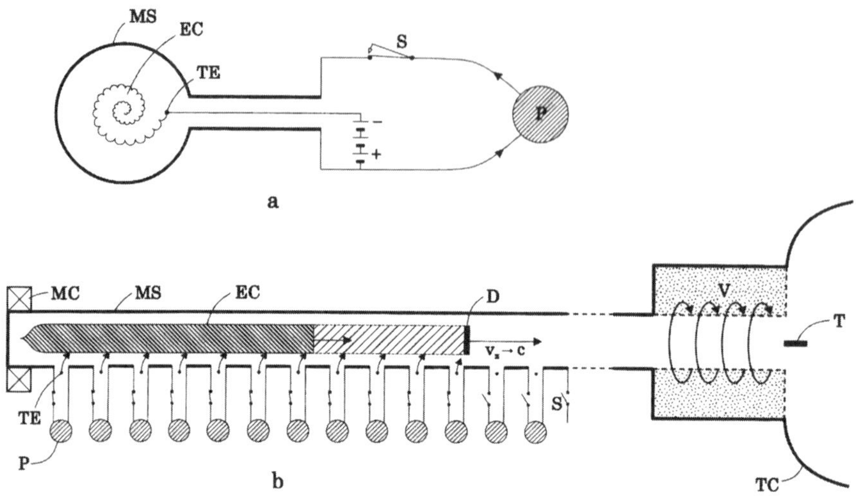

Figure 8.6: The (a) radial and (b) axial cross section: MS, magnetic solenoid; EC, electron cloud; D, dense disk of relativistic electrons; TE, thermionic emmiter; S, switch; P, power supply; MC, magnetic mirror coil; T, thermonuclear target; V, liquid vortex; TC, thermonuclear microexplosion chamber.

The field emission current density is

$$j = 1.55 \times 10^{-6} \left(\frac{E^2}{W}\right) \exp\left[-6.9 \times 10^7 \, \frac{W^{3/2}}{E}\right] \; [\mathrm{A/cm^2}] \tag{8.68}$$

where E is the field strength in V/cm and W the work function of the emitter material in eV. For tungsten, a frequently used material, one has $W = 4.4$ eV. The current drawn from each needle is obtained by multiplying (8.68) with $2\pi r^2$, the semispherical area of each needle. Assuming $r = 0.1$ cm (a rather thick needle), one has for the current emitted by one needle $I \simeq 3.5 \times 10^5$ [A]. For a total current of $\sim 10^8$ [A] one would then need about 300 needles. Distributed over 100 cathode surfaces this amounts to 3 needles per cathode. Increasing the number of needles a thousandfold would reduce the current for each needle to ~ 350 [A], with 3×10^3 needles per cathode surface and a current density of 5×10^3 [A/cm^2] for each needle. For diode voltages in excess of $\sim 10^6$ [V], a rough cathode surface has the same effect and leads to a more uniform current distribution.

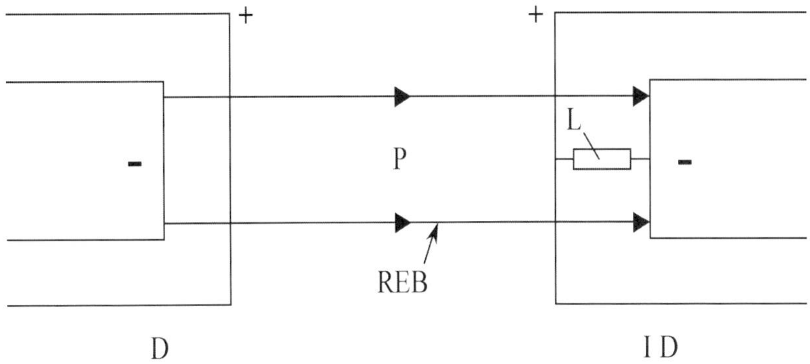

Figure 8.7: Inverse diode ID transforming kinetic energy of an intense relativistic electron beam REB from diode D, propagating through a plasma P (or background gas), back into electromagnetic energy delivered to the load L.

In a "foil-less" diode, the field emission cathode is placed inside a magnetized solenoid in high vacuum. There the beam current is limited by the Alfvén current $I_A \simeq 17\,000\,\beta\gamma$ [A], for $\gamma = 20$ (10 MeV beam) to $I_A \simeq 340\,000$ [A]. But as was explained in chapter 8.3, much larger currents are possible if the beam rotates. One concept promising especially powerful rotating beams is shown in 8.6. There the electrons making up the beam are radially injected by a rapidly rising magnetic field. The electrons released from the thermionic emitters and attached to the rising field, make a spiraling motion towards the center of a common electron beam drift tube, fed along the entire length of the tube by many thermionic emitters and rising magnetic fields. Inside the drift tube, the repulsive electrostatic force accelerates the electrons in the axial direction, producing a powerful beam, while the axial magnetic field radially confines the beam. This configuration, which is conceivably the most powerful concept for the generation of intense relativistic electron beams, does not need a high voltage generator. It is rather driven by many low voltage homopolar flywheel generators, magnetizing a large number of one-turn coils.

The beam can be strongly focused by projecting it into a tenuous background plasma. With the repulsive electric forces neutralized it collapses by the uncompensated magnetic forces. A thermonuclear target, which becomes magnetized by the beam, can then be placed at the entrance of a blast-absorbing liquid vortex tube.

As explained in Fig. 8.7, the kinetic energy of an intense relativistic electron beam well above the Alfvén current and propagating through a space charge and current neutralizing plasma, can be converted back into electromagnetic energy by an inverse diode, and from there to a load. The importance of this concept is that it solves the stand-off problem for electric pulse power driven thermonuclear microexplosions.

8.9 Ion Beam Drivers

They cover the entire range from light to heavy ions, both relativistic and nonrelativistic. According to Fig. 8.5, there are two schemes to make ion beams for thermonuclear microexplosion ignition: (1) With a Marx generator driven magnetically insulated diode; (2) With conventional particle accelerators.

With the first scheme it is relatively easy to make multimegampere - megavolt intense light ion beams but of poor quality (with a large transverse beam temperature resp. large emittance). With the second scheme one can make beams with a much higher particle energy and low emittance, but requiring large particle accelerators. For good beam focusing low emittance is desired, but for better beam energy deposition (small stopping range) a lower beam particle energy is favored.

For both the direct and indirect drive the ions should be stopped in a thin layer. The stopping range is determined by (4.103b) with $\ln \Lambda \simeq 10$, one has

$$\sigma_s \simeq 6 \times 10^{-33} \frac{Z Z_i^2 A_i}{E_0^2} \ [\mathrm{cm}^2] \ . \tag{8.69}$$

The atomic number density in solid matter is $n \simeq 5 \times 10^{22}$ cm^{-3}, and one has for the stopping range $\lambda_i = 1/n\sigma_s$:

$$\lambda_i \simeq 3 \times 10^{9} \frac{E_0^2}{Z Z_i^2 A_i} \ [\mathrm{cm}] \ . \tag{8.70}$$

Short stopping ranges require that the Z-value of the target material is sufficiently large. Let us assume that for a light ion beam $Z_i = 10$, $A_i = 20$, and that $E_0 = 3 \times 10^7$ eV. From (8.70) one finds that for $\lambda_i \sim 10^{-3}$ cm, $Z \simeq 5$ would suffice. To have the same range for a heavy ion with $Z \sim Z_i \sim A_i/2$ would require a particle energy below ~10 GeV.

For light ions it is not difficult to raise the voltage to $\sim 3 \times 10^7$ [V] with available pulse power technology, but it is difficult to lower the particle energy below 10 GeV for heavy ions accelerated in conventional particle accelerators without a drastic loss of the beam power. This can be seen in the nonrelativistic limit of (8.39b), where the beam power scales as

$$P \propto (\varepsilon H_z)\, E_0^2\,. \tag{8.71}$$

Large beam power can be sustained but only at the expense of an increased emittance. Take the example $A = 200$, $Z = 5$, and $E_0 = 40$ GeV, where one finds from (8.39b) that $P \simeq 5 \times 10^{10}\,(\varepsilon H_z)$ [W]. With $H_z \simeq 2 \times 10^4$ G for the bending magnets and requiring that $P = 10^{14}$ [W] one finds that $\varepsilon = 0.1$ cm. But by going down from 40 GeV to 20 GeV, and as before requiring that $P = 10^{14}$ [W], yields $\varepsilon \simeq 2.5$ cm.

For intense ion beams there are two limiting cases of special interest:

1. Momentum-rich low velocity ($\sim 10^8$ cm/s) intense ion beams

2. Magnetic field-rich relativistic ion beams.

In the first case one really has dense nonrelativistic plasma jets. As shown in Fig. 8.8 these jets can be produced as heavy ion beams by a variable voltage magnetically insulated diode axially bunching the beam by radiation cooling. The axial bunching permits large beam power amplification by orders of magnitude. In a momentum-rich beam the particle velocity should ideally be equal to the implosion velocity. With the beam stagnation pressure $p = \frac{1}{2}\rho \mathrm{v}^2$ and the beam energy flux density $\phi = \frac{1}{2}\rho \mathrm{v}^3$, one has $p = \phi/\mathrm{v}$, three times larger than the ablation pressure (8.62). Momentum-rich beams also improve inertial confinement in the direct drive, because the mass of the beam brought to rest at the target acts like a high pressure tamp.

Fig. 8.8 shows an indirect drive where the beam energy is focused onto the target, not by its conversion into soft x-rays, but by a concave ablation shock wave mirror.

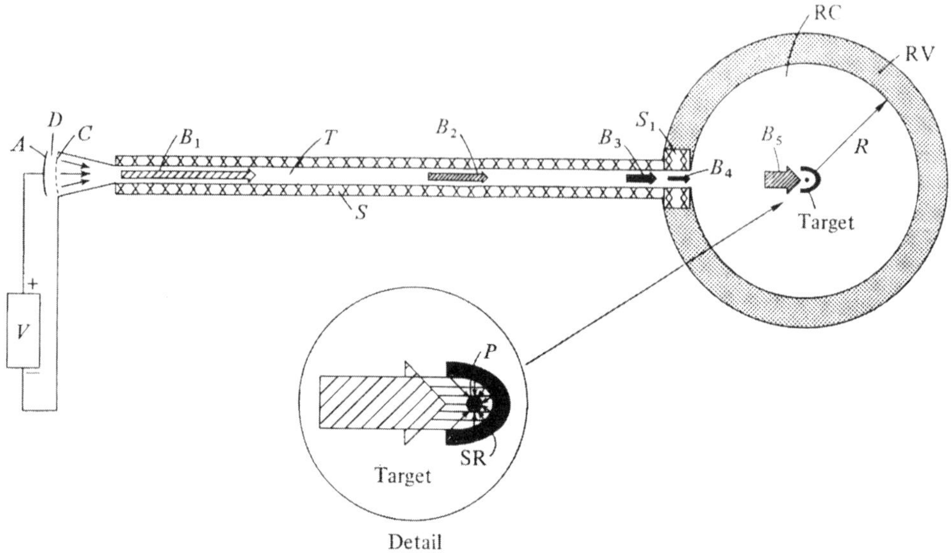

Figure 8.8: Heavy ion beam microexplosion reactor concept: V high voltage source; D ion diode with anode A and cathode grid C; T drift tube; S magnetic solenoid, S_1 pulsed high field magnetic solenoid; B_1, B_2, B_3, B_4, B_5 beam positions; RC reactor chamber; RV reactor vessel of radius R; P thermonuclear target; SR shock wave reflector.

In the other case, one has to consider the Alfvén current for ions (8.31). With $A/Z \sim 2$ and $\beta\gamma \sim 1$ (~ 1 GeV/nucleon), one already has $I_A^i \sim 6 \times 10^7$ [A], large enough to confine the charged fusion products of the reactions listed in table 2.1. At $I = 6 \times 10^7$ [A] a 1 cm radius beam would carry a magnetic field of the order $\sim 10^7$ [G]. Intense relativistic ion beams would for this reason be ideally suited for dense magnetized fusion targets. And like intense relativistic electron beams, they can be transported over large distances, a property of importance for the stand-off problem of thermonuclear microexplosions. With the beam power going in proportion to the square of the beam voltage, the power can be larger by many orders of magnitude if compared with the power of intense relativistic electron beams.

Further possibilities emerge with the prospect of nuclear reaction rates enhanced by the non-Maxwellian beam target environment. This would happen, for example, if an intense relativistic proton beam is shot onto a B^{11} target. As shown in Fig. 8.4 the target for an intense relativistic ion beam would there have the form of a long cylinder.

There remains the problem of generating intense relativistic ion beams. Compared with intense relativistic electron beams requiring megavolt power sources, intense relativistic ion beams require gigavolt power sources. There are two concepts which appear promising. Both make use of magnetic insulation. One is the levitated magnetically insulated torus shown in Fig. 8.1. A way in which it might be realized is shown in Fig.8.9 where the charging of the torus to gigavolt potentials is done by a stream of charged pellets. Another concept shown in Fig. 8.10 uses a staged magnetically insulated transmission line, charged in parallel and, like in a Marx generator, discharged in series.

8.10 Microparticle Beam Drivers

A high velocity microparticle can lead at the location of its impact to thermonuclear temperatures if two conditions are satisfied:

1. The velocity of the microparticle must be large enough for the shock it creates upon impact to have a temperature larger than the ignition temperature of a thermonuclear reaction.

2. The diameter of the microparticle must be larger than the mean free path at the ignition temperature of the thermonuclear reaction.

The impact temperature is obtained from (5.13) with c_v given by (3.6b). For a DT target one finds that

$$T = 5 \times 10^{-9} \mathrm{v}^2 \ [^\circ\mathrm{K}] \,. \tag{8.72}$$

Setting $T = T_{ign} = 5 \times 10^7$ [°K] for the DT reaction, one finds that $\mathrm{v} = 10^8$ cm/s. And according to (4.11) for $T = T_{ign} = 5 \times 10^7$ [°K] = 4.3 keV, the mean free path in a solid DT target with $n = 5 \times 10^{22}$ cm^{-3} is $\lambda \simeq 10^{-6}$ cm. For microparticles smaller than λ, the impact will not lead to a shock wave.

To ignite from a thusly-generated hot spot a thermonuclear deflagration (i.e. burn wave), requires that the particle must have a size larger than the stopping length (6.13) of the charged fusion products. For solid DT this is $\lambda_0 \simeq 0.3$ cm, but for 10^3 fold compressed DT it would be $\lambda_0 \simeq 3 \times 10^{-4}$ cm. If a method can be found to accelerate a microparticle of that size to a velocity

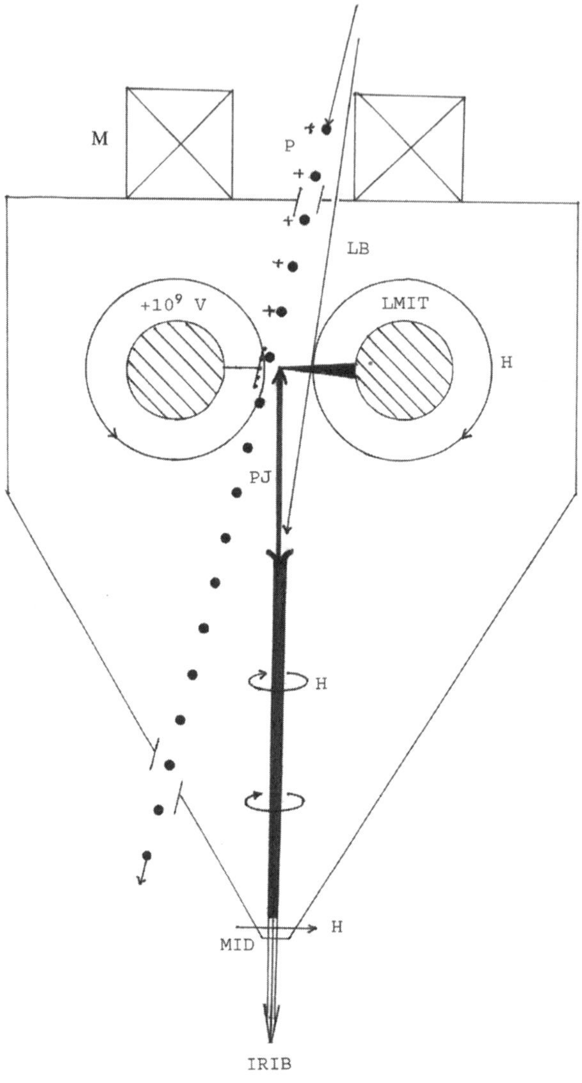

Figure 8.9: *LMIT* levitated magnetically insulated torus: *M* levitation coils: *F* feedback control coils: *PJ* plasma jet; *LB* pulsed laser beam: *P* positively charged pellets moving at high speed: *H* insulation magnetic field: *MID* magnetically insulated diode: *IRIB* intense relativistic ion beam.

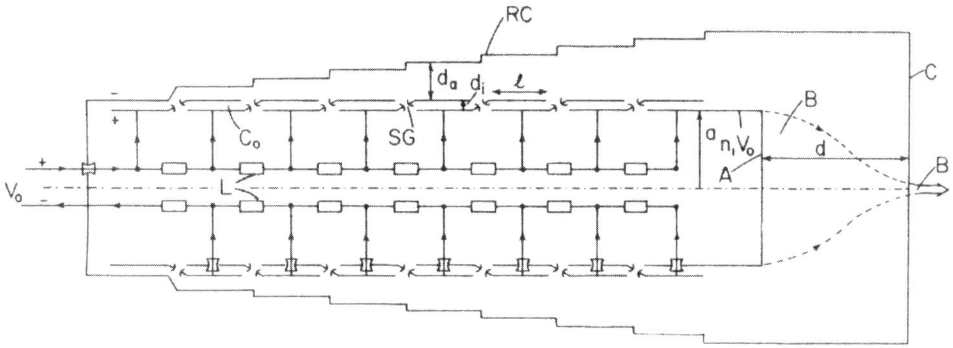

Figure 8.10: Cross-section of a pulsed high-voltage accelerator. The high voltage is obtained using a series of cylindrical capacitors arranged in a multistage transmission line. V_0 is the input from a high-voltage source, C_0 are cylindrical capacitors of length ℓ and inner radius a and separation distance d_i, in between inner and outer conductors. SG are triggered circular spark gap switches, L are inductances, RC is a vacuum vessel also serving as the return current conductor and is separated by distance d_a from cylindrical capacitors, d is the diode gap, A the anode, C the cathode and B the ion beam.

of $\sim 10^8$ cm/s, it could serve as a fast ignitor (see chapter 6.10 and Fig. 6.5). The kinetic energy of such a particle would not be more than 10^7 erg $= 1$ J. Most of the energy needed though would be for compression of the target.

Micron-size particles can be accelerated in linear particle accelerators by positively charging them to the electric field E_0 where the electric stress $E_0^2/8\pi$ becomes equal their tensile strength: One thus has

$$E_0 = \sqrt{8\pi\sigma}\,. \tag{8.73}$$

A typical value is $\sigma \simeq 10^{11}$ dyn/cm^2, whereby $E_0 = 1.6 \times 10^6$ esu $= 5 \times 10^8$ V/cm. The largest electric charge which can be put on a spherical particle of radius r_0 then is

$$q = E_0 r_0^2\,. \tag{8.74}$$

With the density of the particle ρ_0, the charge to mass ratio of the particle is

$$\frac{q}{m} = \frac{E_0 r^2}{(4\pi/3)\,\rho_0 r_0^3} = \left(\frac{3}{4\pi\rho_0}\right)\frac{E_0}{r_0} \,. \tag{8.75}$$

Placed in the electric field E of a linear particle accelerator the acceleration of the microparticle is

$$a = \left(\frac{3}{4\pi\rho_0}\right)\frac{E_0 E}{r_0} \,. \tag{8.76}$$

Expressed in terms of the particle radius the length of the accelerator to reach the velocity v is

$$\frac{L}{r_0} = \left(\frac{2\pi\rho_0}{3E_0 E}\right)\mathrm{v}^2 \,. \tag{8.77}$$

For $\rho_0 = 2\mathrm{g/cm}^3$ (beryllium), $E_0 = 1.6 \times 10^6$ esu, $E \simeq 10^5$ V/cm $\simeq 3.3 \times 10^2$ esu this becomes

$$\frac{L}{r_0} \simeq 0.8 \times 10^{-8}\mathrm{v}^2 \tag{8.78}$$

For $\mathrm{v} = 10^8$ cm/s, one has $L/r_0 \sim 10^8$. If $r_0 \simeq 10^{-4}$ cm, $L \simeq 10^4$ cm $= 100$ meters, but for $r_0 \simeq \lambda_0 = 0.3$ cm (needed to ignite a detonation wave in solid DT) one has $L \simeq 3 \times 10^7$ cm $= 300$ km, with a kinetic particle energy equal to $\sim 4 \times 10^{14}$ erg $= 40$ MJ. Compressing the DT target tenfold would reduce λ_0, and hence r_0, tenfold. Reducing the mass of the particle $\sim 10^3$-fold, reduces its kinetic energy to 4×10^{11} erg, the accelerator though would still be 30 km long. Compressing the DT target 100 fold would reduce the length of the accelerator down to ~ 3 km. The particle size would there be $r_0 \simeq 3 \times 10^{-3}$ cm with a kinetic energy of ~ 40 J.

Instead of accelerating one larger particle, one may accelerate a beam of many small particles, but there the Child-Langmuir law (8.48) sets an upper limit for the number of particles. For the acceleration by the voltage V [esu] in a diode one has

$$\mathrm{v}^2 = \left(\frac{2q}{m}\right)V \tag{8.79}$$

or with the value of q/m (8.75)

$$V = \frac{2\pi}{3}\rho_0 r_0 \frac{\mathrm{v}^2}{E_0} \, . \tag{8.80}$$

For $\rho_0 \simeq 2$ g/cm^3, $E_0 = 1.6 \times 10^6$ esu, and expressing V in volts, this becomes

$$V \simeq 8 \times 10^{-4} \mathrm{v}^2 r_0 \, [\mathrm{V}] \, . \tag{8.81}$$

For $\mathrm{v} = 10^8$ cm/s this is $V \simeq 8 \times 10^{12} r_0$ [V]. If $r_0 \simeq 10^{-4}$ cm, $V \sim 10^9$ [V] would be required. As we have shown this might be attainable with the superconducting levitated ring capacitor. For $r_0 \simeq 10^{-6}$ cm, only $V \sim 10^7$ [V] would be needed which can be reached with Marx generators.

From the Child-Langmuir law (8.48) one has for the electric current of a beam of charged microparticles accelerated in a high voltage diode

$$I = \frac{\sqrt{2}}{9}\left(\frac{q}{m}\right)^{1/2}\left(\frac{r}{d}\right)^2 V^{3/2} \tag{8.82}$$

where r is the beam radius and d the diode gap. Since $q/m = \mathrm{v}^2/2V$, equation (8.82) can be written as follows

$$I = \frac{1}{9}\,\mathrm{v}\left(\frac{r}{d}\right)^2 V \tag{8.83}$$

with the diode impedance

$$Z = \left(\frac{9}{\mathrm{v}}\right)\left(\frac{d}{r}\right)^2 [\mathrm{esu}] = 270\left(\frac{c}{\mathrm{v}}\right)\left(\frac{d}{r}\right)^2 [\Omega] \, . \tag{8.84}$$

For $\mathrm{v} = 10^8$ cm/s this becomes

$$Z = 8.1 \times 10^4 \left(\frac{d}{r}\right)^2 [\Omega] \, . \tag{8.85}$$

If we assume that $d/r \simeq 10^{-1}$, $Z \simeq 10^3$ Ω, the beam power is

$$P = \frac{V^2}{Z} \simeq 10^{-3} V^2 \, [\mathrm{W}] \, . \tag{8.86}$$

For $V = 10^7$ [V] it is only 10^{11} [W], but for $V = 10^9$ [V], one has $P = 10^{15}$ [W]. These two examples demonstrate the importance of going to high voltages. The same is also true for heavy ion fusion, except that there by going to high voltages the stopping range becomes too large. Because of the much smaller particle velocity, the stopping range of a microparticle beam is much shorter. And because their mass is rather large, the thermal velocity of the microparticles is small and the same is true for the beam emittance.

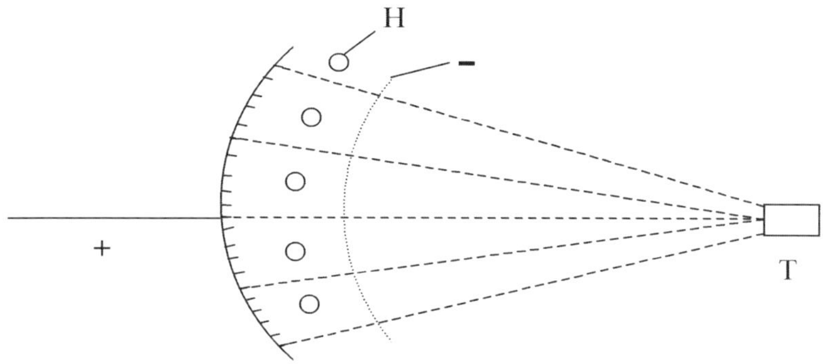

Figure 8.11: Generating a beam of electrically charged microparticles from a concave high voltage diode.

One can launch the microparticles from the concave anode surface of a magnetically insulated diode to which a high voltage pulse is suddenly applied (see Fig. 8.11). In passing through a cathode grid, the microparticles are neutralized by electron field emission from the grid, and behind the grid ballistically focused onto the thermonuclear target.

A problem is that unless all the microparticles have the same q/m ratio, there will be an axial beam spread accompanied by a reduction of the beam power flux. To overcome this problem it is proposed to produce the microparticles by applying a high voltage pulse to a brush of needles attached to the anode as shown in Fig. 8.11. For a repetitive pulsed operation one can replace the needles with tiny jets coming out of many pores in the anode.

In applying the voltage V to the needle brush, each needle tip will have the electric field

$$E \simeq \frac{V}{r} \tag{8.87}$$

where r is the radius of a needle tip, with each needle holding the electric charge

$$q = r^2 E = rV \, . \tag{8.88}$$

If $E > E_0$ the needles disintegrate into n microparticles of radius r_n and charge q_n whereby

$$\frac{q_n}{r_n^2} = E_0 \, . \tag{8.89}$$

If the charge q is evenly distributed between the n microparticles, each of them will have the charge

$$q_n = \frac{q}{n} \tag{8.90}$$

and the radius

$$r_n = rn^{-1/3} \, . \tag{8.91}$$

Inserting (8.90) and (8.91) into (8.89) one has

$$E = n^{1/3} E_0 = \frac{V}{r} \, . \tag{8.92}$$

With the mass of one microparticle

$$m_n = \frac{4\pi}{3} \rho_0 r_n^3 \tag{8.93}$$

where ρ_0 is the density of the microparticles one has for the charge to mass ratio:

$$\frac{q_n}{m_n} = \frac{3}{4\pi\rho_0} \frac{E_0}{r_n} \tag{8.94}$$

or with (8.91) and (8.92):

$$\frac{q_n}{m_n} = \frac{3}{4\pi\rho_0}\frac{E}{r} = \frac{3}{4\pi\rho_0}\frac{V}{r^2} \tag{8.95}$$

the same for all microparticles.

To reach the required velocity one must have

$$\frac{1}{2}m_n \mathrm{v}^2 = q_n V \tag{8.96}$$

or with (8.90), (8.91) and (8.93)

$$\frac{2\pi}{3}\rho_0 \mathrm{v}^2 r^3 = qV \tag{8.97}$$

and by eliminating q with (8.88) and solving for r:

$$r = \frac{1}{(2\pi\rho_0/3)^{1/2}}\frac{V}{\mathrm{v}}\,. \tag{8.98}$$

From (8.92) and (8.98) one can compute n:

$$n = \left(\frac{2\pi\rho_0}{3}\right)^{3/2}\left(\frac{\mathrm{v}}{E_0}\right)^3\,. \tag{8.99}$$

If N is the number of needles, the total beam energy will be

$$E_b = N\frac{4\pi}{3}r^3\rho_0 \mathrm{v}^2 \tag{8.100}$$

and because of (8.98):

$$N = \left(\frac{2\pi\rho_0}{3}\right)^{1/2}\frac{E_b \mathrm{v}}{V^3}\,. \tag{8.101}$$

Assuming $\rho_0 = 2$ g/cm^3, $E_b = 10^{14}$ erg $= 10$ MJ, $\mathrm{v} = 10^8$ cm/s and $V = 10^9$ [V] $= 3.3 \times 10^6$ esu one has $N \simeq 500$. From (8.99) with $E_0 = 1.6 \times 10^6$ esu one has $n \simeq 2 \times 10^6$, and from (8.98) that $r \simeq 1.6 \times 10^{-2}$ cm and hence $r_n \simeq 10^{-4}$ cm.

The focusing of the charge-neutralized microparticles can be improved if they are made from ferromagnetic material with a large saturation field strength, for example gadolinium. The force density of a magnetized medium of (average) density ρ and permeability μ is

$$\mathbf{f} = \frac{1}{8\pi}\nabla\left[H^2\rho\,\frac{\partial\mu}{\partial\rho}\right] - \frac{H^2}{8\pi}\,\nabla\mu\,. \tag{8.102}$$

Since

$$\frac{\mu - 1}{\mu_0 - 1} = \frac{\rho}{\rho_0} \tag{8.103}$$

where μ_0 is the permeability at solid density ρ_0, one has

$$\rho\,\frac{\partial\mu}{\partial\rho} = \mu - 1 \tag{8.104}$$

whereby (8.102) becomes

$$\mathbf{f} = \frac{\mu - 1}{8\pi}\nabla H^2\,. \tag{8.105}$$

The equation of motion of the ferromagnetic fluid made up magnetized microparticles is then

$$\rho\frac{d\mathbf{v}}{dt} = \frac{\mu - 1}{8\pi}\,\nabla H^2 \tag{8.106}$$

or because of (8.103):

$$\frac{d\mathbf{v}}{dt} = \frac{\mu_0 - 1}{8\pi\rho_0}\,\nabla H^2\,. \tag{8.107}$$

If

$$H = H_0\left(\frac{r_0}{r}\right)^2 \tag{8.108}$$

realized in a converging magnetic mirror field, one has for the radial component of (8.107)

$$\frac{d\mathrm{v}_r}{dt} = -\frac{\mu_0 - 1}{2\pi\rho_0}\,\frac{r_0^4 H_0^2}{r^5}\,. \tag{8.109}$$

With $dv_r/dt = \frac{1}{2}dv_r^2/dr$ (8.109) can be integrated. Neglecting smaller terms the result is

$$v_r^2 = (\mu_0 - 1)\frac{H_0^2}{4\pi\rho_0} . \tag{8.110}$$

Let us assume that $H_0 \sim 10^5$ G, $\mu_0 - 1 \sim 10^4$ and $4\pi\rho_0 \sim 10^2$ g/cm^3, one then obtains $v_r \sim 10^6$ cm/s. If by this focusing method the ferromagnetic microparticles come close to each other they will coagulate into one slug, greatly increasing the beam power.

Another intriguing possibility, explained in Fig. 8.12, is to accelerate a thin disc by a magnetically insulated electron cloud inside a long magnetic solenoid. Placed in front of the cylindrical electron cloud, the disc is charged up to the electrical potential V of the cloud. To avoid for the cloud to overcome the disc, the magnetic field of the solenoid is superimposed by a magnetic traveling wave, with the disc placed at the location of the wave. The field of the traveling wave acts as a magnetic mirror for the electrons behind the disc, which in conjunction with the highly charged disc keeps the electron cloud confined in the space behind the disc. The electron cloud is generated by injection of electrons from a foil-less high voltage diode as described in chapter 8.8. If the radius R of the disc is of the order of magnitude of

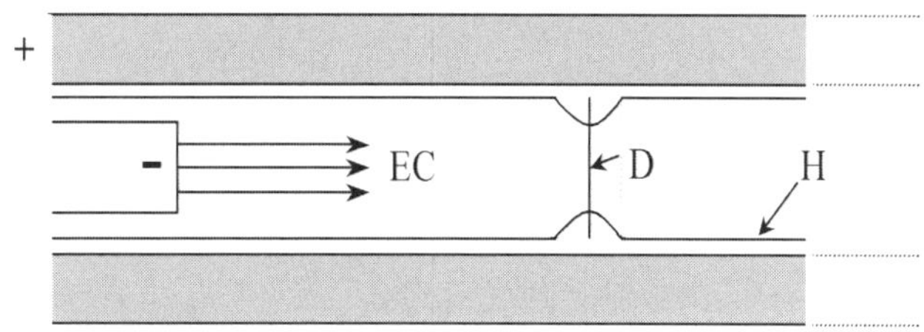

Figure 8.12: Acceleration of a thin disc D by a magnetically insulated electron cloud generated by a foil-less diode.

the inner radius of the solenoid, the electric field at the surface of both the electron cloud and disc is of the order $E \sim V/R$. To prevent the disc from losing its charge by electron field emission, $E \lesssim 10^7$ Volt/cm. Assuming that R is of the order cm, then implies that $V \sim 10^7$ Volts.

With magnetic insulation the electric field at the surface of the cloud is $E < H$. With $H \sim 5 \times 10^4$ G and $E \sim 10^7$ Volt/cm $\sim 3 \times 10^4$ esu, this condition is well satisfied. For $E = 3 \times 10^4$ esu the electric pressure acting on the rear of the cloud is $P_E = E^2/8\pi \simeq 4 \times 10^7$ dyn/cm^2. If the disc has a density $\rho \simeq 4$ g/cm^3 and a thickness δ, its acceleration is $a = P_E/\rho\delta \sim 10^{10}$ cm/s^2. To reach a velocity v [cm/s], the length over which the disc would have to be accelerated is $L = \mathrm{v}^2/2a = 5 \times 10^{-11}\mathrm{v}^2$, for $\mathrm{v} = 10^7$ cm/s one has $L = 50$ m, but for $\mathrm{v} = 10^8$ cm/s, $L = 5$ km. With the mass of the disc of the order $\rho R^2\delta \sim 10^{-2}$ g, the kinetic energy at $\mathrm{v} = 10^7$ cm/s is ~ 50 kJ, and at 10^8 cm/s, 5 MJ.

The current drawn from the diode is $I \sim ne\mathrm{v}R^2$, where $ne \sim E/R$, hence $I \sim E\mathrm{v}R$. For $E \sim 3 \times 10^4$, $\mathrm{v} \sim 10^8$ cm/s and $R \sim 1$ cm one has $I \sim 3 \times 10^{12}$ esu $= 10^3$ Ampere, well below the Alfvén current $I_A = 17\,000\gamma$ Ampere $\sim 400\,000$ Ampere, which can be drawn from a foil-less diode.

We have to add three remarks: First, the pressure acting on the disc must be less than its tensile strength σ. Typically, $\sigma \sim 10^{10}$ dyn/cm^2. This condition is well satisfied. Second, at the thin rim of the disc the electric field would be much larger than $\sim 10^7$ Volt/cm, where it not for the toroidal electron cloud formed by electron field emission from the rim, azimuthally rotating around the rim with the drift velocity $\mathrm{v}_\phi = cE/H < c$, shielding the rim. Third, by making the disc from a ferromagnetic substance, it can be kept in a stable position by magnetic feedback control.

8.11 Magnetic Traveling Wave Macroparticle Accelerator

In the spectrum of particles considered as potential inertial confinement fusion drivers, we finally have reached the end: A single macroscopic particle accelerated to high velocities. As it turns out this was hardly feasible with electrostatic acceleration, but it certainly is possible with magnetic acceleration, even though there, too, the dimensions of the accelerator are not small. With velocities of several 100 km/s needed, the macroparticle is not permitted to come into contact with the wall of the accelerator tube.

The acceleration is possible for a ferromagnetic or superconducting projectile riding on a traveling magnetic wave as shown in Fig. 8.13. As in magnetic plasma confinement, where the Lorentz force acts on the plasma

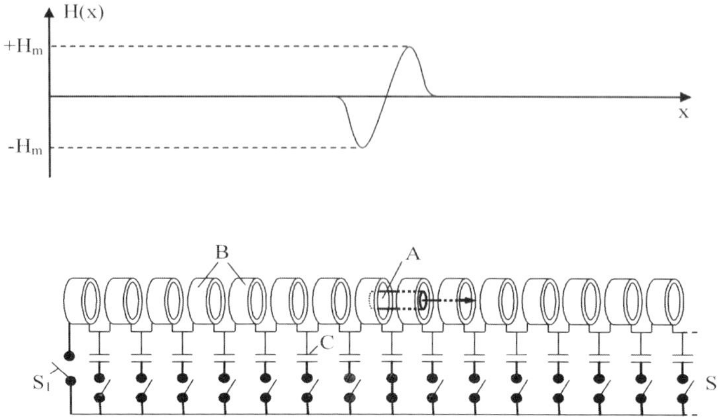

Figure 8.13: Magnetic dipole-type traveling wave accelerator: The projectile A, which can be a small ferromagnetic rod or superconducting solenoid, is magnetically accelerated through external magnetic field coils B. C are capacitors and $S_1, \ldots, S$ are switches to be closed as the projectile moves down the accelerator tube.

electrons, with the plasma ions held by electrostatic forces to the electrons, the same is true here, where the magnetic force acts on the electrons with the ions electrostatically held to the electrons to preserve charge neutrality.

Unlike other particle accelerators, the traveling magnetic wave accelerator is a dipole accelerator with the force on the macroparticle:

$$F = M \nabla H \ . \tag{8.111}$$

where M is the magnetic moment of the macroparticle. If the macroparticle has the shape of a cylinder of length ℓ and cross sectional area A, its magnetic moment is $M = A\ell H_0/4\pi$, where H_0 is the intrinsic magnetic field. For ferromagnets $H_0 \simeq 60$ kG (gandolinium), $H_0 \simeq 80$ kG (holmium), but for superconductors H_0 can conceivably be as large as 300 kG.

If the magnetic field rises for $-H_m$ to $+H_m$ over the length ℓ of the cylindrical macroparticle, one has $\nabla H = 2H_m/\ell$, and hence

$$F = \frac{AH_0H_m}{2\pi} \tag{8.112}$$

With the mass of the cylindrical macroparticle of density ρ, $m = A\ell\rho$, its acceleration is

$$a = \frac{H_0 H_m}{2\pi\rho\ell} \ . \tag{8.113}$$

The length of the accelerator needed to reach the velocity v then is

$$L = \frac{\mathrm{v}^2}{2a} = \frac{\pi\rho\ell}{H_0 H_m}\mathrm{v}^2 \ . \tag{8.114}$$

An upper limit for H_0 and H_m is determined by the tensile strength of the material from which the magnetic field coils are made. With $\sigma \simeq 10^{10}$ dyn/cm^2, one has $H_{max} = \sqrt{8\pi\sigma} \simeq 5 \times 10^5$ G. We therefore may take the following example $H_0 = H_m = 3 \times 10^5$ G, $\rho = 5$ g/cm^3, $\ell = 3$ cm and $\mathrm{v} = 10^8$ cm/s. We find $L \simeq 50$ km, but because of its large momentum, a massive projectile can upon impact not only implode and compress a thermonuclear target, it can also confine the target much longer than otherwise would be possible. This means that a velocity of only $\mathrm{v} = 2 \times 10^7$ cm/s is needed, whereby L is reduced from $L = 50$ km to $L = 2$ km. In reality though, the assumed magnetic field strength $H_0 \sim H_m \sim 3 \times 10^5$ G is too high, with $H_0 \sim H_m \sim 10^5$ G being more likely, whereby $L \simeq 20$km.

A more serious problem is switching. To make the traveling wave, the switching has to be done in the time $\tau \simeq \ell/\mathrm{v} \sim 10^{-7}$ s, not only to turn the current on, but more importantly, to turn it off. A solution to this problem is to abandon the linear traveling wave accelerator altogether, replacing it by a circular accelerator. There the macroscopic particle needs be kept only in a circular orbit by a static magnetic field, while a comparatively weak traveling magnetic wave can leisurely accelerate the macroparticle to the desired final velocity. And there is a second bonus. It is that the accelerator ring needs to consist only of a large circular trough, with the wall of the trough made up from a good conductor. A magnetized macroparticle, be it a ferromagnet or superconductor, is repelled from the conducting wall by the magnetic field of induced mirror image currents. (However, to make full use of this effect the macroparticle should have the shape of a thin plate rather than a sphere).

The analysis of the accelerator is quite similar to that of the linear accelerator, except that here $H_0 = H_m$, with H_m the magnetic field of the induced image currents. With $H_0 = H_m = H$, the radially directed magnetic force on a spherical macroparticle of radius r is

$$F_p = \frac{4\pi}{3} r^2 H^2 \tag{8.115}$$

which has to balance the centrifugal force in a circular orbit of radius R:

$$Z_\rho = \frac{m\mathrm{v}^2}{R} = \frac{4\pi}{3}\rho r^3 \frac{\mathrm{v}^2}{R} \,. \tag{8.116}$$

Equating (8.115) with (8.116) and solving for R one has

$$R = \frac{\rho r}{H^2}\,\mathrm{v}^2 \,. \tag{8.117}$$

For the example $\mathrm{v} = 2 \times 10^7$ cm/s, $r = 1$ cm, $H = 6 \times 10^4$ G, $\rho = 5\mathrm{g/cm}^3$, one finds $R \simeq 5$ km, with a circumference of the accelerator ring $2\pi R \simeq 30$km.

For a kinetic energy of $\sim 10^{14}$ erg $= 10$ MJ, a typical value required for impact fusion at a velocity of 200 km/s the macroparticle mass is ~ 0.5 g. Fig. 8.14a shows the macroparticle in the trough with its magnetic mirror image, and Fig. 8.14b shows how the macroparticle is accelerated in essentially the same way as in the linear accelerator, except that the traveling wave can here be much weaker, greatly reducing the switching problem. Fig. 8.14c shows the injection-ejection macroparticle switchyard, with fresh macroparticles injected into the accelerator, while fast ones leave the accelerator ring and are delivered to a thermonuclear reactor for ignition. Finally, Fig. 8.15 shows the arrangement of four thermonuclear power plants positioned along the periphery of the accelerator, each one of them receiving fast moving macroparticles.

A substantial reduction in the length of the magnetic linear macroparticle accelerator seems possible, if the projectile to be accelerated carries with it a coolant, evaporated during the acceleration. This allows us to replace superconductors or ferromagnets with ordinary conductors, driving the macroparticles with magnetic fields at the tensile strength limit $H_{max} = \sqrt{8\pi\sigma} \simeq 5 \times 10^5$ G. At this limit the maximum acceleration possible is

$$a_{max} = \frac{\sigma}{\rho r} \,. \tag{8.118}$$

For steel with $\sigma \sim 10^{10}$ dyn/cm^2, $\rho \simeq 7$ g/cm^3, and $r \simeq 1$ cm, one has $a_{max} \simeq 1.4 \times 10^9$ cm/s^2. For $\mathrm{v} = 2 \times 10^7$ cm/s this would require an accelerator length of $L = \mathrm{v}^2/2a_{max} = 1.4$ km.

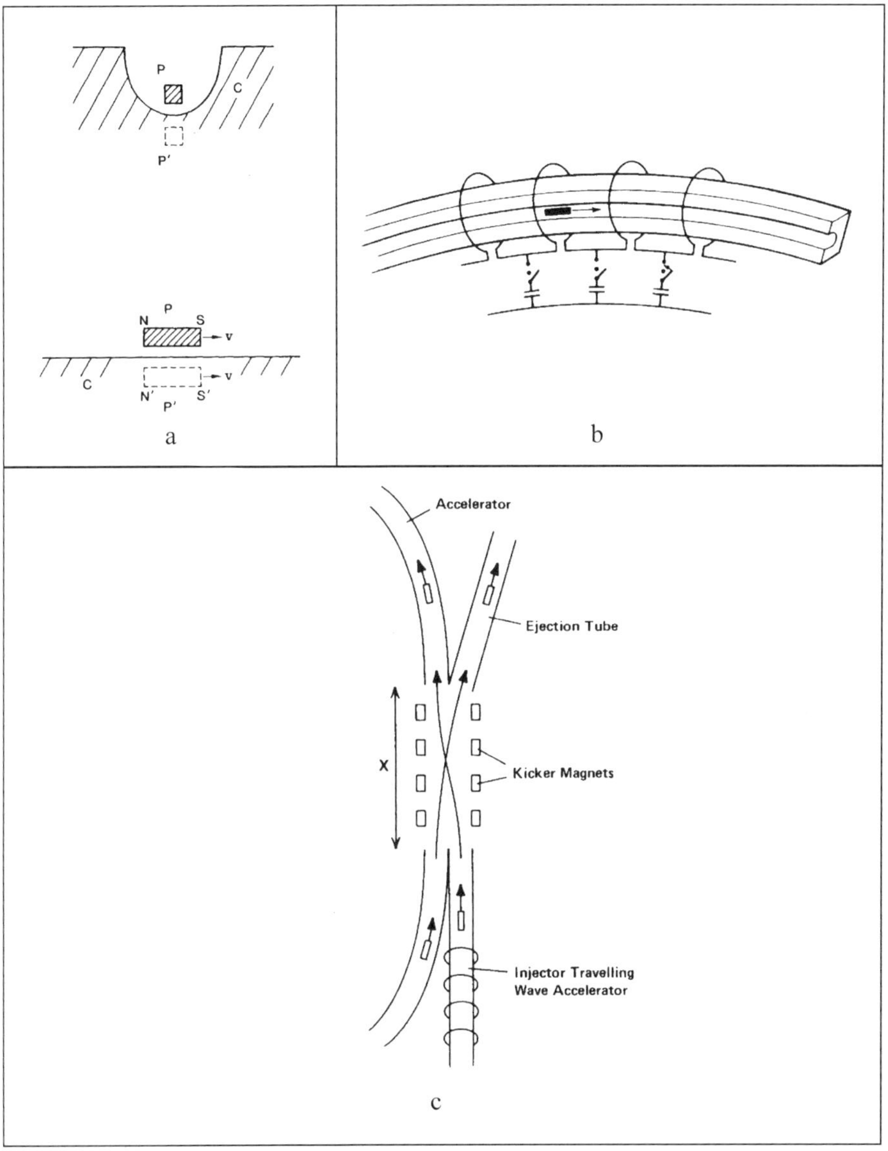

Figure 8.14: (a) Perpendicular and parallel cut through accelerator at the position of the projectile: P projectile; C conductor; N, P north and south pole of magnetic dipole projectile; and P' virtual mirror image of projectile, with N' , S' virtual north and south pole. (b) The generation of the traveling magnetic wave accelerating the projectile inside the ring. (c) Injection-ejection switchyard.

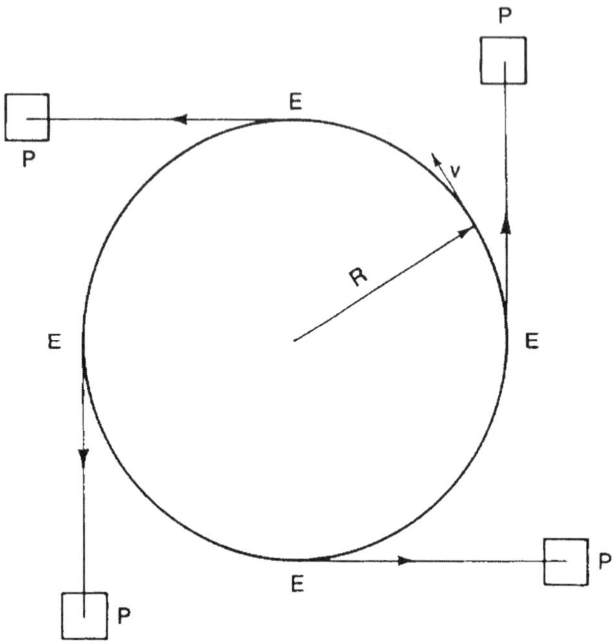

Figure 8.15: General layout of the accelerator and power plants; P power plants; E ejection points for macroparticles.

To compute the maximum velocity for a macroparticle carrying along a coolant, we start from the magnetic four density

$$\mathbf{f} = \frac{1}{c}\mathbf{j} \times \mathbf{H} \tag{8.119}$$

which with Maxwell's equation $4\pi\mathbf{j}/c = \operatorname{curl}\mathbf{H}$, resp. $H \simeq 4\pi jr/c$ one has

$$f = \frac{4\pi r}{c^2}\, j^2\,. \tag{8.120}$$

The rate of resistive energy dissipation into heat (σ is here the electrical conductivity) is

$$\varepsilon = \frac{j^2}{\sigma} \tag{8.121}$$

and hence

$$\frac{f}{\varepsilon} = 4\pi r \, \frac{\sigma}{c^2} \; . \tag{8.122}$$

The electrically conducting part of the macroparticle shall have the density ρ_0 and occupy the volume V_0, while the nonconducting coolant shall have the density ρ_1 and occupy the volume V_1. The mass of the macroparticle then is

$$M = \rho_0 V_0 + \rho_1 V_1 \tag{8.123}$$

and the force acting on V_0 only is

$$R = f V_0 \tag{8.124}$$

hence the acceleration

$$a = \frac{F}{M} = \frac{f}{\rho_0}\left[1 + \frac{\rho_1 V_1}{\rho_0 V_0}\right]^{-1} \; . \tag{8.125}$$

During the time t_0, the heat removed by the coolant is

$$E_{out} = c_{\mathrm{v}} T_0 \rho_1 V_1 \tag{8.126}$$

where T_0 is the evaporation temperature of the coolant with c_{v} the specific heat of the coolant including the evaporation energy. The heat released by resistive dissipation is in the same time t_0

$$E_{in} = \varepsilon V_0 t_0 \; . \tag{8.127}$$

Setting $E_{out} = E_{in}$ and solving for t_0 one has

$$t_0 = c_{\mathrm{v}} T_0 \rho_1 \, \frac{V_1}{\varepsilon V_0} \; . \tag{8.128}$$

At the time t_0 all the coolant is evaporated, and one has for the velocity of the macroparticle in this instant

$$\mathrm{v} = a t_0 = \frac{4\pi\sigma r c_{\mathrm{v}} T_0}{c^2}\left[1 + \frac{M_0}{M_1}\right]^{-1} \tag{8.129}$$

where $M_0/M_1 = \rho_0 V_0/\rho_1 V_1$ is the projectile to coolant mass ratio. The importance of this result is that the coolant keeps the conductor at a low temperature and thereby σ high. This, of course, requires that $M_1 > M_0$. With σ remaining large we may put $\sigma \simeq 10^{18}\,\mathrm{s}^{-1}$, furthermore $c_v T_0 \sim 10^{10}$ erg/g, and find v $\lesssim$ 500 km/s.

One might realize this idea by a sponge-like projectile, where the sponge is filled with a coolant (for example liquid hydrogen). During its acceleration the evaporated coolant is squeezed out of the sponge by the pressure of the magnetic traveling wave.

A more efficient way is the concept of the electromagnetic rocket gun shown in Fig. 8.16. There the evaporated coolant is heated by the traveling magnetic wave and contributes to the thrust on the projectile-like macroparticle.

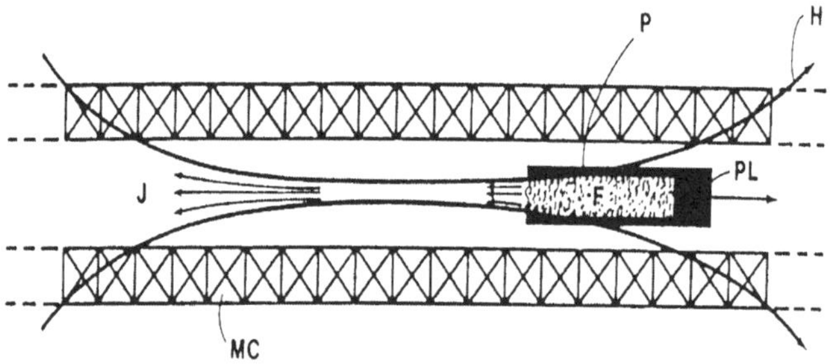

Figure 8.16: Electromagnetic rocket gun principle. *P* part of the projectile holding the propellant *F* which vaporizes, and together with the propellant becomes part of the jet *J*. *PL* projectile payload, *MC* magnetic field coils, *H* magnetic lines of force.

8.12 Magnetic Acceleration of Magnetically Confined Dense Matter

Magnetic acceleration of ordinary conductors can lead to much higher velocities if the conductor to be accelerated is held together by magnetic forces rather than by the internal cohesive forces of the solid state. If the time t_0 to accelerate the conductor is equal to the time for its evaporation by resistive heating, one has

$$\frac{j^2}{\sigma}\, t_0 = \rho c_{\rm v} T_0 \tag{8.130}$$

where $c_{\rm v}T_0$ is the evaporation energy per unit mass and ρ the density of the conductor. In accordance with (8.120) the acceleration of the conductor is

$$a = \frac{f}{\rho} = \frac{4\pi r}{\rho c^2}\, j^2 \tag{8.131}$$

and the final velocity up to the moment of its vaporization

$$\mathrm{v} = a t_0 = 4\pi\sigma r \frac{c_{\rm v} T_0}{c^2}\ . \tag{8.132}$$

If one assumes that $\sigma \sim 10^{16}\ \mathrm{s}^{-1}$ (hot conductor), $c_{\rm v}T_0 \sim 10^{10}$ erg/g, $r \sim 1$ cm, one has $\mathrm{v} \sim 10$ km/s. But if the conductor is held together by magnetic pressure forces, one has for the confinement time t_0^H:

$$\frac{j^2}{\sigma}\, t_0^H = \frac{H^2}{8\pi} \tag{8.133}$$

and hence

$$\mathrm{v} = a t_0^H = \sigma r \frac{H^2}{2\rho c^2}\ . \tag{8.134}$$

With magnetic confinement larger velocities are possible if $t_0^H > t_0$, or if

$$\frac{H^2}{8\pi} > \rho c_{\rm v} T_0\ . \tag{8.135}$$

With $c_{\rm v}T_0 \sim 10^{10}$ erg/g, $\rho \simeq 10$ g/cm^3 one needs $H > 2 \times 10^6$ G, requiring megagauss fields.

That in fact the conductor is held together by the magnetic pressure forces for the time t_0^H, follows from comparing t_0^H with the diffusion time (3.86) for the magnetic field to penetrate the conductor. For times smaller than the diffusion time, the magnetic field can there confine a conductor by magnetic pressure forces acting on its surface. With $j \simeq Hc/4\pi r$ one obtains from (8.133) that $t_0^H \simeq 2\pi r^2/c^2$, which up to a factor 2, (setting $t_0 \to t_0^H$ and $R \to r$), is the same as (3.86).

If we assume that $H = 10^7$ G, attainable with magnetic flux compression, $\sigma \sim 10^{16}$ s^{-1}, $\rho \sim 10$ g/cm^2 and $r \sim 1$ cm, one finds from (8.134) that v $\sim$ 500 km/s.

With thin wires multimegagauss magnetic fields can be produced by letting large currents flow through the wires. For two parallel wires of radius r_0, separated by the distance r, and with an equal current I [esu] flowing through both of them, the magnetic force density on each of the wires is

$$f = \frac{1}{c}\, jH\,. \tag{8.136}$$

With $H = 2I/rc$ and $j = I/\pi r_0^2$ this becomes

$$f = \frac{2}{\pi r_0^2 c^2}\frac{I^2}{r} \tag{8.137}$$

and with the mass of each wire $m = \pi r_0^2 \ell \rho$ where ρ is the density of the wire and ℓ its length one has for their mutual acceleration

$$a = \frac{f}{\rho} = 2\frac{\ell I^2}{mc^2 r}\,. \tag{8.138}$$

Replacing m with the reduced mass $m/2$ of the two wires, their mutual equation of motion for the relative velocity v is:

$$\frac{d\mathrm{v}}{dt} = -\frac{4\ell I^2}{mc^2}\frac{1}{r}\,. \tag{8.139}$$

With $d\mathrm{v}/dt = \frac{1}{2}d\mathrm{v}^2/dr$ one can integrate (8.139) to obtain the mutual impact velocity

$$\mathrm{v} = \sqrt{\frac{8\ell I^2}{mc^2}\log\left(\frac{r}{r_0}\right)}\,. \tag{8.140}$$

or with

$$m = \pi r_0^2 l \rho$$
$$\mathrm{v} = \frac{I}{c r_0}\sqrt{\frac{8}{\pi\rho}\log\left(\frac{r}{r_0}\right)}\,. \tag{8.141}$$

Expressing I in Ampere this is

$$\mathrm{v} = 0.1\frac{I}{r_0}\sqrt{\left(\frac{8}{\pi\rho}\right)\log\left(\frac{r}{r_0}\right)}\,. \tag{8.142}$$

As a first example we take two wires with a radius $r_0 = 10^{-3}$ cm, separated by $r = 4$ cm, and with a current $I = 10^6$ [A] flowing through each wire. From (8.142) one finds for the relative velocity $\mathrm{v} \simeq 1.4 \times 10^8$ cm/s and hence for the absolute velocity $\mathrm{v} \simeq 700$ km/s. We have to check if this velocity is less than the upper limit given by (8.134), where we have to set $r = r_0 = 10^{-3}$ cm. For a current $I = 10^6$ [A] and wire radius $r_0 = 10^{-3}$ cm one has $H = 2 \times 10^8$ G. One can rewrite (8.134) as a condition for the conductivity requiring that $\sigma \geq (2\rho c^2 \mathrm{v}) / (r_0 H^2)$. For $\rho \simeq 10$ g/cm^3 one finds that $\sigma \gtrsim 4 \times 10^{17}$ s^{-1}. This compares well with good metallic conductors for which $\sigma \sim 10^{18}$ s^{-1}.

If the two wires collide, their kinetic energy is transformed into heat given off as blackbody radiation. To estimate the temperature of the blackbody radiation one has to equate the kinetic energy density $\frac{1}{2}\rho \mathrm{v}^2$ of the wires with the energy density aT^4 of the blackbody radiation. One finds that $T \simeq 6 \times 10^7$ °K.

For two wires with a length $\ell = 4$ cm and having a velocity of $\mathrm{v} = 700$ km/s, the kinetic energy is $E_{kin} \simeq 10^{12} = 100$ kJ, converted into heat in the inelastic collision time $\tau_c \simeq 2r_0/\mathrm{v} \simeq 2.5 \times 10^{-11}$ s. The time τ_r this energy is released as radiation can be obtained with the help of (4.72) and one finds that $\tau_r = 2\pi r_0 j_r / \pi r_0^2 a T^4 \simeq \frac{2}{3}(c/r_0)(\lambda_{opt}/r_0)$, where $\lambda_{opt} = (\kappa\rho)^{-1}$, with κ given by (4.71). For $\rho = 10$ g/cm^3, $T \simeq 10^7$ °K one finds that $\tau_r \simeq 2 \times 10^{-13}$ s. Quite obviously, the radiation cannot be released in a time shorter than the inelastic collision time τ_c. With this in mind the keV-x-ray pulse has the power $E_{kin}/\tau_c \simeq 4 \times 10^{15}$ [W]. The diode impedance is of the order $1\,\Omega$. For wires with a good conductivity, a radius $r_0 \sim 10^{-3}$ cm and length ℓ their resistance is of the same order of magnitude. Hence for a total current of 2×10^6 [A], with $I = 10^6$ [A] through each wire, the diode voltage must be $\simeq 2 \times 10^6$ [V] with an input power $\sim 4 \times 10^{12}$ [W]. With an output power of $\sim 4 \times 10^{15}$ [W], this implies a $\sim 10^3$-fold pulse power compression.

For an input energy of $\sim$ 100 kJ, the high voltage pulse must last $\sim 2 \times 10^{-8}$ s. This time compares well with the time needed to accelerate the wires to $\sim$ 700 km/s over a distance of $\sim$ 4 cm. Of course, not all energy goes into kinetic energy, with a comparable fraction going into magnetic field energy.

A velocity of $\sim$ 700 km/s is sufficient to ignite upon impact a DT target, provided the fusion α-particles are trapped in the target. Since this happens for a pinch current in excess of I_c given by (6.23), suggests that we should go to a current of $\sim 10^7$ [A] and to place a cylindrical DT target in between the colliding wires. For a current of $\sim 10^7$ [A] and a wire radius $r_0 \simeq 10^{-2}$ cm almost the same velocity is reached. At a velocity of $\sim$ 700 km/s two 4-cm long wires would there acquire a kinetic energy of $\sim$ 10 MJ, sufficient to ignite a DT cylinder containing $\sim 10^{21}$ DT nuclei. With the impact pressure $\frac{1}{2}\rho \mathrm{v}^2 \sim 3 \times 10^{16}$ dyn/cm^2 set equal the plasma pressure $\sim nkT$ at $kT \sim 10$ keV $\sim 10^{-8}$ erg, one finds that $n \sim 10^{26}$ cm^{-3}. From the Lawson criterion $n\tau \gtrsim 10^{14}$ cm^{-3}s^{-1} then it follows that $\tau \gtrsim 10^{-12}$ s, about 100 times shorter than the inertial confinement time $\tau = 10^{-9}$ s of the DT cylinder. This implies a large burn-up which for $\sim 10^{21}$ DT nuclei releases about 10^{16} erg, with a gain of $\sim$ 100.

For a diode impedance $\sim 1\ \Omega$ and a current of 2×10^7 A the voltage must be $\sim 2 \times 10^7$ V, with a power of $\sim 4 \times 10^{14}$ W. To deliver an energy of $\sim$ 10 MJ to the load requires, as before, a pulse length larger than 2×10^{-8} s.

8.13 Multiple Wire Implosions

Instead of two wires one may take a cylindrical assembly of n wires imploding onto their cylindrical axis. If the implosion velocity is v, the mutual collision velocity of the wires is $\mathrm{v}_0 = 2 \times (2\pi/n)\,\mathrm{v} = (4\pi/n)\,\mathrm{v}$. To compute the temperature upon impact one has to equate the kinetic energy density with the energy density of the black body radiation:

$$\frac{1}{2}\rho \mathrm{v}_0^2 = aT^4 \tag{8.143}$$

or

$$\frac{8\pi^2}{n^2}\rho v^2 = aT^4 \ . \tag{8.144}$$

For the example $n = 100, \rho = 19.3\text{g/cm}^3$ (tungsten wires), $v = 10^7$ cm/s, one finds $T \simeq 6.3 \times 10^6$ °K $= 540$ eV, which is in the soft x-ray range.

It has been proposed to use the intense x-ray pulse from the colliding wires to drive an ablation implosion thermonuclear target. A configuration for this purpose is shown in Fig. 8.17, where the x-ray pulse enters a cavity (hohlraum) into which a spherical thermonuclear target is placed. A limitation of this scheme is the Rayleigh-Taylor instability of the imploding cylindrical wire assembly.

A different configuration is shown in Fig. 8.18, which is conceivably better. It is the same as the plasma focus configuration, if the plasma is replaced with wire spokes. The observed high energy accumulation of the plasma focus results from shear flow stabilization (see chapter 8.18) of the pinch focus. The same can be expected to happen for the imploding wire focus.

The acceleration of the wires to a velocity of $\sim 10^7$ cm/s happens over some distance, whereas the x-rays are released the moment the wires collide. The imploding wires are, for this reason, a pulse compression scheme, with the energy first more slowly transformed into kinetic energy of the wires and thereafter rapidly released as a soft x-ray pulse.

In the configuration shown in Fig. 8.17, the acceleration takes place over a shorter distance than possible with the focus configuration shown in Fig. 8.18. In the latter, the wires are first accelerated axially before they implode radially inward. Therefore, apart from the much better stability of the second configuration, it promises larger pulse compression.

In a typical example for a multiple wire implosion the input energy is ~ 10 MJ delivered to the wires in $\sim 2 \times 10^{-7}$ s with an input power of 50 TW, amplified in the course of the implosion to ~ 200 TW. By order of magnitude an ~ 8 g solid projectile with a velocity of ~ 50 km/s and kinetic energy of ~ 10 MJ, could upon impact deliver an input power of 50 TW. Letting the macroparticle pass through an induction coil it could drive an array of exploding wires inside a cavity into which the macroparticle is shot (Fig. 8.19). Because the wires attract each other magnetically, they collide with high velocities, resulting in a pulse power amplification of the soft X-rays released.

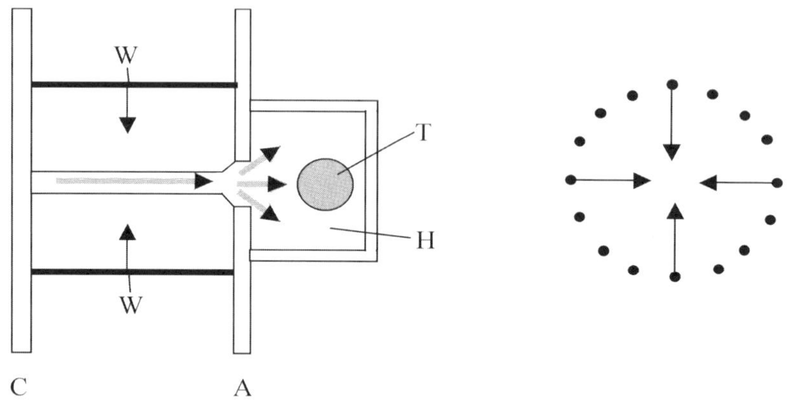

Figure 8.17: Cylindrical imploding multiple wire configuration. *C* cathode, *A* anode, *W* wires, *H* hohlraum, *T* thermonuclear target.

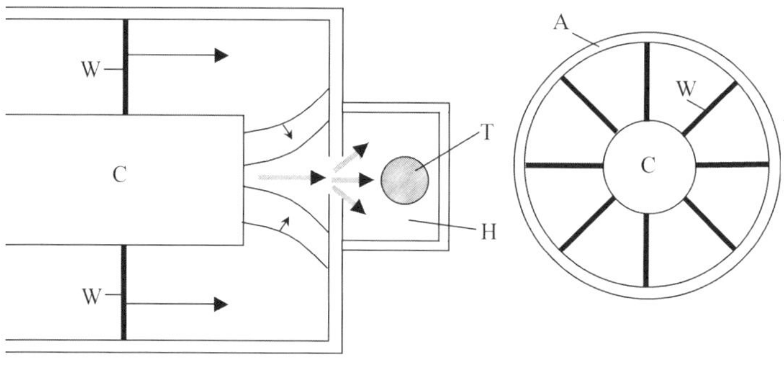

Figure 8.18: Multiple wire plasma focus configuration. *C* cathode, *A* anode, *W* wires, *H* hohlraum, *T* thermonuclear target.

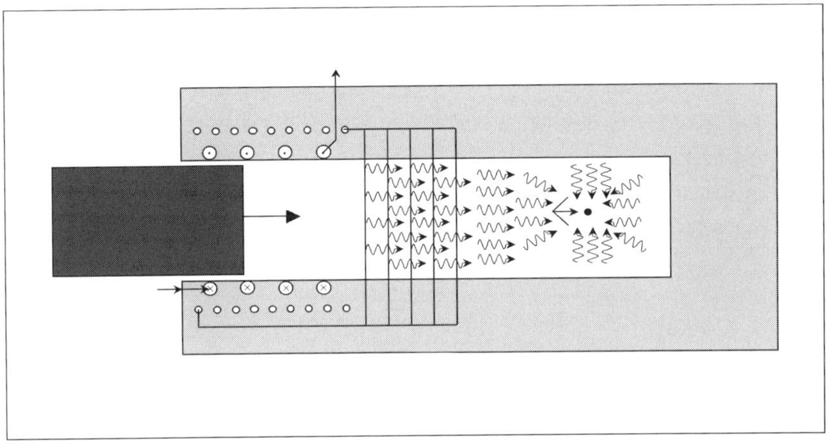

Figure 8.19: Exploding wires driven by fast moving projectile, and with the soft X-ray released, to compress and ignite a thermonuclear target.

Placing a high gain target inside the cavity, with a conical shield placed in front of it (as shown in Fig. 8.19), the target can be precompressed to high densities and ignited by a fast jet from the collapsing cone. To reach the high required ignition velocities, the cone is ablatively accelerated towards the target, with the shaped charge effect of the collapsing cone directing a jet serving as the ignition pulse onto the target.

Experiments have shown that the kinetic energy into heat upon impact model cannot be completely correct, in particular for the collision of thin wires were the impact velocity is highest, because there much more x-rays are emitted than this simple model predicts. A possible explanation for this behavior is given in chapter 11.6.

8.14 Some General Comments on Pulse Power Compression

The example of pulse power compression by the kinetic energy conversion into a soft x-ray pulse with an array of imploding wires is just one example of a more general method of achieving pulse power compression by slow kinetic energy accumulation, followed by a rapid transformation of the kinetic energy into a high power pulse. The factor by which the pulse power is compressed is equal to the ratio of the longer time needed for the acceleration, and the shorter time the kinetic energy is set free as radiation. More generally, in a macroparticle accelerator energy slowly accumulated is rapidly delivered onto a target upon impact. The power to drive the macroparticle can be kept there comparatively low, at the expense of a long macroparticle accelerator. If the length of the accelerator is L, the average particle velocity v, and the energy accumulated into the particle as kinetic energy E, the power to accelerate the particle is

$$P \sim \frac{\mathrm{v}}{L} E \, . \tag{8.145}$$

For the example $\mathrm{v} \sim 10^7$ cm/s, $L \sim 10^6$ cm, $E \sim 10^7$ J one has $P \sim 2 \times 10^7$ W. For a cm-size macroparticle the pulse power compression ratio is here 5×10^6, raising P from $P \sim 2 \times 10^7$ W to $P \sim 10^{14}$ W.

In chapter 8.2 we have shown that flywheel generators can deliver a power $P \sim 10^{10}$ W. There a $\sim 10^4$-fold pulse power compression would be needed to reach a power of 10^{14} W.

Three concepts are promising pulse power compression candidates powered by a $\sim 10^{10}$ W homopolar flywheel generator:

1. The cumulative magnetically driven electron beam accelerator shown in Fig. 8.6.

2. The heavy ion beam plasma jet accelerator shown in Fig. 8.7.

3. The magnetic macroparticle accelerator for an ordinary conductor carrying a coolant as described in chapter 8.11.

One may also attain large pulse power compression by magnetizing a large storage coil, followed by the interruption of the current through a sequence of faster opening switches, with the load serving as the fastest last opening switch. How this idea might be realized is explained in chapter 8.20.

Pulse power compression can be described in a power-time, P-t diagram, where the energy $E = Pt = const.$ As shown in Fig. 8.20 pulse power compression requires that a low-power long-duration pulse be inverted into a high-power, short-duration pulse.

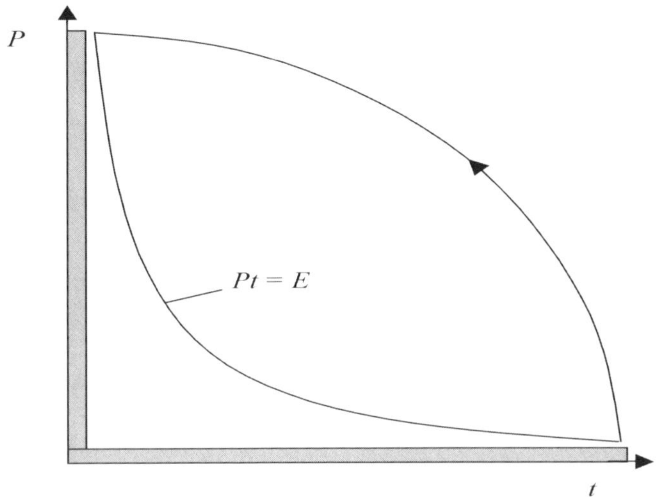

Figure 8.20: Pulse power compression by inverting a low-power, long-duration pulse into a high-power, short-duration pulse.

8.15 The Magnetic Booster Impact Fusion Concept

In chapter 6.10 we have shown that the velocity of $\sim$ 200 km/s required for impact fusion can be lowered by about one order of magnitude for magnetized fusion targets. A $\sim$10-fold reduction in the projectile velocity would imply a $\sim$ 100-fold reduction in the length of the magnetic macroparticle accelerator down to a few 100 meters. However, for magnetized targets the expected thermonuclear gain G is rather small, in the given example $G \sim 30$, while for commercial thermonuclear energy release one should have $G \sim 10^3$. (The same demand is made for thermonuclear rocket propulsion.) We will now show how this limitation can be overcome by the magnetic booster impact fusion concept. Shown in Fig. 8.21a-d, it proceeds as follows:

1. A high velocity projectile P having a conical depression at its front, implodes the cylindrical hollow booster target chamber T placed into the conical depression of an "anvil" A. The booster target chamber is filled with DT gas which is permeated by a magnetic field of the order

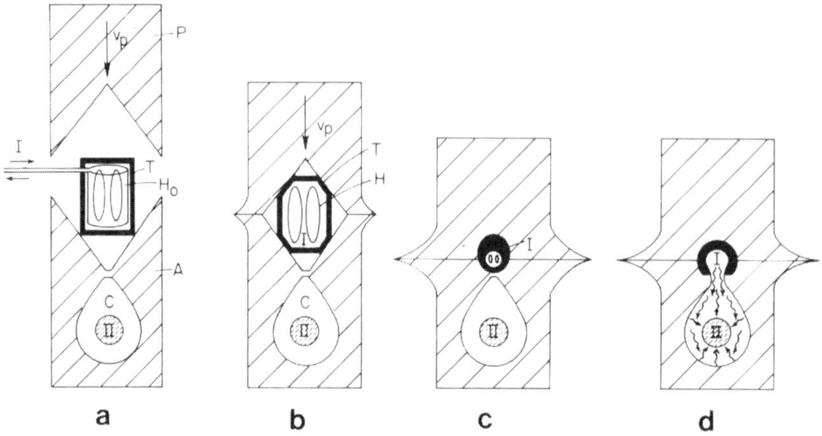

Figure 8.21: In the two-stage magnetic booster impact-fusion target, the first stage target I is the low gain booster target and the second stage target II is the high-gain target. P is the incoming projectile moving with velocity v_p and B a laser or charged-particle beam that passes through the opening O into T, the booster target chamber. V is the conical vertex position, and H_0 is the initial magnetic field inside T. (a) An incoming projectile implodes the booster low density DT gas which has been magnetized and preheated by the laser or charged-particle beam. (b) Magnetic field reversal closes the field lines with the target I highly compressed. (c) The magnetic field rises to maximum compression where the target chamber is at its minimum diameter. (d) In reaching its ignition temperature the DT plasma ruptures the cavity wall at V, releasing a large amount of energy into the chamber C. The radiation and hot plasma released into C ablatively implodes and ignites the high-gain target II.

$H_0 \sim 10^5$ G. The magnetic field in this first stage can be produced by a pulsed one-turn magnetic field coil, or, alternatively, by a small superconducting coil or a ferromagnet with a large saturation field strength like gadolinium or holmium.

2. Just a short moment before the incoming projectile strikes the target chamber, a short laser pulse of comparatively low energy passes through a hole into the target chamber preheating the DT gas to a temperature of $T_0 \sim 10^6$ °K, transforming it into a magnetized plasma.

3. The magnetized DT plasma in the target chamber is imploded by the incoming projectile and highly compressed, leading to a rapid rise of the magnetic field trapped inside the chamber. Currents induced in the plasma and the target chamber wall lead to magnetic field reversal closing the lines of force within the plasma. After field reversal has taken place, the heat conduction losses of the DT plasma into the target chamber wall are only those perpendicular to the magnetic lines of force, greatly increasing the energy confinement time. Without energy losses, the plasma temperature and magnetic field in the target chamber rise as

$$\frac{T}{T_0} = \frac{H}{H_0} = \left(\frac{\ell_0}{\ell}\right)^2 , \tag{8.146}$$

where ℓ_0 and ℓ, respectively, are the linear dimensions of the target chamber at the beginning of the implosion process and some time later. Field reversal occurs when $H \gtrsim 2H_0$, that is, for

$$\ell \lesssim \frac{\ell_0}{\sqrt{2}} .$$

Because a temperature of $T_i \simeq 10^8$ °K is needed to reach thermonuclear ignition, the cavity must be imploded to a minimum diameter

$$\ell_{min} \simeq \frac{\ell_0}{10} .$$

Therefore, apart from the initial phase of the implosion process, the magnetized plasma is in a state of complete field reversal most of the time. At the final minimum diameter of the target chamber, the magnetic field has risen to its maximum value $H_{max} \simeq 10^7$ G.

4. After reaching the ignition temperature T_i the DT plasma confined inside the target chamber makes a thermonuclear excursion. The rapid rise in the internal energy caused by the thermonuclear reactions leads to a rapid rise in the plasma pressure, whereby the target chamber wall is ruptured at the weakest point. If this weakest point is chosen to be at the vertex point V of the conical cavity formed by the anvil and projectile, a large amount of energy in the form of radiation and hot

plasma is released into the adjacent chamber C, into which a second-stage thermonuclear target II made up of liquid or solid DT is placed.

5. The radiative energy flowing with the velocity of light into the chamber C ignites the second stage II by ablative implosion with a $\sim 10^3$-fold gain of this stage. It is this second stage were most of the energy is released, with the booster stage I serving as a trigger for the high gain second stage.

With this two-stage target concept one can work with much smaller impact velocities and still have a high gain.

These general ideas will now be supported by some more quantitative estimates:

1. If we assume that the projectile has a density $\rho_p \simeq 10$ g/cm^3 and moves with $\mathrm{v}_p = 2 \times 10^6$ cm/s, leading upon impact to a stagnation pressure

$$p_s = \frac{1}{2}\rho_p \mathrm{v}_p^2 = 2 \times 10^{13} \text{ dyn/cm}^2 ,$$

to be set equal to the final plasma pressure in the target chamber $p = 2nkT_i$ at $T_i \simeq 10^8$ °K, we then find that $n \simeq 10^{21}$ cm^{-3}. The initial density of the DT gas inside the target chamber is smaller by the factor

$$\left(\frac{\ell_{min}}{\ell_0}\right)^3 \simeq 10^{-3} .$$

Therefore, the initial number density of the gaseous DT target is $n_0 \simeq 10^{18}$ cm^{-3}.

2. Let us assume that the initial and final diameters of the target chamber are $\ell_0 \simeq 4$ cm and $\ell_{min} \simeq 0.4$ cm, respectively. The volume at maximum compression is thus

$$\ell_{\min}^3 \simeq 6 \times 10^{-2} \text{cm}^3 ,$$

and the total number of atoms in the chamber is $N \sim 6 \times 10^{19}$.

3. Because the implosion of the chamber is three-dimensional, the time in which the DT gas is heated from $T_0 \simeq 10^6$ °K up to $T_i \simeq 10^8$ °K, under

the assumption that the compression is completely isentropic, is given by

$$\tau_A \simeq \frac{\ell_0}{2\mathrm{v}_p} \,. \qquad (8.147)$$

For our example we find $\tau_A \simeq 10^{-6}$ s. As we had shown in chapter 6.10, at these velocities and particle number densities, the energy losses of a magnetized DT plasma can be neglected in comparison to its isentropic heating by implosive compression. The assumption of isentropic compression is therefore reasonably well satisfied.

4. To heat a plasma composed of $N \simeq 6 \times 10^{19}$ ions to $T = 10^6$ °K requires the energy

$$E_0 = 3NkT \simeq 2.4 \times 10^{10} \text{ erg} = 2.4 \text{ kJ} \,.$$

This relatively small energy required for preheating can be easily supplied by a short-pulse laser or charged-particle beam. Because the beam pulse has to enter the chamber through a small opening, a laser beam seems to be better suited for this purpose. Furthermore, since the initial density of the DT gas is rather low, an infrared gas laser of high efficiency can be used.

5. To heat the DT gas by isentropic compression to the ignition temperature $T_i \simeq 10^8$ °K requires that its internal energy be raised to

$$E_i = 3nkT \simeq 2.4 \times 10^{12} \text{ erg} = 240 \text{ kJ} \,.$$

Assuming pessimistically that only about 1% of the kinetic projectile energy goes into this internal energy, the projectile energy would have to be 2.4×10^{14} erg. With a projectile velocity of 2×10^6 cm/s, the projectile mass is $m_p = 120$ g. The remaining 99% of the projectile energy would not be lost but would serve to inertially confine the target plasma. Under this assumption most of the energy is used for inertial confinement and not for ignition, as in laser or charged-particle beam fusion.

6. After the DT plasma has reached the thermonuclear ignition temperature $T_i \simeq 10^8$ °K, the part of the thermonuclear energy set free in

the form of α-particles is dissipated within the DT plasma because the Larmor radius of these α-particles at $H = 10^7$ G is $r_L \simeq 0.03$ cm, more than 10 times smaller than the diameter of the imploded chamber, which is $\ell_{min} \simeq 0.4$ cm. As a result the DT plasma undergoes a thermonuclear excursion, greatly raising its temperature as long as the inertial confinement lasts. The inertial confinement time is of the order

$$\tau_i \simeq \frac{h}{\mathrm{v}_p} , \tag{8.148}$$

where h is the thickness of the material made up from the projectile and the anvil. The value of h can be estimated putting $h^3 \rho_p = m_p$, which in our example, with $\rho_p = 10$ g/cm^3 and $m_p = 120$ g, gives $h = 2.3$ cm. It thus follows that $\tau_i \simeq 10^{-6}$ s. The fuel burnup time, on the other hand, is given by

$$\tau_b \simeq \frac{1}{n\langle\sigma \mathrm{v}\rangle} . \tag{8.149}$$

In a thermonuclear excursion the temperature rises until $\langle\sigma \mathrm{v}\rangle$ has reached its maximum, for the DT reaction

$$\langle\sigma \mathrm{v}\rangle_{max} \simeq 10^{-15} \text{ cm}^3/\text{s}$$

is reached at a temperature of $\sim 8 \times 10^8$ °K. For $n = 10^{21}$ cm^{-3}, one has $\tau_b \simeq 10^{-6}$ s.

7. Since $\tau_i \simeq \tau_b$ we may assume a large fuel burnup, for example, 50%. The total energy released into α-particles, each having a kinetic energy of 2.8 MeV, for a DT plasma of 6×10^{19} ions, thus is

$$E_\alpha = \frac{1}{2}\left[\frac{1}{2}\left(6 \times 10^{19}\right)\right]\left(4.5 \times 10^{-6}\right) = 3.4 \times 10^{14} \text{ erg} = 34 \text{ MJ} . \tag{8.150}$$

$$p \sim \frac{E_0}{\ell_{min}^3} \simeq 5 \times 10^{15} \text{ dyn/cm}^2 ,$$

which is about 100 times larger than the magnetic pressure at 10^7 G. As a result the hot plasma will convectively mix with the wall material. Because of this mixing effect most of the energy will go into blackbody radiation. The temperature T_b of this blackbody radiation is determined by

$$aT_b^4 = \frac{E_\alpha}{\ell_{min}^3}\,, \tag{8.151}$$

where $a = 7.67 \times 10^{-15}$ erg/cm^{3}K^4, and one finds that $T_b \simeq 3 \times 10^7$ °K. If we assume that the chamber is permitted to expand approximately 3-fold from the high pressure before it breaks at the vertex, the temperature would go down to $T'_b = T_b/3 \simeq 10^7$ °K.

8. After rupture at the vertex point, the photon energy flux into the cavity C, inside which the high-yield thermonuclear target is placed, is given by

$$P = \sigma T_b'^4\,, \tag{8.152}$$

where $\sigma = ac/4 = 5.75 \times 10^{-5}$ erg/cm^{2}s°K^4. With $T'_b = 10^7$ °K, we find that

$$P = 5.75 \times 10^{23}\ \text{erg/cm}^2\text{s} = 5.75 \times 10^{16}\ \text{W/cm}^2\,.$$

If the cross section of the opening formed at the breaking point through which energy can flow is of the order $\ell_{min}^2 \sim 10^{-1}$ cm^2, the power flux through this opening is $\sim 5 \times 10^3$ TW. Of the α-particle energy released in the booster stage, equal to about 30 MJ, only one-third, that is ~10MJ, is available as blackbody radiation. The remaining ~20 MJ goes into work expanding the target chamber ~3-fold in its diameter, but the remaining ~10 MJ is more than enough to implode a high-density, high-gain thermonuclear target. The wavelength of blackbody radiation at $T'_b \simeq 10^7$ °K is sufficiently short to ensure good coupling to the target for its compression up to $\sim 10^4$ times solid densities.

Finally, we would like to mention that for the magnetized booster stage one may use instead a fast z-pinch discharge as shown in Fig. 8.22. There the z-pinch can be stabilized by axial shear flow described in chapter 3.8. In the configuration shown in Fig. 8.22 fast jets are generated by the shape charge effect described in chapter 5.8, through the collision of the conical depressions in projectile and anvil. With the shear flow produced by these jets, the z-pinch can be stabilized, provided the kinetic energy density of the jet is of the same order of magnitude as the magnetic energy density of the pinch.

Figure 8.22: Fast z-pinch magnetic booster high gain target.

8.16 Laser Ignition of the Dense Z-Pinch

The linear z-pinch discharge, the first plasma magnetic confinement configuration considered for the controlled release of thermonuclear energy, was abandoned a long time ago, primarily because of the $m = 0$ (sausage) and $m = 1$ (kink) instabilities. More recently the fast z-pinch has been revived by discharging a high voltage pulse power source over a thin solid filament, which may be a thin wire or even a DT fiber (actually only deuterium fibers have so far been used). To avoid the $m = 0$ and $m = 1$ instabilities one could superimpose an axial magnetic field which must be by order of magnitude as strong as the azimuthal magnetic field of the discharge current, but with the magnetic field of the discharge current many megagauss, this stabilization method can be ruled out. The stabilization by axial shear flow described in chapter 3.8 is more promising.

An axial shear flow pattern can be realized if a fast jet is shot through the core of the pinch, or by a hollow jet flowing along its surface. In chapter 3.8

it was assumed that the density of the shear flow is the same as for the pinch. If the density of the jet is ρ_s, stabilization requires that

$$\frac{1}{2}\rho_s \mathrm{v}_z^2 \geq \frac{H^2}{8\pi} \, . \tag{8.153}$$

For $\rho = \rho_s$ (8.153) is the same as (3.157), and for $\rho = \rho_s$ (8.153) can be written as follows

$$\mathrm{v}_z \geq \mathrm{v}_A \tag{8.154}$$

where v_A is the Alfvén speed. Computer simulations show that $\mathrm{v}_z/\mathrm{v}_A \gtrsim 3$ is required to reach stabilization. With jet densities larger than the plasma density, smaller values of v_z can stabilize the pinch discharge.

At first sight it appears that this kind of stabilization would make the attainment of thermonuclear temperatures difficult if not impossible, with a large axial thermal energy loss along the pinch channel. But these losses are insignificant if a thermonuclear detonation wave is ignited at one point from where it propagates supersonically along the pinch discharge channel. There the discharge channel should ideally be at absolute zero, where for a given magnetic pressure its density would be highest. Therefore, a cool rather than a hot pinch is here desired.

For currents above the Pease-Braginskii current (4.87), the pinch column shrinks down to a small radius until it becomes optically opaque. This happens for a pinch radius about equal to a photon path length $\lambda_p = 1/n\sigma_{opt}$, obtained from (4.70). For a DT plasma at $T \sim 10^7$ °K, $n \sim 5 \times 10^{24}$ cm^{-3}, ($\sim$ 100-fold compression above solid state density), one finds $\lambda \sim 5 \times 10^{-2}$ cm. To reach this high density the plasma would have to collapse from say solid density to $\sim$100 fold solid density. For $T \simeq 10^7$ °K and $n = 5 \times 10^{24}$ cm^{-3} one finds that the radiative collapse time (1.26) is $\tau_R \sim 10^{-8}$ s. For the given temperature and particle number density one has $p = 2nkT \sim 10^{14}$ dyn/cm^2, and with $p = H^2/8\pi$ one has $H \sim 5 \times 10^7$ G. For $r \sim 10^{-2}$ cm this requires that $I = 5Hr \sim 2 \times 10^6$ A, just above the Pease-Braginskii current for DT, which is $I_{PB} \simeq 1.7 \times 10^6$ A.

For a fast radiative collapse, the current should be well above the Pease-Braginskii current, that means $I \gtrsim 10^7$ A.

In order for a thermonuclear detonation wave to propagate along the pinch discharge channel, the discharge current must be larger than I_c, listed in table 6.1 for a number of thermonuclear reactions. For the DT reaction this current is $I_c \simeq 1.35 \times 10^6$ A. This again means that $I \gtrsim 10^7$ A.

To launch a thermonuclear detonation wave, the length of the pinch discharge channel must be larger than the α-particle range λ_0 (4.25), resp. (6.13), at $T \sim 10^9$ °K where $\langle \sigma v \rangle$ is largest. For $T = 10^9$ ° K and $n = 5 \times 10^{24}\text{cm}^{-3}$ on has $\lambda_0 \simeq 0.2$ cm. For this temperature the thermal DT ion particle velocity (i.e. the velocity of sound) is $a_0 \sim 2 \times 10^8$ cm/s.

If a segment of the pinch discharge channel of length λ_0 is heated to $T \sim 10^9$ °K, its expansion time is $\tau \sim \lambda_0/a_0 \sim 10^{-9}$ s. Hence, to ignite a thermonuclear detonation wave the energy

$$E_{ign} = 3nkT\pi r^2 \lambda_0 \tag{8.155}$$

must be supplied in less than $\sim 10^{-9}$ s to a segment of length $\lambda_0 \simeq 0.2$ cm of the pinch discharge channel. For the given example one has $E_{ign} \sim 3 \times 10^{11}$ erg $= 30$ kJ, with a power $P \sim E_{ign}/\tau \sim 3 \times 10^{13}$ W. This energy and power focused onto a $\sim$ 0.01-cm thick 0.2-cm long filament can be supplied by a pulsed laser. Because of the large atomic number density (5×10^{24} cm^{-3}), the ignition must follow the recipe of the fast ignitor, except that no hole has here to be drilled through a plasma corona, absent in the magnetically confined z-pinch plasma.

The jet needed to stabilize the pinch discharge can be produced by an auxiliary discharge as shown in Fig. 8.23. To stabilize the pinch discharge, the energy of the jet must be of the same order of magnitude as the energy of the pinch discharge. Because the discharge channel cannot be made too long, the gain is limited. However, much larger gains are possible, even with a small pinch discharge channel, by staging. There, the thermonuclear energy released in the pinch discharge channel is set free as a magnetically confined jet, facilitating staging by the bombardment of the second stage with the jet.

8.17 Laser Ignition of an Isentropically Compressed Dense Z-Pinch

Radiative collapse of a pinch discharge above the Pease-Braginskii current is not the most efficient means to reach high densities. The isentropic compression used in laser fusion is more effective. As we had remarked, for the

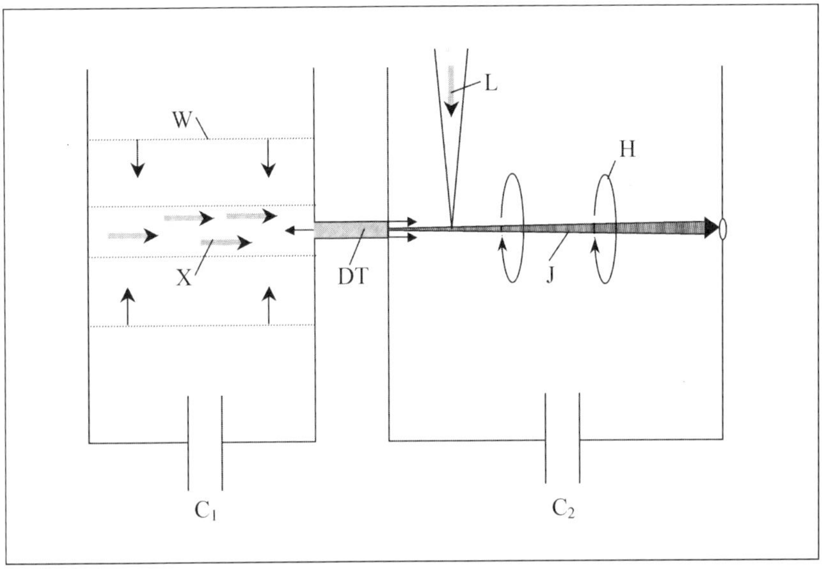

Figure 8.23: Laser ignited shear flow stabilized dense z-pinch. C_1 , C_2 Marx capacitor banks; W wire array imploded by discharge of C_1; soft X-rays X heat solid or liquid DT contained in thin tube, resulting in an ablatively driven jet J over which C_2 is discharged forming a dense z-pinch with azimuthal magnetic field H. L is a pulsed laser beam igniting the dense pinch channel.

ignition of a thermonuclear detonation wave the density, not the temperature, should be as high as possible. High densities can hardly be reached with a large plasma current, in excess of the Pease-Braginskii current, leading to a hot, not a cold plasma. But only in a cold, not hot, plasma can the highest densities be reached. To reach high densities the DT (or other thermonuclear material) is placed inside a metallic capillary tube. The current from an electric pulse power driven discharge over the tube generates a large magnetic pressure on the surface of the tube compressing the DT inside the tube. Futhermore, with a properly chosen time dependence of the current $I = I(t)$, the compression can be made isentropic, as it is required to reach the highest densities by keeping the DT on its lowest adiabat. The magnetic pressure acting on the capillar goes in proportion to I^2 and the power of the discharge in proportion to I^2Z, where Z is the impedance of the discharge channel. Ignoring the weak logarithmic dependence of Z on the radius of the discharge channel, the power for the isentropic compression of a cylindrical assembly, must have according to (5.40) the time dependence

$$\frac{P}{P_0} = \left(1 - \frac{t}{t_0}\right)^{-9/5} . \tag{8.156}$$

If, for example, $I = 10^7$ A with the final capillary tube radius $r = 2 \times 10^{-2}$ cm, the magnetic field has risen to $H = 0.2I/r = 10^8$ G, with a magnetic pressure $H^2/8\pi = 4 \times 10^{14}$ dyn/cm^2. At this pressure the particle number density of cold hydrogen, according to (3.170), is $n = 5 \times 10^{24}$ cm^{-3} (100 times solid state density). And with a current of 10^7 A, a thermonuclear detonation wave, (ignited by a laser pulse at one end of the capillar), can propagate along the capillar.

During its implosion the capillar is subject to the Rayleigh-Taylor instability (chapter 5.7), in addition to the $m = 0$ and $m = 1$ magnetohydrodynamic pinch instabilities. As explained in chapter 3.8 the magnetohydrodynamic instabilities can be suppressed by axial shear flow and the Rayleigh-Taylor instability by rapid rotation.

We now show that by a helical corrugation of the capillar both the magnetohydrodynamic and Rayleigh-Taylor instability can be suppressed. The idea is explained in Fig. 8.24.

If the electrical conductivity of the capillary tube is sufficiently high, and the implosion short compared to the time needed for the magnetic field to diffuse into the tube, the magnetic field cannot penetrate into the tube. With the electric current j flowing along the surface of the capillary tube, the magnetic body force $(1/c)\mathbf{j} \times \mathbf{H}$ is directed perpendicularly onto the surface of the sawtooth. The magnetic field then acts like a piston on the surface of the capillary tube, and one has to set the wedge implosion velocity v_0 equal the Alfvén velocity $\mathrm{v}_A = H/\sqrt{4\pi\rho}$, where $H = 0.2I/r_0$ is the magnetic field of the current I in Ampere, with r_0 the initial radius of the capillary tube and ρ its density. As a result, sheet-like jets are ejected along the imploding sawtooth. The recoil from these jets generates a massive "slug" moving in a direction opposite to the jet. Both the jet and slug generate shear and implode the capillary tube. According to chapter 5.8, eq. (5.104),

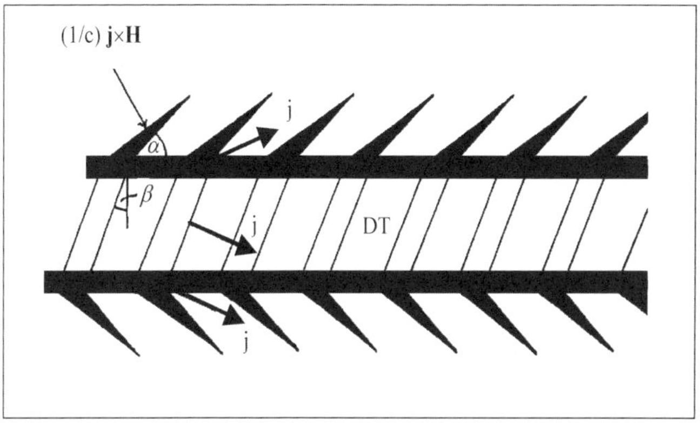

Figure 8.24: Corrugated capillary tube filled with solid DT: α wedge angle, β pitch angle of corrugated surface, J jet, $(1/c)\,\mathbf{j}\times\mathbf{H}$ magnetic body force.

the velocity of jet and slug are

$$\left.\begin{aligned} \mathrm{v}_j &= \left(\frac{\mathrm{v}_0}{\sin\alpha}\right)(1+\cos\alpha) \\ \mathrm{v}_s &= \left(\frac{\mathrm{v}_0}{\sin\alpha}\right)(1-\cos\alpha) \end{aligned}\right\} \tag{8.157}$$

where α is the sawtooth angle as shown in Fig. 8.24. The relative fraction of jet and slug mass are given by (5.103), in our case by

$$\left.\begin{aligned} \frac{m_j}{m} &= \frac{1}{2}(1-\cos\alpha) \\ \frac{m_s}{m} &= \frac{1}{2}(1+\cos\alpha) \end{aligned}\right\} \tag{8.158}$$

The stagnation pressure of jet and slug are

$$\left.\begin{aligned} p_j &= \frac{1}{4}(1-\cos\alpha)\,\rho\mathrm{v}_j^2 = \frac{1}{4}\rho\mathrm{v}_A^2\,(1+\cos\alpha) \\ p_s &= \frac{1}{4}(1+\cos\alpha)\,\rho\mathrm{v}_s^2 = \frac{1}{4}\rho\mathrm{v}_A^2\,(1-\cos\alpha) \end{aligned}\right\} \tag{8.159}$$

with the sum

$$p_j + p_s = \frac{1}{2}\rho\mathrm{v}_A^2 = \frac{H^2}{8\pi} \tag{8.160}$$

just the necessary condition for shear flow stabilization.

The momentum density of the slug, which gives the capillary tube its spin, but also the shear flow and radial implosion, is obtained from (8.157) and (8.158):

$$I_s = \frac{1}{2}\rho \mathrm{v}_A \sin\alpha \,. \tag{8.161}$$

It is equal and opposite to the momentum density of the ablated material. The tangential and radial components of I_s are:

$$\left.\begin{aligned} I_s \cos\alpha &= \frac{1}{4}\rho \mathrm{v}_A \sin 2\alpha \\ I_s \sin\alpha &= \frac{1}{2}\rho \mathrm{v}_A \sin^2\alpha \end{aligned}\right\} \tag{8.162}$$

with the tangential and radial velocity component

$$\left.\begin{aligned} \mathrm{v}_t &= \frac{1}{4}\mathrm{v}_A \sin 2\alpha \\ \mathrm{v}_r &= \frac{1}{2}\mathrm{v}_A \sin^2\alpha \end{aligned}\right\} \tag{8.163}$$

The tangential velocity component has a maximum for $\alpha = 45°$, where one has

$$\left.\begin{aligned} \mathrm{v}_t &= \frac{1}{4}\mathrm{v}_A \\ \mathrm{v}_r &= \frac{1}{2}\mathrm{v}_A \end{aligned}\right\} \tag{8.164}$$

The tangential velocity component can be further decomposed into an axial and azimuthal component, the first generating shear, the second rotation:

$$\left.\begin{aligned} \mathrm{v}_z &= \frac{1}{4}\mathrm{v}_A \cos\beta \\ \mathrm{v}_\phi &= \frac{1}{4}\mathrm{v}_A \sin\beta \end{aligned}\right\} \tag{8.165}$$

where β is the pitch angle of the helical corrugation as shown in Fig. 8.24.

For spherical implosions the Rayleigh-Taylor instability poses a serious problem limiting the maximum attainable compression. For a cylindrical implosion the situation is much better, because there a superimposed rotational motion can suppress this instability. A cylindrical implosion, though making use of this effect, requires a cylinder long compared to its diameter, but this is just realized in a sufficiently long z-pinch. To suppress the Rayleigh-Taylor instability, the capillary tube is brought into fast rotational motion generated by the helical winding of its corrugated surface with still to be determined pitch angle β.

To compute β we assume that initially all the energy deposited is kinetic energy, with part of it going into the radial implosion and part of it into the rotation of the capillary tube. If r_0 is the initial and r_1 the final implosion radius, further $\mathrm{v}_r^{(0)}, \mathrm{v}_\phi^{(0)}$ and $\mathrm{v}_r^{(1)}, \mathrm{v}_\phi^{(1)}$, the initial and final velocity components, where $\mathrm{v}_r^{(1)} = 0$, energy and angular momentum conservation require that

$$\left(\mathrm{v}_\phi^{(1)}\right)^2 = \left(\mathrm{v}_r^{(0)}\right)^2 + \left(\mathrm{v}_\phi^{(0)}\right)^2 \tag{8.166}$$

$$r_1 \mathrm{v}_\phi^{(1)} = r_0 \mathrm{v}_\phi^{(0)} . \tag{8.167}$$

Eliminating $\mathrm{v}_\phi^{(1)}$ from (8.166) and (8.167) one has

$$\frac{\mathrm{v}_r^{(0)}}{\mathrm{v}_\phi^{(0)}} = \left[\left(\frac{r_0}{r_1}\right)^2 - 1\right]^{1/2} \tag{8.168a}$$

which for $r_0 \ll r_1$ is

$$\frac{\mathrm{v}_r^{(0)}}{\mathrm{v}_\phi^{(0)}} \simeq \frac{r_0}{r_1} . \tag{8.168b}$$

The radial deceleration

$$a_1 = \frac{\left(\mathrm{v}_r^{(0)}\right)^2}{r_0} \tag{8.169}$$

leads to Rayleigh-Taylor instability, but the centrifugal acceleration

$$a_2 = \frac{\left(\mathrm{v}_\phi^{(0)}\right)^2}{r_1} \tag{8.170}$$

is counteracting and stabilizing. The Rayleigh-Taylor instability is suppressed if

$$\frac{a_2}{a_1} \gg 1 \, . \tag{8.171}$$

With (8.167) and (8.168a) this means that

$$\frac{a_2}{a_1} = \frac{r_0}{r_1} \gg 1 \, . \tag{8.172}$$

For a 100 fold compression $r_0/r_1 = 10$ inequality (8.172) is well satisfied.

Inserting $\mathrm{v}_r = \mathrm{v}_r^{(0)} = \frac{1}{4}\mathrm{v}_A$, $\mathrm{v}_\phi = \mathrm{v}_\phi^{(0)} = \frac{1}{4}\mathrm{v}_A \sin\beta$ from (8.164), (8.165) into (8.168a) one has

$$\sin\beta = \frac{r_1}{r_0} = 0.1 \tag{8.173}$$

and hence $\beta = 6°$.

Apart from stabilizing the Rayleigh-Taylor instability, the rapid rotation has the benefit that it separates by centrifugal forces the DT inside the capillar from the high atomic number material of the capillar, which is driven radially outward away from the DT.

8.18 Laser Cutting the Dense Z-Pinch and Inductive Energy Storage

As was shown in chapter 8.1, about 10^3 more energy per unit volume can be stored inductively in magnetic field coils than electrostatically in capacitors, but the problem of inductive energy storage is the need to open rather than to close a switch as in electrostatic energy storage. If the resistance of the switch is R, the discharge time for a capacitor C is

$$\tau_c = RC \tag{8.174}$$

whereas for an inductor L it is

$$\tau_L = \frac{L}{R} \, . \tag{8.175}$$

To draw a high power, either from a capacitor or an inductor, requires short discharge times. For a capacitor it requires a small resistance of the switch. For an inductor the resistance of the switch should instead be large, but then most of the energy is dissipated in the switch. The fast z-pinch offers a solution to this problem by making the pinch both the switch and the load into which the inductively stored energy is dissipated.

As shown in Fig. 8.25 this can be done in three stages:

1. An inexpensive low power d.c. source magnetizes a coil storing an energy of several MJ. (If done on a time scale of ~ 0.1 sec, the d.c. power is $\sim 10^9$ Watt, attainable with homopolar generators. With a d.c. voltage of ~ 100 V the current would have to be $\sim 10^7$ A).

2. A mechanically moved switch opens the coil with the current diverted to pass through a tamped wire exploding in $\sim 10^{-7}$ sec. (For a ~ 10-cm long wire the voltage rises to ~ 100 kV).

3. The high voltage generated by the exploding wire triggers the pinch discharge subsequently cut by a laser beam further increasing the voltage and pulse power.

The problem of one large current passing through one switch can be solved with the concept of the Xram (chapter 8.2, and Fig. 8.1) where a bank of coils is charged in series and discharged in parallel, adding the currents of all coils. There the current is equally divided between many switches. For a bank of 100 coils, for example the current in each switch would be reduced 100-fold, for a current of $\sim 10^7$ A, down to $\sim 10^5$ A.

The pulse power compression scenario then assumes the following sequence of events:

1. A 10^7 Watt, 10^2 V, 10^5 A d.c. source magnetizes a bank of 100 coils in series.

2. By mechanically opening the switches connecting the coils and switching them in parallel, their currents are added up, increasing the pulse power to 10^9 Watt.

3. The added up current of 10^7 A passes through a tamped wire which explodes in $\sim 10^{-7}$ sec, raising the voltage from 10^2 V to 10^5 V, and the pulse power from 10^9 Watt to 10^{12} Watt.

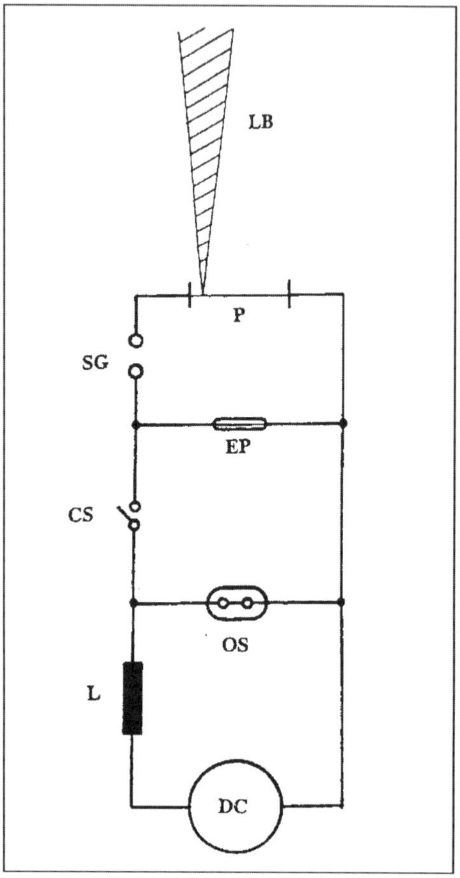

Figure 8.25: DC homopolar generator. L storage coil; OS mechanical opening switch; CS mechanical closing switch; EP exploding wire opening switch; SG spark gap closing switch; P pinch discharge; LB laser beam.

4. The pinch discharge ignited by the 10^{12} Watt pulse is cut by a laser beam raising the voltage from 10^5 to 10^8 V, and the power to 10^{15} Watt.

The high voltage pulse in the last step occurs on a time scale of 10^{-8} sec, short enough for the insulating material to prevent breakdown by streamer formation, while the inertia of the magnetic field energy stored in the space surrounding the pinch keeps the current constant.

We estimate the dimensions of the magnetic storage coil, needed to store an energy of ~ 10 MJ to be discharged in $\sim 3 \times 10^{-8}$ sec, as follows: In electrostatic units the self-inductance of a coil with a wire of length ℓ and height h is approximately given by $L \sim \ell^2/h$, and its capacitance is approximately $C = h/\log[R/r]^2$, where r is the coil radius and R the radius of the wall "containing" the coil. One thus has $LC \sim \ell^2/\log[R/r]^2$, and for the discharge time

$$\tau \sim \sqrt{LC}/c = (\ell/c)\sqrt{\log{(R/r)}^2}\,. \tag{8.176}$$

The impedance of the coil expressed in practical units is

$$Z = 30\,(\ell/h)\sqrt{\log{(R/r)}^2}\;[\Omega] \tag{8.177}$$

for the example $\ell \sim h$, $Z \simeq 10\ \Omega$. For a current of 3×10^6 A, the voltage pulse would thus rise to $\sim 10^8$ V.

To store an energy of ~ 10 MJ in a magnetic field of $\sim 5 \times 10^4$ G (possessing a magnetic pressure of $\sim$100 atm), requires a volume of $\sim 1\ \text{m}^3$. By comparison, the volume of a capacitor bank storing the same amount of energy would be of the order $10^3\ \text{m}^3$. Ignoring the logarithmic factor of the order unity, the discharge time is simply given by ℓ/c, the time needed for an electromagnetic pulse propagating along the coil wire with the velocity of light. For a discharge time of $\sim 3 \times 10^{-8}$ s the length of the coil wire would have to be $\ell = 10$ m. In the Xram circuit, this is the wire length for each coil, because the time to discharge all coils does not change if the coils are switched in parallel.

According to (6.96) the laser intensity to make a cut is for yellow light $I_L \sim 3 \times 10^{19}\ \text{W/cm}^2$. For a pinch of radius r, the laser power required to make a cut of width d, then is

$$P \sim I_L r d\,. \tag{8.178}$$

The width d of the cut cannot be made smaller than the laser wave length which for yellow light is $\lambda = 6 \times 10^{-5}$ cm. Choosing $r \sim 10^{-2}$ cm and $d \sim 10^{-4}$ cm, one has $P \sim 3 \times 10^{13}$ W, and if the cut shall be kept open for the time τ, the laser energy required is

$$E \sim I_L r d \tau \tag{8.179}$$

for $\tau \sim 10^{-8}$ s, this is $E \sim 3 \times 10^5$ J.

We now show that if a cut can be made as small as $d \sim 10^{-4}$ cm, it can become magnetically self-insulating in favor of an ion current crossing the gap at the expense of an electron current.

The concept of magnetic insulation applied to the cut is as follows: If V is the voltage across the cut, where e and m are the charge and mass of an electron, v the electron velocity, and $\gamma = (1 - \mathrm{v}^2/c^2)^{-1/2}$, the relativistic energy equation is

$$\frac{eV}{mc^2} = \gamma - 1 \, . \tag{8.180}$$

Magnetic insulation then requires that the width d of the cut is larger than the electron Larmor radius

$$r_L = \frac{\gamma m \mathrm{v} c}{eH} = \frac{mc^2}{eH}\sqrt{\gamma^2 - 1} < d \, . \tag{8.181}$$

By eliminating γ from (8.180) and (8.181) one obtains

$$\left(\frac{eHd}{mc^2}\right)^2 > \left(1 + \frac{eV}{mc^2}\right)^2 - 1 \, . \tag{8.182}$$

With the magnetic field of the pinch (in Gaussian units) given by

$$H = \frac{2I}{rc} \tag{8.183}$$

one obtains from (8.182)

$$4\left(\frac{I}{I_A^e}\right)^2 \left(\frac{d}{r}\right)^2 > \left(1 + \frac{eV}{mc^2}\right)^2 - 1 \, , \tag{8.184}$$

where $I_A^e = mc^3/e = 17\,000$ A is the electron Alfvén current. The ion current across the cut is given by the Child-Langmuir law where M is the ion mass:

$$I = \frac{\sqrt{2}}{9}\left(\frac{e}{M}\right)^{1/2}\left(\frac{r}{d}\right)^2 V^{3/2} \,. \tag{8.185}$$

From (8.184) and (8.185) one obtains

$$\frac{I}{I_A^e} \geq \frac{9}{4\sqrt{2}}\left(\frac{M}{m}\right)^{1/2}\left[\left(1+\frac{eV}{mc^2}\right)^2 - 1\right]\left(\frac{eV}{mc^2}\right)^{-3/2} , \tag{8.186}$$

$$\frac{d}{r} = \frac{2\sqrt{2}}{9}\left(\frac{m}{M}\right)^{1/2}\left[\left(1+\frac{eV}{mc^2}\right)^2 - 1\right]^{-1/2}\left(\frac{eV}{mc^2}\right)^{3/2} . \tag{8.187}$$

With the magnetic field decreasing towards the axis of the pinch discharge channel the magnetic insulation is imperfect, with some electron current crossing the cut. For this reason the estimates made are somewhat too optimistic. The laser can make the cut in less than 10^{-10} s, short enough to lead to the formation of a magnetically insulated cut possessing the parameters given by (8.186) and (8.187).

Instead of (8.186) one can also write

$$I \geq \frac{9}{4\sqrt{2}}\sqrt{I_A^e I_A^i}\left[\left(1+\frac{eV}{mc^2}\right)^2 - 1\right]\left(\frac{eV}{mc^2}\right)^{-3/2} , \tag{8.188}$$

where $I_A^i = Mc^3/e$ is the ion Alfvén current. For protons it is $I_A^i = 3.1\times 10^7$ A, about two thousand times larger than the electron Alfvén current. As a function of eV/mc^2, (8.188) has a minimum for $eV/mc^2 = 2$, where $I \geq \frac{9}{2}\sqrt{I_A^e I_A^i}$, for protons $\sim 3 \times 10^6$ A. From the Bennett relation $I^2 = 400NkT$ (I in Ampere, and $kT \sim 10^{-8}$ erg for $T \sim 10^8$ °K), one finds for $I \sim 3\times 10^6$ A that $N \sim 2 \times 10^{18}$ cm^{-1}. To make $n_c = 3 \times 10^{21}$ cm^{-3}, the critical density for yellow laser light, requires that $r \leq 10^{-2}$ cm.

According to (8.188) one has

$$I_A^e < I < I_A^i \,, \tag{8.189}$$

where $I \sim \sqrt{I_A^e I_A^i}$. For $I \gg I_A^i$ the pinch is unstable or "soft", which is the case if the current is carried by electrons, but it is stable or "stiff" if $I \ll I_A^i$, if the current is carried by ions. Since this condition cannot be completely satisfied, the instabilities are only reduced, not eliminated.

For the minimum current at $eV/mc^2 = 2$, one obtains from (8.187) $d/r = \frac{2\sqrt{2}}{9}(m/M)^{1/2}$, for protons $d/r \sim 7 \times 10^{-3}$. If the current is given in Gaussian units one must have $\pi r^2 \leq I/nec$, where n is the number density of the electrons. For $I \sim 3 \times 10^6$ A $\sim 10^{16}$ esu and $n \sim 10^{21}$ cm^{-3} one finds that $r \geq 10^{-3}$ cm, and $d \sim 10^{-5}$ cm.

Because of the smallness of the cut one may go to higher voltages where the cut becomes larger.

In the limit $eV/mc^2 \gg 1$ one has

$$I \geq \left(\frac{9}{4\sqrt{2}}\right)\sqrt{I_A^e I_A^i}\left(\frac{eV}{mc^2}\right)^{1/2}, \tag{8.190}$$

$$\frac{d}{r} = \left(\frac{2\sqrt{2}}{9}\right)\left(\frac{m}{M}\right)^{1/2}\left(\frac{eV}{mc^2}\right)^{1/2}. \tag{8.191}$$

In practical units, where $I_A^i = 1.7 \times 10^4$ A and $mc^2 \approx 0.5 \times 10^6$ eV, one has

$$I \geq 38.0\left(\frac{M}{m}\right)^{1/2} V^{1/2}. \tag{8.192}$$

For protons this is

$$I \geq 1.6 \times 10^3 V^{1/2} \tag{8.193}$$

or

$$V \leq 4 \times 10^{-7} I^2. \tag{8.194}$$

The gap impedance is given by

$$Z = \frac{V}{I} = 7.5 \times 10^{-4} V^{1/2} = 4 \times 10^{-7} I. \tag{8.195}$$

In practical units (8.191) is

$$\frac{d}{r} = 4.5 \times 10^{-4}\left(\frac{m}{M}\right)^{1/2} V^{1/2} \tag{8.196}$$

for protons

$$\frac{d}{r} = 10^{-5} V^{1/2}. \tag{8.197}$$

For $V = 10^8$ V one would have $d/r = 0.1$, that is for $r = 10^{-2}$ cm, $d = 10^{-3}$ cm.

For $r > d$, the largest laser light wavelength should be $\lambda \sim r \sim 10^{-3}$ cm with a polarization of the electric field vector in the direction of the cut, and one would have to use laser light with a wavelength distribution ranging from the infrared with $\lambda \sim 10^{-3}$ cm down to the ultraviolet with $\lambda \sim 10^{-5}$ cm, covering the plasma density range from $n_c^{min} \sim 10^{19}$ cm^{-3} to $n_c^{max} \sim 10^{23}$ cm^{-3}. The intensity distribution is given by (6.97), but for $d < \lambda$ it is proportional to $1/\lambda$. The residual current carried across the gap is then by order of magnitude equal to

$$I \sim n_c^{min} r^2 ec \approx 10nr^2 \text{ [esu]} = 3 \times 10^{-9} nr^2 \text{ [A]} , \tag{8.198}$$

for the given example about $\sim 3 \times 10^4$ A, well below megampere pinch currents.

The large axial electric field E_z, set up in the cut, leads to a large inward radially directed Poynting vector $S = (c/4\pi)\, E_z H_\phi$, where H_ϕ is the magnetic field in the space surrounding the pinch.

The range of the fast ions crossing the cut for protons is given by (4.25):

$$\lambda_i = \frac{3}{8\sqrt{\pi}\log\Lambda} \frac{(kT)^{3/2}}{e^4 n} \left(\frac{M}{m}\right)^{1/2} \sqrt{E_{ion}} , \tag{8.199}$$

where E_{ion} is here the proton energy, $\log\Lambda \sim 10$ the Coulomb logarithm, and n the plasma particle number density. Expressing E_{ion} in eV one finds

$$\lambda_i = 3.5 \times 10^7 \frac{T^{3/2}}{n} \sqrt{E_{ion}} . \tag{8.200}$$

For the example $T = 10^8$ °K , $n = 5 \times 10^{22}$ cm^{-3} (corresponding to solid state density), one finds for $E_{ion} = 3 \times 10^6$ eV that $\lambda_i \sim 1$ cm, and for $E_{ion} = 10^8$ eV that $\lambda_i \sim 7$ cm. For deuterons the range is twice as large. As expected, the high voltage leads to a larger range, and hence to a larger length over which the stability of the pinch is increased. Over this same length there is a departure from a Maxwellian velocity distribution by a fast ion component. This, of course, increases the $\langle \sigma v \rangle$ value for a fusion reaction, possibly enough to reach burn for the neutronless HB^{11} reaction.

Under radiative collapse of the pinch discharge channel the highest densities are reached if the plasma temperature is as low as possible. But even then, the cut leads to a burst of energetic ions. If, for example, $T \sim 10^6$ °K and $n = 5 \times 10^{23}$ cm^{-3} (corresponding to 10 times solid density), one finds that now $\lambda_i \sim 10^{-3}$ cm, which means that a hot spot is created at the location of the cut, from which a thermonuclear detonation wave may be launched propagating along the pinch channel.

For the example $I = 3 \times 10^6$ A, one obtains from (8.194) that $V \sim 3 \times 10^6$ V, and from (8.195) that $Z \sim 1\,\Omega$. But for $V = 10^8$ V one would have $I = 1.67 \times 10^7$ A and $Z = 6.4\,\Omega$. By comparison, a 10-cm long plasma column with a cross section $\sim 10^{-4}$ cm^2 and a temperature of $\sim 10^8$ °K would have a resistance equal to $R = 6 \times 10^{-3}\,\Omega$, with the resistive losses about 10^3 times smaller. Therefore, a magnetically insulated cut increases the power dissipated into the pinch $\sim 10^3$-fold, leading to a large pulse power compression. With the pinch discharge as the load, the cut acts as a fast switch dissipating at the location of the switch the magnetically stored energy into the load.

The magnetic energy stored in the space surrounding the pinch of length ℓ and return current conductor radius R is

$$\varepsilon_M = 10^{-9} I^2\,\ell \log\left(\frac{R}{r}\right)\ [\mathrm{J}]\,. \tag{8.201}$$

For the example for $I = 3 \times 10^6$ A, $\ell = R = 10$ cm, $r \sim 10^{-2}$ cm, one has $\varepsilon_M \sim 30$ MJ.

The inductance of the discharge channel is

$$L = 2 \times 10^{-9}\,\ell \log\left(\frac{R}{r}\right)\ [\mathrm{H}] \tag{8.202}$$

for the given example $\ell = R = 10$ cm, $r \sim 10^{-2}$ cm equal to $L \sim 10^{-7}$ H.

The discharge time is

$$\tau = \frac{L}{Z}\ [\mathrm{s}]\,. \tag{8.203}$$

For $Z = 1\,\Omega$, where $I = 3 \times 10^6$ A, $V \sim 3 \times 10^6$ V, one has $\tau \sim 10^{-7}$ s, and for $Z = 6\,\Omega$, where $I \sim 1.7 \times 10^7$ A, $V \sim 10^8$ V, one has $\tau \sim 2 \times 10^{-8}$ s.

If the cut is made larger and violates the magnetic insulation criterion, the cut is bridged by an intense relativistic electron beam. In the limit of high relativistic electron energies, the repulsive space charge is compensated for by attractive magnetic forces. As a result, the beam current remains equal to the current carried by the pinch. The voltage across the cut is given by

$$V \sim \frac{LI}{\tau}\,. \tag{8.204}$$

For $L \sim 10^{-7}$ H, $I \sim 3 \times 10^{6}$ A and $\tau \sim 3 \times 10^{-8}$ s, one has $V \sim 10^{7}$ V.

After reentering the plasma on the other side of the cut, the electron beam can propagate inside the plasma only as long as $I \leq I_A^e \beta\gamma$. For an electron energy of 3×10^7 eV one has $\beta = 1$ and $\gamma = 20$, hence $I_A^e \beta\gamma = 5 \times 10^5$ A and thus $I = 3 \times 10^6$ A $> I_A^e \beta\gamma$. In this case the electrons are forced into Larmor motion around the self-magnetic beam field, preventing the beam from propagating. Losing their energy by magnetic bremsstrahlung (see chapter 4.12), they are brought to rest over a distance given by ($\gamma \gg 1$)

$$\lambda_e \sim \frac{(mc^2)^4}{e^4 H^2}\frac{1}{E}\,, \tag{8.205}$$

where E is the electron energy. With $E_0 = mc^2 \sim 5 \times 10^5$ eV and $e^2/r_0 = mc^2$, where r_0 is the classical electron radius, this can be written as

$$\lambda_e \sim \left(\frac{mc^2}{r_0^2}\right)\left(\frac{1}{H^2}\right)\left(\frac{E_0}{E}\right)\,. \tag{8.206}$$

For $\gamma = 20$, ($E/E_0 \approx 20$) and $H \sim 10^8$ G (valid for $I = 5 \times 10^6$ A, $r = 10^{-2}$ cm), one finds $\lambda_e \sim 10^{-5}$ cm. Because of this short range, the side of the cut where the electron beam reenters the plasma becomes an intense x-ray point source with a power equal to

$$P = IV \sim \frac{LI^2}{\tau}\,. \tag{8.207}$$

For $L \sim 10^{-7}$ H, $I \sim 10^7$ A, $\tau \sim 10^{-8}$ s one has $P \sim 10^{14}$ Watt. The maximum of this radiation occurs at the frequency

$$\omega_{max} \approx \left(\frac{eH}{mc}\right)\gamma^2 \tag{8.208}$$

with the photon energy

$$E_{max} = \hbar\omega_{max} = e\left(\frac{\hbar}{mc}\right)H\gamma^2 \approx 10^{-8}H\gamma^2\,[\mathrm{eV}]\,. \tag{8.209}$$

For the example $H \sim 10^8$ G, $\gamma \approx 20$ one finds $E_{max} \sim 400$ eV. This is a very interesting result, because it shows that the x-ray point source can be used to drive a thermonuclear hohlraum target, with the hohlraum placed near the cut, for example on the axis of the pinch. It should be emphasized that this configuration is possible with an exploding wire pinch discharge.

8.19 Ignition of a Thermonuclear Detonation Wave in the Focus of Two Concentric Magnetically Insulated Transmission Lines

As I have shown the combination of different concepts, like the two-stage magnetic booster impact fusion concept, or the laser ignition of a dense z-pinch, promise large gains without the compression of the fusion fuel to ultrahigh densities, of the order $\sim 10^3$ times solid density. In the two-stage magnetic booster concept this large compression is achieved only in the second stage by the burn of the much lower density first stage. The achievement of such high densities will in general always be difficult.

I now present a configuration in which two inertial confinement fusion drivers are used, operating in different regimes, a smaller one with a high voltage and lower current, and a larger one with a high current but lower voltage. Both transmit their energy to the thermonuclear target by magnetically insulated transmission lines.

As shown in Fig. 8.26, the transmission lines are nested, with the inner-higher voltage lower current-transmission line ending in a cone, and with the tip of the cone serving as the cathode for a field emitted intense relativistic electron beam. The return current conductor of the inner transmission line is, at its smallest diameter, connected to a slender cone of solid DT, which also serves as the anode of the inner transmission line. An intense relativistic electron beam is focused by self-magnetic beam forces onto the DT cone, heating the tip of the cone to thermonuclear temperature by electrostatic two-stream instability. At the same time a large electric current is discharged through the outer transmission line, with the current passing over the DT cone, and the outer side of the conical segment of the inner transmission line. This current must be large enough to generate a magnetic field with a magnetic pressure which can balance the pressure of the DT plasma at thermonuclear temperatures. In addition to heating the DT plasma, the relativistic electron beam emitted from the end of the inner transmission line must compensate the axial expansion losses of the hot plasma blown off in the opposite direction. If this condition is met, a shock wave moves to the right into the DT cone, and if the charged fusion reaction α-particles are confined within the cone by the magnetic field of the current flowing over the cone, the shock wave goes over into a thermonuclear detonation wave, supersonically moving down the DT cone. With the exception of the small region near the vertex of the cone, no magnetic plasma confinement is required, with the magnetic field serving only to entrap the charged fusion reaction α-particles in the cone. Therefore, very large fusion gains can here be reached.

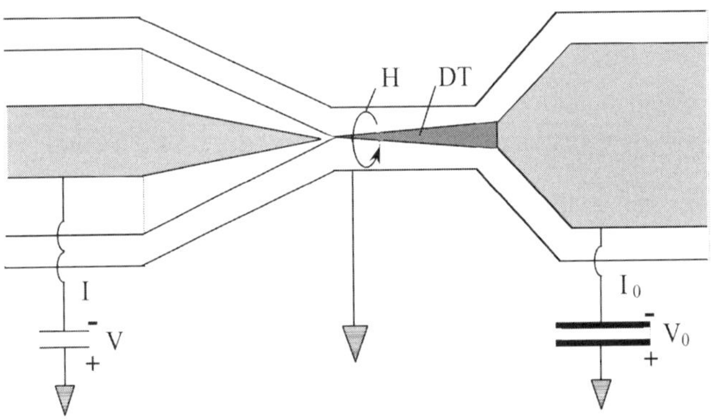

Figure 8.26: Ignition of a thermonuclear detonation wave in the focus of two nested magnetically insulated transmission lines.

The current density of field-emitted electrons was given by (8.68). For $V = 10^7$ V, $r = 0.1$ cm and $W = 4.4$ eV (valid for tungsten) we find $I = 3.5\times\ 10^5$ A. By the self-magnetic field and the repulsive image currents in the convergent conical return conductor, the beam can be focused down to an even smaller diameter, with the axial electric field from the cathode tip to the anode preventing the beam from being reflected back to the cathode. Assuming that the beam can be focused down to a radius $r_0 = 10^{-2}$ cm, the current density would be $\sim 10^9$ A/cm^2, with an electron number density in the beam[2] $n_b \simeq 0.7 \times 10^{17}$ cm^{-3}.

[2]Alternatively, one may let the beam be emitted from a smaller cathode tip. With $E = V/r$ this would not reduce the total current, which is proportional to $2\pi r^2 E^2$

The e-fold beam stopping length by the collective two-stream instability was given by (4.57a), and with $\varepsilon = n_b/n_0$, ($n_0 = 5 \times 10^{22}$ cm^{-3} valid for solid DT), $\omega_p = \sqrt{4\pi n_0 e^2/m} = 1.3 \times 10^{16}$ s^{-1}, and $\gamma \simeq 20$ (~ 10 MeV electrons), one finds $\lambda_c = 6 \times 10^{-3}$ cm. This means that just the tip of the DT is heated.

Under steady state conditions the power flux density of the incoming electron beam

$$\Phi_{in} = \frac{IV}{\pi r_0^2} \tag{8.210}$$

must balance the power flux density of the ablated DT

$$\Phi_{out} = 2n_0 \, \frac{M\text{v}^2}{2} \frac{\text{v}}{6} = \frac{1}{6} \rho \text{v}^3 \tag{8.211}$$

where v is the nondirectional ablation velocity, with the fraction (1/6) going in one direction, M the mass of the DT nuclei and $\rho = 0.21$ g/cm^3 the density of solid DT. For $V = 10^7$ V, $I = 10^5$ A, $r_0 = 10^{-2}$ cm, one finds $\Phi_{in} = 3 \times 10^{22}$ erg/cm^{2}s. Equating Φ_{out} with Φ_{in} one finds that $\text{v} \simeq 10^8$ cm/s.

From

$$\frac{M\text{v}^2}{2} = \frac{3kT}{2} \tag{8.212}$$

one finds that for $\text{v} = 10^8$ cm/s, $T \simeq 10^8$ °K, about the ignition temperature of the DT thermonuclear reaction.

Equating the plasma pressure with the magnetic pressure $H_0^2/8\pi$ at $r = r_0$, by the current I_0 passing through the outer transmission line, one finds from

$$2n_0 kT = \frac{H_0^2}{8\pi} \tag{8.213}$$

that $H_0 = 1.9 \times 10^8$ G and $I_0 = 5r_0 H_0 \simeq 10^7$ A.

According to (6.23) and table 6.1, a current in excess of 1.35×10^6 A is needed to entrap the charged fusion α-particles of the DT reaction. With a current $I_0 = 10^7$ A this condition is well satisfied.

Approximating the magnetically confined tip of the DT cone by a cylinder of radius and height $2r_L$ (the effective range of the magnetically trapped fusion α-particles) with r_L given by (6.21) the condition for ignition is

$$E_{ign} > 3nkT\pi r_0^2 \cdot 2r_L \simeq 2 \times 10^9 \text{ erg} = 200 \text{ J} . \tag{8.214}$$

To overcome the bremsstrahlungs losses, this energy has to be supplied by the relativistic electron beam in less than 10^{-8} s. The e-fold stopping length of the electron beam equal to 6×10^{-3} cm, is about twice as large as $2r_L \simeq 3 \times 10^{-3}$ cm, requiring that $E_{ign} > 400$ J. Existing beam technology enables us to generate a 10^7 V, 10^5 A, 10^{12} W beam, lasting 10^{-9} s and delivering ~ 1 kJ, according to the estimate of what is required for ignition.

If the DT cone has the height h, and the radius at its base is R, then there are $(\pi R^2 h/3)(n/2)$ pairs of DT nuclei in the cone, releasing an energy equal to

$$E_{out} = \frac{\pi R^2 h}{3} \frac{n}{2} \varepsilon_f \tag{8.215}$$

where $\varepsilon_f = 17.4$ MeV $\simeq 2.8 \times 10^{-5}$ erg. As an example with a cone of height $h = 1$ cm and base radius $R = 0.1$ cm, we find that $E_{out} = 7.3 \times 10^{15}$ erg. For a slightly larger cone one would have $E_{out} \simeq 1$ GJ. The input energy E_{in} is essentially determined by the energy of the high current pulse, with the ignition energy of the high voltage pulse small in comparison. For $I_0 = 10^7$ A at a voltage $V_0 = 10^6$ V, with a power of 10^{13} W lasting 10^{-8} s, the input energy is $E_{in} = 10^5$ J, and the gain $E_{out}/E_{in} = 10^4$.

The impedance of a coaxial transmission line was given by (8.50), and if its impedance is matched to a current pulse passing through the line, it is magnetically self-insulated. For the inner transmission line we have $I = 10^5$ A, $V = 10^7$ V requiring that $Z = 100\,\Omega$. Therefore, if $a = 0.5$ cm is the radius of the inner conductor, the radius of the outer conductor must be $b = 2.1$ cm, keeping the ratio b/a constant in the conical section. For the outer transmission line $I = 10^7$ A, $V = 10^6$ V requiring that $Z = 0.1\,\Omega$. Such a small impedance is possible with several lines in parallel.

8.20 Chemical Ignition

The chemical energy density of condensed high explosives is of the order $\sim 10^{11}$ erg/cm^3. With the ignition energy for thermonuclear reaction of the order 10^{14} erg (10^7 J), a few kilograms of high explosives should be sufficient to ignite a thermonuclear reaction. The problem is that this energy has to be focused in space and time onto an area less than 1 cm^2, and in a time less than 10^{-8} s. With detonation velocities of the order $\sim 10^6$ cm/s, the speed chemical energy can be released is fast, but not fast enough for thermonuclear ignition. Chemical ignition thus requires some kind of energy cumulation concept. It would be ideal if a high percentage of the chemical energy could be directly converted into a short pulse laser beam. If this does not work, one could use the explosively released chemical energy to generate a powerful light flash, optically pumping a laser. One may also think of an explosively driven magnetohydrodynamic generator, with the electric current drawn from the generator pumping a laser. To release the laser energy in the short time required, one can use as lasing material a substance with self-induced transparency, where the firing of the laser is delayed until a high population inversion has been reached.

A totally different approach for the chemical ignition of thermonuclear reactions, is by high explosive driven fluid dynamic energy focusing, to reach the high energy flux of $\sim 10^{15}$ W/cm^2 (10^7 J in 10^{-8} s onto ~ 1 cm^2) needed for ignition. This can be done by convergent shock waves or imploding shells, ultimately in combination with the magnetic booster target concept described in chapter 8.15.

Driver	**Direct/Indirect**	**Concept**	**Remarks**
1. Laser	Direct	Azides (and other) High Explosives	Self-induced Transparency
2. Laser	Indirect	Shock Wave Pumping	Argon Bomb
3. Laser	Indirect	High Explosive Generator Pumping	Promising
4. Impact	Direct	Convergent Shock Wave	Too Large
5. Impact	Direct	Imploding Shells	Rayleigh-Taylor Instability
6. Impact	Indirect	Magnetized Target	Small Gain
7. Impact	Indirect	Magnetic Booster Target	Large Gain, Promising

Table 8.2: Chemical ignition concepts.

In table 8.2 a number of conceivable thermonuclear ignition concepts using chemical high explosives are summarized. They are now described in some detail.

1. For the direct conversion of the chemical energy into a laser pulse, the high explosive must be simultaneously ignited throughout the entire volume it occupies. Otherwise detonation waves will develop, leaving behind a medium in thermodynamic equilibrium without a population inversion. This problem may be overcome by giving the high explosive a layered structure, with the thickness of the layers larger than the optical path length at low laser radiation intensities, but transparent at high laser radiation intensities (condition for self-induced transparency). The volume ignition of the high explosive could be done by an intense flash of light passing through the gap space separating the layers from each other.

Azides of dioxitanes are high explosives which might be suitable for this purpose. Both are the source of intense luminescent radiation. Ideally, the explosive would be the triplett state of helium, with the upper laser level in the far ultraviolet, but until now it has not been possible to stabilize triplett helium for atomic number densities of interest. Other interesting candidates are noble gas compounds. As explosives they too have higher energy densities compared with conventional explosives.

Figure 8.27 shows how the direct conversion of chemical energy into laser energy can be implemented.

The high explosive is arranged in a layered zig-zag pattern, ignited from both sides with triggered flash lamps. The expansion of the population inverted detonation products helps to sustain a high population inversion until the last moment where an auxiliary short pulse laser triggers a photon avalanche in the population inverted medium made up from the detonation products.

2. Glass lasers can be efficiently pumped with argon flash lamps, because at a temperature of a few eV argon is a brilliant light source. In the argon flash lamp the argon has a low density. A much more powerful light source is the argon bomb where a detonation wave from a high explosive goes into solid argon. Such an argon bomb can pump a laser which is powerful enough to ignite a thermonuclear microexplosion. Because the hot argon emits light in a broad spectrum, a dye laser with a large number of upper laser level states can be used.

One possible configuration for an argon bomb dye laser is shown in Fig. 8.28a, with the laser rod in the center of a cylindrical configuration surrounded by the solid argon and high explosive. One problem of this configuration is that the thickness of the solid argon must be less than the optical

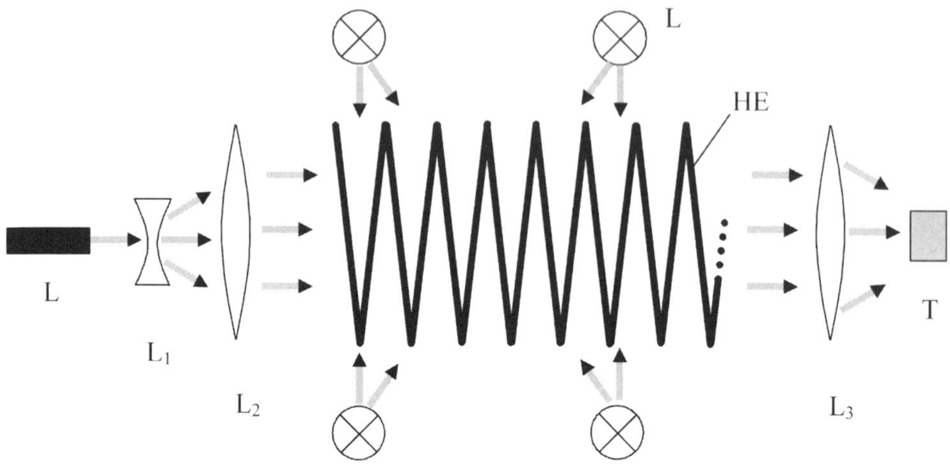

Figure 8.27: HE layers of high explosives, FL flash lamps, L trigger laser, L_1 dispersion, $L_{2,3}$ collection lenses, T thermonuclear target.

path length of the solid argon at the temperature of a few eV where the light emission is strongest. For this one needs a rather thin layer of solid argon, which is insufficient to release the many photons required to pump the laser rod to an energy of $\sim 10^7$ J. This problem though can be solved by the layered structure shown in Fig. 8.28b, where each argon layer is sufficiently thin to make it transparent. There are then many laser beamlets emitted from all the layers, which are focused onto the thermonuclear target by a Fresnel lens.

With a population inversion in a dye laser of the order 10^{-2}, an atomic number density of the order $n \sim 10^{22}$ cm^{-3}, an upper laser level energy of ~ 0.1 eV $\sim 10^{-13}$ erg, the energy density of the pumped laser rod would be $\sim 10^7$ erg/cm^3, requiring a volume of cubic meters to store $\sim 10^{14}$ erg $= 10^7$ J. If the fraction f of the explosively released energy is converted into photons emitted by the hot argon, and if the laser efficiency is η, the volume of the high explosive with an energy density of $\sim 10^{11}$ erg/cm^3 needed to pump the laser to $\sim 10^{14}$ erg, would be $V \sim (10^{14}/10^{11}) f$ cm^3, with an energy of $V \times 10^{11}$ erg/cm$^3 = 10^{16}$ erg, assuming that $f = 0.1$. This is equivalent to 100 kg of TNT.

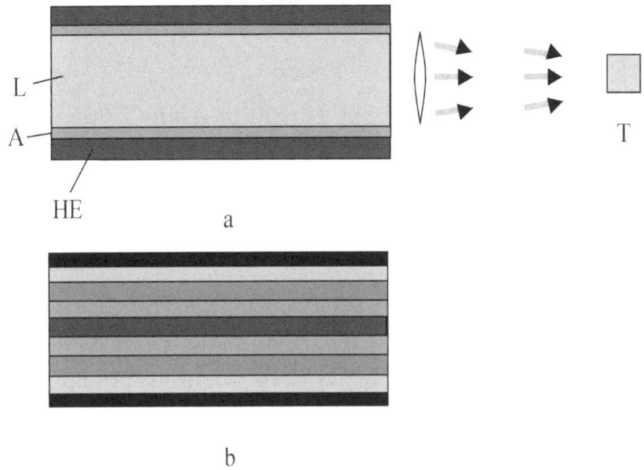

Figure 8.28: a, b Argon bomb pumped dye laser: L laser rod, A solid argon, HE high explosive, L focusing lens, T thermonuclear target.

3. High explosives have a detonation velocity of the order $\mathrm{v} \sim 10^6$ cm/s. Good electric conductors propelled by high velocities to these velocities can by magnetic flux compression reach magnetic fields of the order $H = \sqrt{4\pi\rho}\,\mathrm{v}$, where ρ is the density of the conductor. For $\rho \simeq 10$ g/cm^3 and $\mathrm{v} = 10^6$ cm/s one finds that $H \simeq 10^7$ G. In a one-turn magnetic solenoid of length ℓ the current is $I \sim H\ell$ A, for the example $H \simeq 10^7$ G, $\ell \sim 10$ cm, $I \sim 10^8$ A. The electric power which can be drawn from such a magnetic flux compression generator is determined by the Poynting vector $\mathbf{S} = (c/4\pi)\,\mathbf{E} \times \mathbf{H}$, where $\mathbf{E} = (\mathbf{v}/c)\,\mathbf{H}$, hence $|S| = (\mathrm{v}/4\pi)\,H^2$, or because of $H = \sqrt{4\pi\rho}\,\mathrm{v}$, $|S| = \rho \mathrm{v}^3$. For the example $\rho \simeq 10$ g/cm^3, $\mathrm{v} = 10^6$ cm/s, one has $|S| = 10^{19}$ erg/cm^{2}s $= 10^{12}$ W/cm^2. To compress the magnetic flux in a conductor of linear dimension ℓ requires the time $\tau \sim \ell/\mathrm{v}$, for the given example $\tau \sim 10^{-5}$ s. Over an area of 1 cm^2 the energy delivered in 10^{-5} s thus is $\sim 10^{12}$ [W/cm^2] $\times 10^{-5}$ s $= 10^7$ J. The electric field [EMF] of the magnetic flux compression generator is $E = (\mathrm{v}/c)\,H$ [esu] $= 10^{-8}\mathrm{v}H$ [V/cm] $= 10^5$ [V/cm], large enough to drive a diode laser. The problem though remains to compress the

laser pulse to $\sim 10^{-8}$ s. A conceivable solution might be to pump a second stage laser with self-induced transparency with the output drawn from the diode laser.

4. Using convergent shock waves for thermonuclear ignition is probably the oldest non-fission ignition idea. In a spherical convergent shock the temperature and pressure rise as $r^{-0.9}$ (see chapter 5.3 eq. (5.29)), and in the center of convergence liquid DT would be compressed $\sim$ 30-fold to a density of ~ 3 g/cm^3. Therefore, to satisfy the $\rho r \simeq 1$ g/cm^2 condition for thermonuclear ignition and burn, would require that the ignition temperature $T \sim 10^8$ °K is reached at $r \sim 0.3$ cm. Assuming $T_0 \sim 3 \times 10^3$ °K as the initial temperature of the high explosive driven convergent shock wave, would then have to launch the shock from a radius $r_0 \sim 300$ meters. Therefore, this idea is really not feasible.

5. A much faster rise in temperature, compared to the rise in convergent shock waves, is achieved through the implosion of metallic shells described in chapter 5.5. There, for megabar pressures, the implosion velocity rises as $\sim 1/R$, and hence temperature upon shell impact as $1/R^2$. To reach a velocity of ~ 200 km/s needed for impact fusion, an initial shell radius of meter size dimension propelled by high explosives to a velocity of a few km/s would be needed. Instead of an impact fusion target placed in the center of the imploding shells, one may place there a smaller shell which upon impact by the imploding large shell would generate a burst of soft x-rays to implode a high gain target of the kind used in laser fusion (see Fig. 8.29).

In reality the overall implosion of a shell is much more complex than described by the self-similar solution presented in chapter 5.5. This solution is valid for an ideal gas, and ignores energy losses by plastic deformation. Fig. 8.30 supplements the idealized analytical solution given in chapter 5.5, showing the energy flow diagram for the different processes involved.

An important problem is that in contrast to a convergent shock wave an imploding shell is subject to Rayleigh-Taylor instability. This problem though can be alleviated by a multishell configuration analyzed in chapter 5.6, standing somehow in between the convergent shock wave (limit of an infinite number of shells), and the single shell. It is for this reason that for the multiple shell configuration the Rayleigh-Taylor instability is less serious, but the price paid is in more energy losses. In the limit of an infinite number of shells, realized in the convergent shock wave, this loss is expressed by the

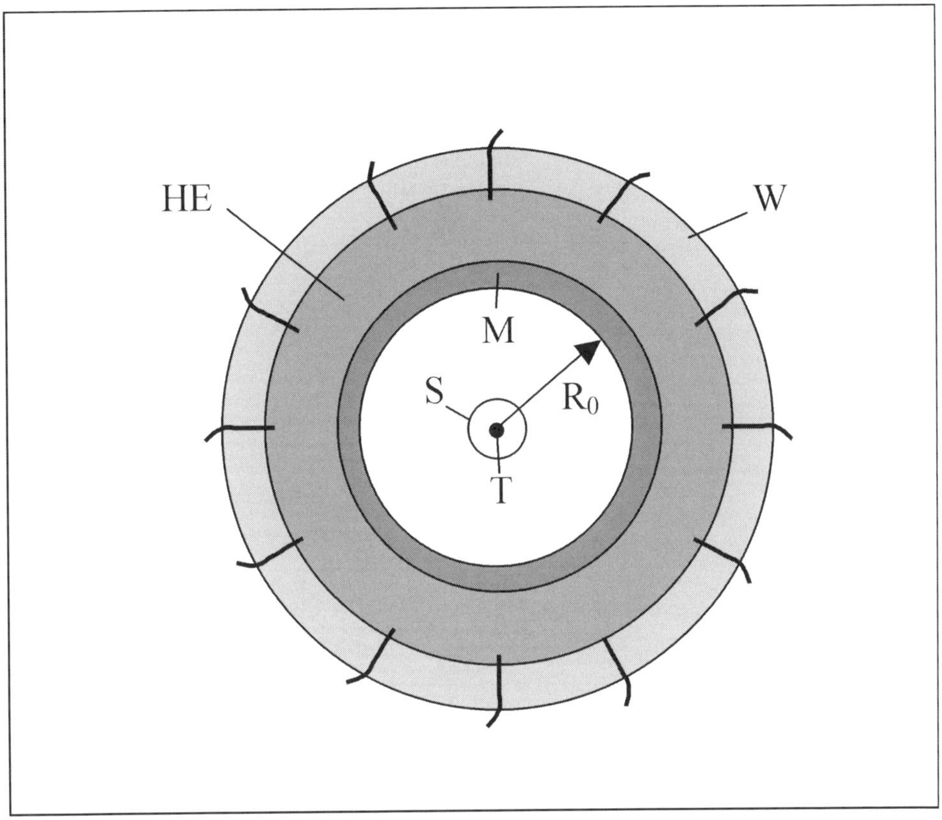

Figure 8.29: Shell driven by chemical implosion for the generation and compression of blackbody radiation. W outer tamp, HE high explosive, I ignition cables, M shell of initial inner radius R_0, S inner shell, T thermonuclear target.

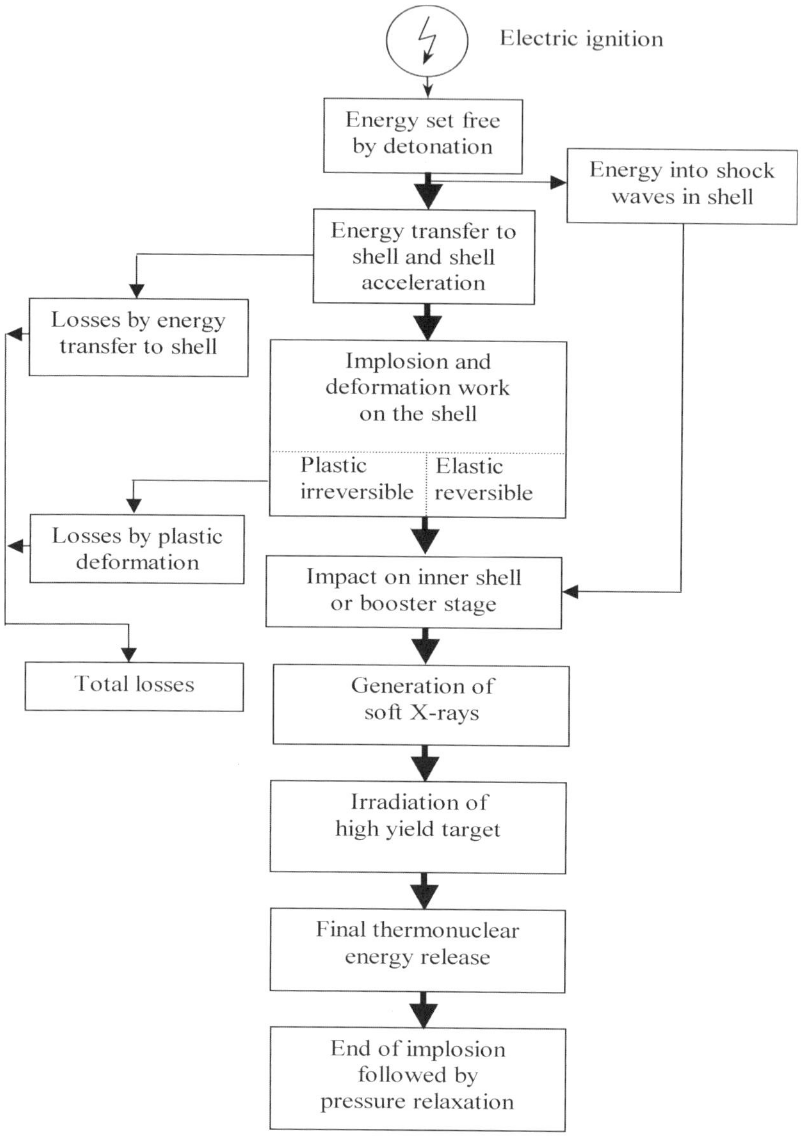

Figure 8.30: Energy flow diagram for imploding shell ignition.

irreversible nonisentropic change made by a shock wave. For a multiple shell configuration with inelastic energy losses we have from (5.83a) and (5.83b), putting $T \propto \mathrm{v}^2$, that there $T \propto r^{-1.4}$.

In summary we have with decreasing stability but increasing temperature rise:

1. Convergent shock wave $T \propto r^{-0.9}$.

2. Multiple shell implosion $T \propto r^{-1.4}$.

3. Single shell implosion $T \propto r^{-2}$.

6. In chapter 6.10 we have shown that for magnetized fusion targets a $\sim$ 10 times smaller implosion velocity than for impact fusion is sufficient for ignition. If driven by an imploding shell, this would, for a single shell, reduce the implosion velocity from $\sim$ 200 km/s down to $\sim$ 20 km/s, with a $\sim$100-fold reduction in the initial shell radius. For the more stable multiple shell configuration the radius would be reduced by the factor $10^{1.4} = 25$, from say $\sim$1 meter down to $\sim$5 cm. However, as we also have pointed out in chapter 6.10 the thermonuclear gain for magnetized fusion targets is not very large, and this is also true for the gain of a magnetized fusion target driven by a high explosive propelled shell.

7. Large thermonuclear gains are possible with the magnetic booster target concept described in chapter 8.15. A configuration combining this concept with the ignition by an imploding shell is shown in Fig. 8.31.

There the booster ignites a high gain target T by ablative implosion with soft X-rays. If this target launches a thermonuclear detonation wave into thermonuclear explosive placed inside the central current carrying rod, very large gains (and yields) are possible. If the current in the central rod exceeds the critical current I_c (6.23), a thermonuclear detonation wave can propagate there even without compression of the thermonuclear explosive.

8.21 The Goal Towards Low-Yield High-Gain Thermonuclear Explosive Devices

The importance of low-yield high-gain thermonuclear microexplosion is for the commercial release of fusion energy, but also for nuclear rocket propulsion.

(a)

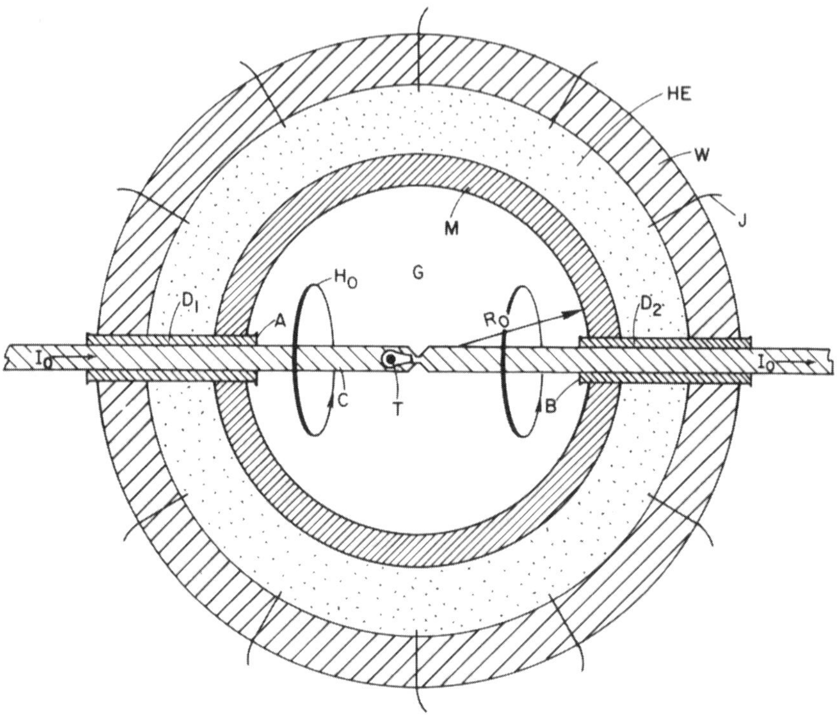

(b)

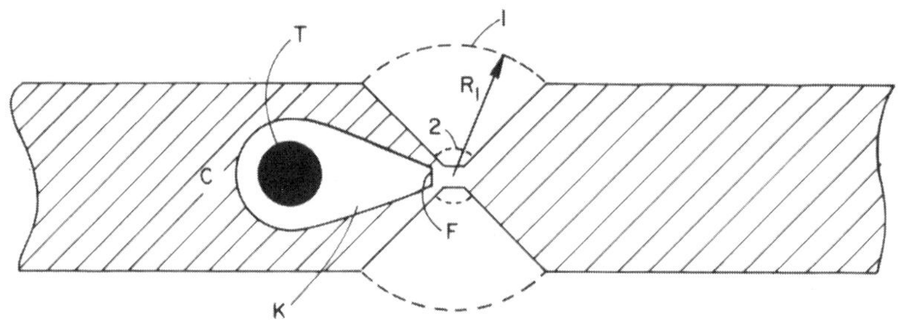

Figure 8.31: The imploding shell driven by chemical explosive in combination with the magnetic booster stage concept. HE high explosive; M shell of initial radius R_0; J ignition cables; W tamp; G gas composed of DT; C conduction rod carrying current I_0; D_1 , D_2 insulators; K chamber containing thermonuclear target T; H_0 initial magnetic field; F window.

For power production the yield should not exceed the energy set free in one ton of TNT with a gain of $\sim 10^3$. For nuclear rocket propulsion the yield can be larger, but not more than $\sim$ 100 tons of TNT equivalent, with a large gain implying a small fusion trigger.

For fission ignited thermonuclear explosive devices, the trigger energy cannot be smaller than the energy released in a fission explosion, which is much larger than what is actually needed for ignition. It is for this reason that fission triggered thermonuclear explosive devices below a kiloton (the energy released in a small fission bomb) become very extravagant. Thermonuclear explosive devices with a yield below a kiloton would not only be of military interest but also for nuclear rocket propulsion. This is where fissionless ignition would fill a void. Unfortunately this also carries the specter of fissionless ignited large thermonuclear explosive devices, opening a fissionless shortcut to the most destructive thermonuclear weapons.

Apart from pure fissionless trigger devices there are the hybrid fission-fusion ignition concepts. To avoid the "tyranny of the critical mass" (F. Dyson), one can reduce the critical mass by compressing a fissile pellet using the same kind of techniques used for the compression of DT targets. As was shown in chapter 2.3, the critical radius is inversely proportional to the number density of fissile nuclei. At a pressure of $\sim 10^{16}$ dyn/cm^2, DT can be compressed $\sim 10^3$-fold, but uranium (or plutonium) only $\sim$10 fold. A $\sim$10 fold compression would imply a $\sim$100 fold reduction in the critical mass, from $\sim$10 kg down to $\sim$100 g. This is still quite large, but a further reduction is possible with the fission-fusion chain reaction described in chapter 2.6. There the critical radius can be reduced $\sim$30 fold, reducing the critical mass to $\sim$10 g. To obtain these small critical masses one though has to pay a price in energy needed to compress the fission-fusion material to high densities. Nevertheless, in the form of the autocatalytic fission-fusion implosion described in chapter 2.6, this energy may be comparatively small. It appears that the greatest hopes for a low-yield high-gain thermonuclear microexplosion device are the shear flow stabilized laser ignited dense z-pinch, and the novel concept described in chapter 8.19.

8.22 Bibliography for Chapter 8

W. H. Bostick, V. Nardi and O. S. F. Zucker, Energy Storage, Compression and Switching, Plenum Press, New York 1976.
Proceedings IEE International Pulsed Power Conference, Nov. 9-11, Lubbock, Texas, Institute of Electrical and Electronics Engineers, New York 1976.
4th IEE Pulsed Power Conference, Albuquerque, New Mexico 1983; Library of Congress Catalog Number 83-80951; IEE Catalog Number 83CH1908-3.

F. Winterberg, Physical Review **174**, 212 (1968).

R. E. Kidder, in "Physics of High Energy Density", Academic Press, New York 1971, p. 306 ff.

F. Winterberg, in "Physics of High Energy Density", Academic Press 1971, p. 370 ff.

J. M. Dawson, The Physics of Fluids **7**, 981 (1964).

N. G. Basov and O. N. Krokhin, 3rd International Conference on Quantum Electronics Paris, 1963, Vol. 2, Dunod, Paris 1964, p. 1373 ff.

F. Winterberg, Physics of Plasmas **2**, 733 (1995).

R. Martin, IEEEE Trans. Nucl. Sci. **22**, 1763 (1975).

W. Maschke, IEEE Trans. Nucl. Sci. **22**, 1825 (1975).

F. Winterberg, J. Plasma Physics **21**, 301 (1979).

F. Winterberg, Physics of Plasmas **7**, 2654 (2000).

F. Winterberg, J. Plasma Physics **24**, 1 (1980).

F. Winterberg, Z. f. Naturforschung **19a**, 231 (1964).

E. R. Harrison, Plasma Physics **9**, 183 (1976).

C. Maisonnier, Nuovo Cimento **42B**, 232 (1966).

F. Winterberg, J. of Nuclear Energy Part C, Plasma Physics, Accelerators, Thermonuclear Research, **8**, 541 (1966).

Proceeding Impact Fusion Workshop Los Alamos New Mexico USA 1979 (LA-8000-C Los Alamos Scientific Laboratory).

F. Winterberg, Nuclear Fusion **30**, 447 (1990).

F. Winterberg, Physics of Fluids **B4**, 3350 (1992).

F. Winterberg, Z. f. Naturforschung **41a**, 495 (1986).

F. Winterberg, Atomkernenergie/Kerntechnik **41**, 291 (1982).

F. Winterberg, Beiträge Plasmaphysik **25**, 117 (1985).

F. Winterberg, Plasma Physics **10**, 55 (1968).

C. Stallings, K. Nielsen and R. Schneider, Applied Physics Letters **29**, 404 (1976).

T. W. L. Sanford et al. Physics of Plasmas **4**, 2188 (1997).

F. Winterberg, Z. f. Naturforschung **53a**, 933 (1998).

F. Winterberg, Z. f. Naturforschung **54a**, 443 (1999).

F. Winterberg, Z. f. Naturforschung **54a**, 459 (1999).

E. Sänger, Z. f. Naturforschung **6a**, 302 (1951).

F. Winterberg, Atomkernenergie **40**, 56 (1982).

F. Winterberg, Acta Astronautica **10**, 443 (1983).

K.-J. Boller, A. Imamoglu and S. E. Harris, Physical Review Letters **66**, 2593 (1991).

F. Winterberg, Z. f. Naturforschung **58a**, 197 (2003).

Chapter 9

Thermonuclear Lenses and Shaped Charges

9.1 Thermonuclear Lenses

For conventional high explosives detonation wave-shaping techniques have been developed into an art. In the so-called explosive-lens technique, it is possible to obtain plane, (convergent) cylindrical and spherical detonation waves. In one of these wave-shaping techniques, explosives with two different detonation velocities are used. This technique is not applicable under circumstances where thermonuclear explosives are involved because of the greatly differing temperatures for essentially all thermonuclear explosives, especially true if DT and D are used as the two explosives.

Another technique for wave shaping is to place holes or inert bodies in the path of the detonation front around which the wave has to pass. This technique also works with just one explosive. It is therefore ideally suited for wave shaping of thermonuclear detonation fronts. Wave shaping of thermonuclear detonations may have great importance not only for certain technical applications, but also for basic science.

The wave-shaping technique is most easily explained by the example of a thermonuclear plane-wave lens, shown in Figure 9.1. A thermonuclear detonation front starting from the ignition point IP propagates into the thermonuclear explosive TF, conical in shape and surrounded by the tamp T. In the path of the detonation front, hollow bubbles or solid inert bodies B are placed as shown. The density $\rho(B)$ of these obstructions as a function

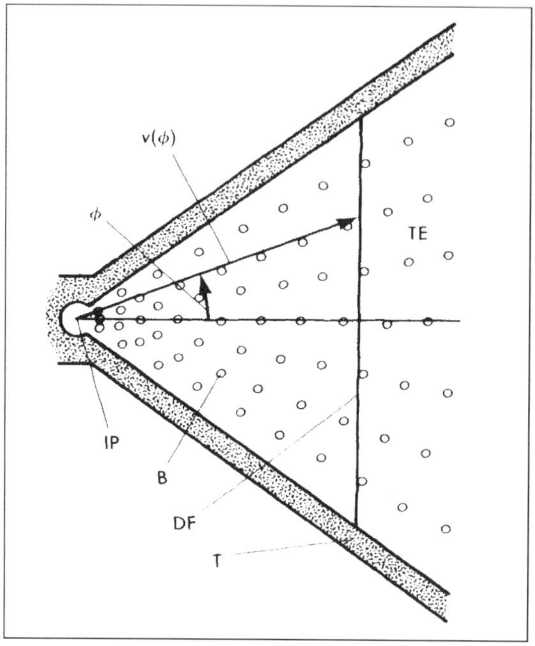

Figure 9.1: Thermonuclear plane wave lens where the detonation wave is shaped by bubbles of inert obstacles B placed in the path of the detonation front DF moving from the ignition point IP of the thermonuclear explosive TE. T is a tamp.

of the radial distance r from the axis of the conical assembly is chosen in such a way to give the detonation front velocity v as a function of the polar angle ϕ the following dependence:

$$\mathrm{v}\,(\phi) = const./\cos\phi\ . \tag{9.1}$$

This dependence leads to a plane wave emerging from the ignition point IP. Since it is reasonable to assume that the wave speed is inversely proportional to the density of the obstacles, we have to put

$$\rho\,(B) = const.\ \cos\phi\ . \tag{9.2}$$

Other distributions $\rho\,(B) = f\,(r,\phi)$ obviously permit the attainment of other wave shapes. In Figure 9.2, for example, a wave-shaping lens is shown that is designed to make a conical implosion.

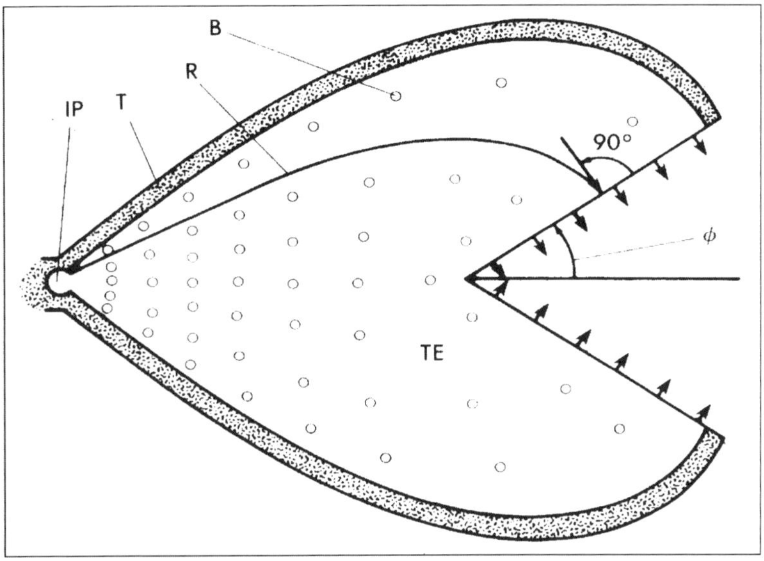

Figure 9.2: A wave shaping lens for conical implosion producing a convergent conical wave. IP is the ignition point of the thermonuclear explosive TE, B are the bubbles placed in the wave path; T is a tamp, and R is a ray of the detonation wave.

In Figure 9.3, finally, it is shown how a spherical implosion can be obtained by wave shaping. To obtain a cylindrical implosion a more complicated three-dimensional lens configuration is needed if the detonation wave starts from a point.

The spatial dependence of the detonation velocity, and, by implication, the density of the obstacles standing in the way of the detonation wave, can be obtained from the requirement of a constant travel time for the detonation wave rays launched from the point of ignition. This time is given by

$$t = \int \frac{ds}{\mathrm{v}} \tag{9.3}$$

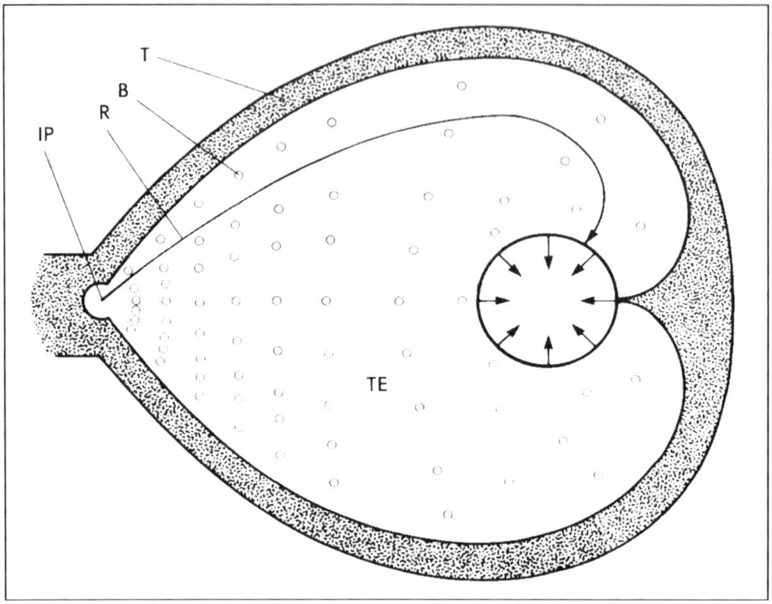

Figure 9.3: A spherical implosion obtained by wave shaping. The detonation lens here produces convergent spherical waves. IP is the ignition point; TE is the thermonuclear explosive; T is the tamp; R is a ray of the detonation wave; and B are bubbles.

In (9.3) $\mathrm{v} = \mathrm{v}(x, y, z)$ is the variable detonation velocity. For rotational symmetry realized in all three configurations described in Figures 9.1-9.3, one can introduce a cylindrical r-z coordinate system whereby (9.3) becomes

$$t = \int \frac{\sqrt{1 + (dr/dz)^2}}{\mathrm{v}(r, z)} \, dz \, . \tag{9.4}$$

The constant time condition for all rays requires that $\delta t = 0$. It leads to the Euler-Lagrange equation

$$\frac{d}{dz} \frac{dr/dz}{\mathrm{v}(r, z)\sqrt{1 + (dr/dz)^2}} + \frac{\sqrt{1 + (dr/dz)^2}}{\mathrm{v}^2(r, z)} \frac{\partial \mathrm{v}(r, z)}{\partial r} = 0 \, . \tag{9.5}$$

The rays are given by an equation of the form $r = r(z)$, whereby (9.5) is a first order partial differential equation for the unknown function $\mathrm{v}(r, z)$, which can be solved by the methods of characteristics.

9.2 Thermonuclear Shaped Charges

Next we discuss the concept of shaped charges in the context of thermonuclear explosives. A good shaped charge requires two things: (1), a well-shaped wave and (2), a liner propelled by the wave. A simple shaped charge is shown in Figure 9.4. A thermonuclear detonation starting from the igni-

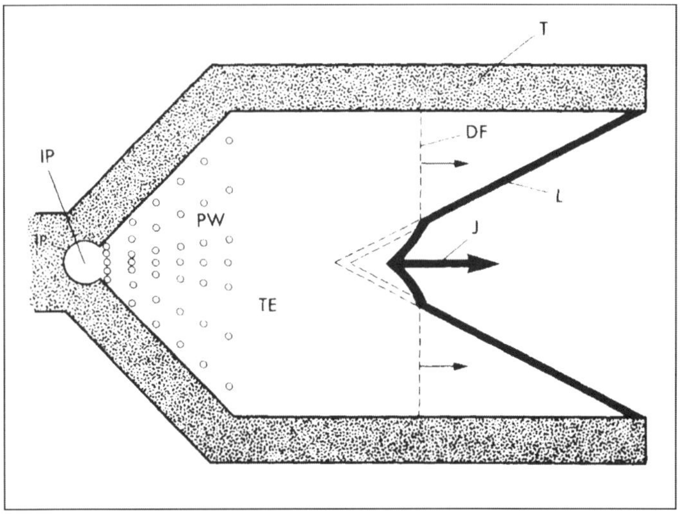

Figure 9.4: Simple thermonuclear shaped charge: IP ignition point; PW plane wave lens; TE thermonuclear explosive; DF detonation front; L metallic liner; J jet of the collapsing liner; and T tamp.

tion point IP enters the conical plane wave lens PW. After leaving the lens the plane wave thusly produced propagates into the thermonuclear explosive TE, collapsing the conical metal liner L toward its axis, and resulting in a fast forward metal jet J. Figure 9.4 shows the partially collapsed liner in the moment the detonation wave has reached the position indicated by the line DF. This simple shaped charge leads to maximum jet velocities only twice as large as the detonation velocity. Since the density of the metal liner can be roughly 10 times larger than the density of the thermonuclear explosive, the energy density of the jet is roughly 40 times larger than in the exploding fusion material.

A shaped charge leading to much higher velocities uses a conical implosion onto a conical liner. How a conical implosion can be made by wave shaping was shown in Figure 9.2. In this case the resulting jet velocity is inversely proportional to $\sin\phi$, where ϕ is the angle shown in Figure 9.2. From this it seems to follow that by decreasing the cone angle ϕ, the jet velocity could become arbitrarily large. However, a lower practical cutoff for ϕ has been found to be $arc\phi \simeq 1$. Therefore, the maximum attainable jet velocity is about 10 times larger than the detonation velocity. For deuterium the detonation velocity is about 1/100 the velocity of light, and it therefore follows that a jet with roughly 10% the velocity of light could be achieved.

9.3 Some Applications of Thermonuclear Lenses and Shaped Charges

Several potential applications of the thermonuclear explosive lens and shaped charge techniques shall now be presented:

1. In the "Orion" nuclear pulse rocket propulsion concept a chain of small fission explosions ($\sim$ 1 kiloton TNT) is set off behind a pusher plate as shown in Figure 9.5a. The Orion nuclear propulsion concept has both a large thrust and specific impulse and is for this reason well suited for large manned missions within the solar system requiring short transit times.

Much larger payloads can, of course, be propelled with larger nuclear explosions, replacing fission with fusion explosions. In comparison to fission explosions, fusion explosions, with their thermonuclear lens and shaped charge techniques, are much more flexible. Lacking a nozzle as in chemical propulsion, most of the energy in the fission bomb propulsion is dissipated into space. Fusion bomb propulsion using a thermonuclear shaped charge as shown in Figure 9.5, can, in part compensate for this inefficiency.

2. The same thermonuclear shaped charge technique which can be used for rocket propulsion, can also be used for the deflection of comets or asteroids to prevent their impact on the earth. Even though these impact events are rare, they do happen, and a comet of the size of Halley's comet would upon

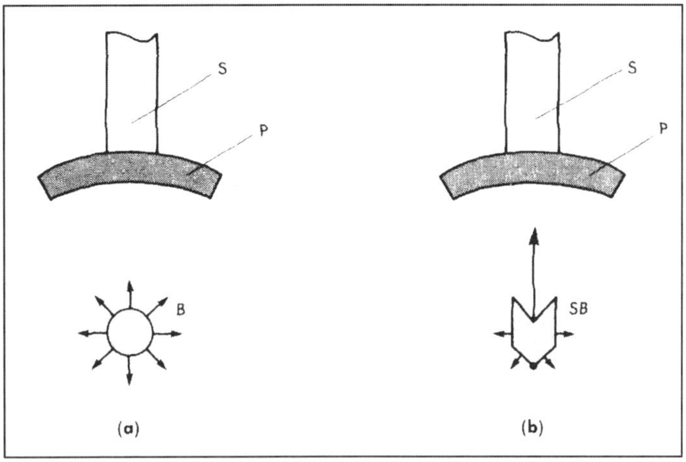

Figure 9.5: Simple Orion type pusher plate configuration for spherically exploding fission bomb (a). The same pusher plate configuration in (b) with asymmetrically exploding thermonuclear shaped charge bomb SB. S is a shaft connecting the pusher plate to the payload.

impact cause a disaster of unimaginable proportions. Before the advent of nuclear explosive and space rocket techniques there was nothing which could have been done to prevent such a disaster, but now the technology exists, awaiting development.

Simply detonating a nuclear explosive attached to the surface of a comet or asteroid would not be advisable, since this would lead to the disintegration of the comet or asteroid into a "buckshot" of many small bullets with an even more destructive shotgun-like effect upon impact. The most sensible technique rather is to ablate the surface of the comet or asteroid with the jet from a thermonuclear shape charge in the same manner as would be done for thermonuclear pulse rocket propulsion shown in Figure 9.5.

Using the staging and autocatalytic detonation wave techniques explained in Chapters 7.9 and 7.10, very large thermonuclear explosive devices can be made which may be needed to deflect large comets and asteroids.

3. Thermonuclear shaped charges can also become an important tool for deep mining by blasting a shaft to great depths. Calculations show that

with this technique one may even blast a tunnel through the center of the moon (see Figure 9.6).

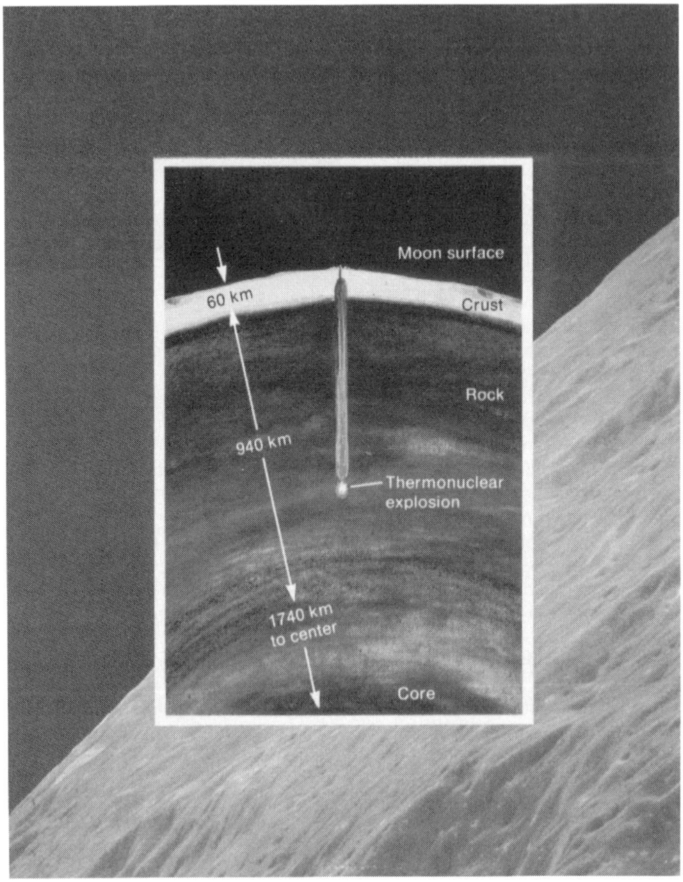

Figure 9.6: Making a tunnel through the moon.

A modern civilization depends on the availability of heavy elements like nickel and chromium to make steel. Because on earth all the mineral deposits will eventually be exhausted, the mining of planetary bodies can one day become very important. Heavy elements should be concentrated in the center of planetary bodies, making the deep mining of these bodies of great interest. The planet Mercury, with its high density appears to be especially promising candidate for this kind of nuclear mining. The "pencil-type" thermonuclear bomb shown in Figure 7.9, in combination with a plane wave lens would be well suited to blast a narrow bore hole into rock.

4. For particle densities $n \gtrsim 10^{27}$ cm^{-3} nuclear fusion chain reactions occur with ease even in exotic fusion fuels like HB^{11}. These densities are reached in spherical convergent shock waves with the thermonuclear lens shown in Figure 9.3. In a plane thermonuclear detonation wave the temperature is of the order $\sim 10^8$ °K. With a detonation velocity of $\sim 10^9$ cm/s, the pressure is $p \sim \rho \mathrm{v}^2 \sim 10^{18}$ dyn/cm^2, rising in a spherical convergent wave as $\sim r^{-0.9}$. If the radius of the wave decreases $\sim$ 10-fold, the pressure rises to $\sim 10^{19}$ dyn/cm^2. At this pressure the cold particle number density rises to $\sim 10^{27}$ cm^{-3}, sufficiently high to ignite a fusion chain reaction. To prevent substantial heating of the material to be compressed, it would have to be put inside a capsule isentropically imploded by the convergent detonation wave.

5. Let us assume that at a radius $r = 10$ cm, one has $T \sim 10^9$ °K. In a convergent shock wave it would have risen to $T \sim 10^{10}$ °K $\sim$ 1 MeV at $r \sim 1$ cm. This example shows that the convergent detonation wave technique may become an important tool in the study of matter under extreme conditions like those found in exotic stars.

6. With a pressure of $p \sim 10^{19}$ dyn/cm^2, magnetic fields can be compressed to $H \simeq \sqrt{8\pi p} \sim 10^{10}$ G, as may occur in magnetic stars.

7. Finally, there are military applications, in particular for antiballistic missile defense. With a thermonuclear shaped charge one can make a shotgun aimed against a swarm of objects released from a ballistic missile, both warheads and decoys.

9.4 Bibliography for Chapter 9

Schall, in "Physics of High Energy Density" edited by P. Caldirola and H. Knoepfel, Academic Press New York 1971, p. 230 ff.

F. Winterberg, in "The Physical Principles of Thermonuclear Explosive Devices", chapter 16, p. 117 ff. Fusion Energy Foundation, New York 1981.

F. Winterberg, Kerntechnik, **63**, 202 (1998).

F. Winterberg, Fusion, August 1981, May-June 1986.

Chapter 10

The Significance of Thermonuclear Microexplosions for Fundamental Research

10.1 Synopsis

There are two great frontiers of science, the quest to reach ever greater distances in space, and the study of ever smaller subnuclear regions. Both quests, towards large and small distances, the first done with space rockets and the second with high energy particle accelerators, require ever larger energies. At the frontier of space research we have the capability to explore the physical properties of all the bodies in our solar system. By placing large telescopes outside the disturbing influence of the earth's atmosphere we can probe deeper into space than ever before, even getting a glimpse of that momentous initial event which we call the big bang. At the frontier of high energy physics, mile-long particle accelerators have permitted us to explore the microscopic world down to $\sim 10^{-15}$ cm, with 10^{-18} cm soon be reached. Because of the stupendous advances made, both in space exploration and high energy physics, it is reasonable to contemplate where the limits might be. Less than 100 years ago it was still possible to make a fundamental discovery in a private laboratory. Today this is still possible in biology but much less likely in physics and impossible in space research. The very large expenditures, both for space research and high energy physics, have completely outgrown the budgets of ordinary university physics departments. Nowadays both space research and high energy physics are big business, depending crucially on the good will of the government and hence society for its support. The required funds can stress the financial capabilities of an entire nation. In the Apollo moon landing project the expenditures were running into the billions. The same is true for particle accelerators, needed to explore the region above which the electromagnetic and weak interaction are unified at center of mass energies in excess of 100 GeV. It has been speculated that something exciting and new is always going to happen if the energy is increased by a factor ~ 137, the inverse value of the fine structure constant. If this should be true, then the next breakthrough would occur at center of mass energies in excess of 10^4 GeV. To reach these energies with presently available accelerator techniques would be so expensive that no government is likely to be willing or able to provide the needed funds.

Turning to the other frontier, space exploration, there too are ever growing funding problems, especially if only presently available propulsion technologies are used. In particular this is true for a manned exploration of Mars.

Apart from the scientific interest to explore other celestial bodies, space research also has a direct impact on high energy physics. If the universe was created out of a space-time singularity about 2×10^{10} years ago, energies all the way up to the Planck energy of $\sim 10^{19}$ GeV must have occurred in the distant past. Assuming the big bang hypothesis is correct, the universe can serve as an ultrahigh energy accelerator, provided we can read the very weak impressions left over after 2×10^{10} years. This can be done with space telescopes far away from the disturbing influence of this earth. High energy physics making use of particle accelerators can also tell us something about the beginning of the universe. We therefore see that a bridge can be drawn between the two great frontiers of science. The exploration of the world on a large scale can give us answers about the world in the small and vice versa. The energies which can be reached with present accelerators are many orders of magnitude below the Planck energy. The gap in between what can be experimentally explored with high energy particle accelerators, and observationally deduced from cosmology may then be closed by clever theoreticians.

Present experiments suggest a unification of all forces at $\sim 10^{16}$ GeV, $\sim 10^3$ times smaller than the Planck energy, far too high to be attainable with any accelerator available now or in the foreseeable future. Less ambitiously, it would be important to see if our present ideas can be confirmed up to center-of-mass energies of $\gtrsim 10^4$ GeV.

Let us first have a closer look at the frontier of space exploration. To make a manned expedition to the moon was marginally feasible with chemical rocket propulsion. Going to Mars, a manned expedition using chemical propulsion would take very long and be very expensive. What is problematic for a manned expedition to Mars using chemical propulsion is, of course, even more problematic for manned expeditions to Jupiter and beyond.

If however, chemical fuels are replaced by nuclear fuels the situation is completely different. Compared with chemical fuels the energy density of nuclear fuels is many million times larger. From the first moment uranium fission was discovered scientists already wondered if it could be used for rocket propulsion. *Wernher von Braun* discussed this question with *Heisenberg* around 1942. *Heisenberg* told *von Braun* that submarines could possibly be propelled with a uranium pile, but that there was no simple way in sight to drive a rocket engine. This negative outlook, still valid for nuclear fission, does not apply to thermonuclear microexplosions by inertial confinement. Non-fission triggered mini-H bombs would change all that.

With fusion microexplosion propulsion the manned exploration and the industrialization of our entire solar system becomes possible, even making visits to nearby solar systems a distant but real possibility. This implies that in the course of less than 10^7 years our entire galaxy could be colonized in this way. A time of 10^7 years is small compared to the age of the galaxy. Incidentally, this fact has cast doubt on the popular belief that there are millions of other human-like civilizations. A large number of suns in our galaxy are billions of years older than our sun, and a civilization which would have arisen in one of these solar systems could have colonized the entire galaxy a long time ago. With microexplosion rocket propulsion we may also launch large research observatories into space, using the solar system as a laboratory to test Einstein's theory of gravitation. The rapidly spinning Jupiter, for example, could be used to test the Lense-Thirring effect predicted by general relativity, the gravitational analog of the magnetic field generated by a spinning electric charge in electrodynamics.

Turning to the exploration of the world in the small, thermonuclear microexplosions again promise a breakthrough if we use them to drive particle accelerators to reach energies in excess of 10^4 GeV.

In going to nearby solar systems our knowledge in the large is extended $\sim 10^3$-fold. And likewise, according to Heisenberg's uncertainty principle an energy of $\sim 10^5$ GeV likewise corresponds to a 10^3-fold extension of our knowledge into the small.

In addition to particle acceleration and rocket propulsion there are other important research applications for thermonuclear microexplosions.

10.2 Thermonuclear Microexplosion Reactors

The energy released in a thermonuclear microexplosion is likely to be 10^9 Joules, equivalent to 1 ton of chemical energy, but can be much higher. Much smaller energies are only economical if the thermonuclear fuel is compressed more than what is presently believed possible. For thermonuclear microexplosion reactors explosive energies very much larger than $\sim 10^9$ Joules pose confinement problems for the exploding fireball which has to be placed inside a cavity serving as the reactor combustion chamber.

The typical expansion velocity of the fireball is about 10^8 cm/s. For a reactor cavity with a radius of ~ 300 cm, the fireball could convert a large part of its energy into an electric pulse by magnetohydrodynamic power conversion within 3×10^{-6} s. Therefore, in each microexplosion in which an energy of say 3×10^9 Joules is released, the power would be 10^{15} Watts.

An early 1969 version by the author of such a microexplosion reactor is shown in Figure 10.1, where a magnetic field permeates the reactor cavity. The rapidly expanding highly conducting fireball of each microexplosion pushes the magnetic field aside creating a field-free space. The magnetic energy displaced from this space is $E = (H^2/8\pi)V$, where $E = H^2/8\pi$ is the magnetic energy density and V the volume of this space. This energy is converted into electromagnetic energy in the time it took the fireball to push the field aside.

A very different reactor concept is suggested by the scheme explained in Chapter 8.21. Because this configuration promises a high gain at a modest yield, one can replace the DT cone in Figure 8.26 with a long tamped DT cylinder positioned in the center of a liquid lithium vortex as shown in Figure 10.2. With 80% of the DT nuclear reaction energy released in fast neutrons, the neutrons dissipate their energy in the liquid lithium, breeding there tritium at the same time. Making the tamp from U238 or TH232, substantial additional concurrent fast fission burn is also here possible.

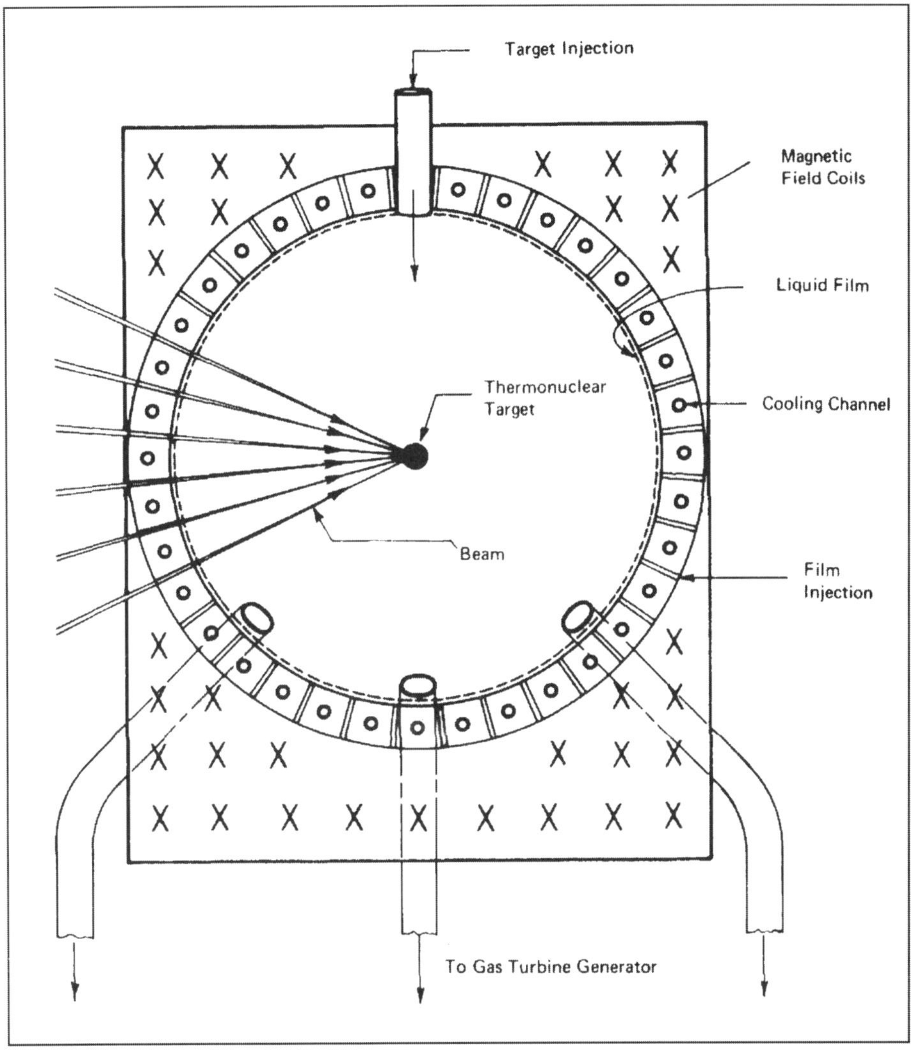

Figure 10.1: Schematic drawing of a thermonuclear reactor based on the confinement of a chain of microexplosions inside a spherical chamber, ignited by laser or particle beams.

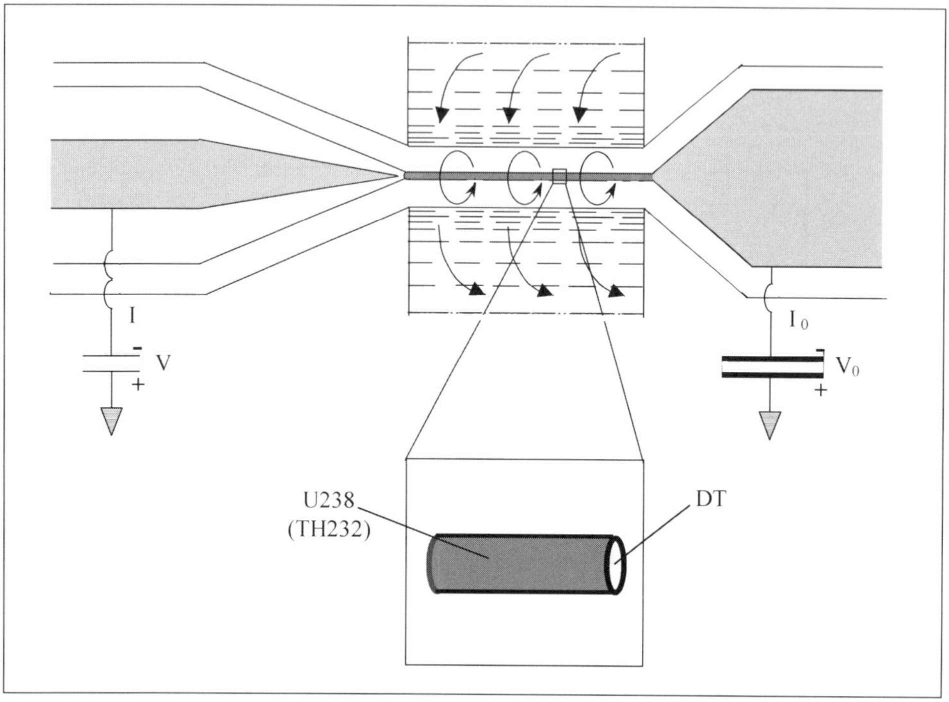

Figure 10.2: Reactor configuration with U238 (TH232) tamped DT cylinder in the core of a liquid lithium vortex.

This configuration can even be used to burn D, if a small amount of DT is placed at the end of the cylinder where the electron beam dissipates its energy. There then the DT burn can ignite the D in the remaining portion of the cylinder.

Because the current flowing over the tamped DT cylinder mus be very large, exceeding $\sim 10^7$ A, a convoluted feed of the outer transmission line as explained in Figure 10.3 is needed. There the outer transmission line is a circular strip line, with the liquid lithium vortex placed inside the strip line.

As shown in Figure 10.4, the electron beam can be precisely focused on the anode by laser initiated breakdown in a tenuous background gas inside the diode, where the electron beam is established through run-away electron in large electric fields, as described in Chapter 4.15.

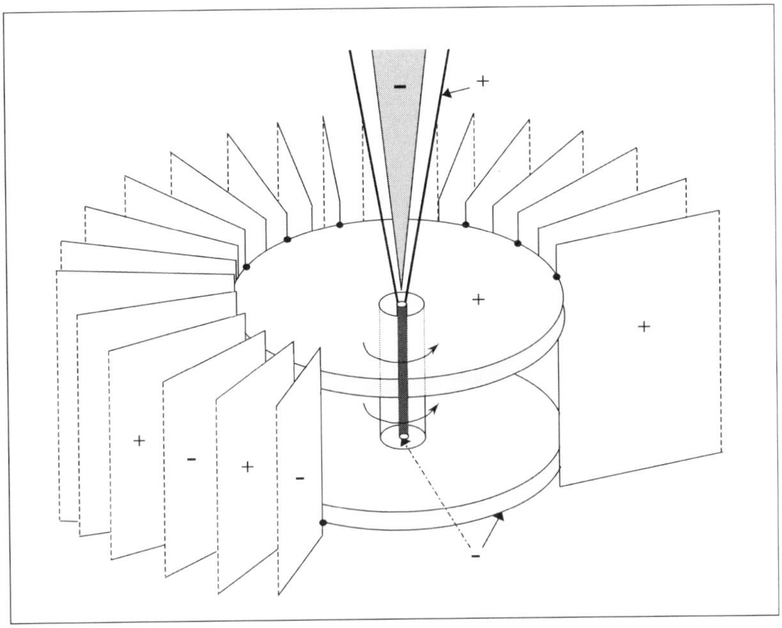

Figure 10.3: Convoluted high current feed for the outer transmission line positioned below the inner high voltage transmission line with liquid lithium vortex in the center.

10.3 Thermonuclear Microexplosion Rocket Propulsion

The concept of a thermonuclear rocket propulsion engine is shown in Figure 10.5 (taken from a 1969 paper by the author). A thermonuclear microbomb injected into the focus of a magnetic mirror is ignited by an intense beam. The magnetic field of the mirror deflects the fireball in a preferred direction, with the field produced by superconducting magnets. To prevent these magnets from going normal during the rapid change of the magnetic field by the expansion of the fireball, they have to be protected by a conducting shield with a high melting point, for example tungsten. The eddy currents induced in this shield then give rise to a $\mathbf{j} \times \mathbf{H}$ body force resulting in thrust.

A fraction of the energy released in the expanding fireball is recovered through the magnetohydrodynamic loop, used to recharge the trigger appa-

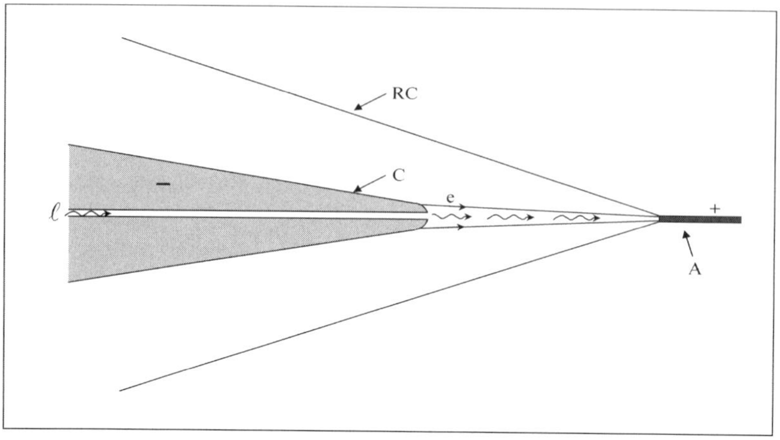

Figure 10.4: Laser-guided run-away electron beam in low-density space charge neutralizing background gas with the laser beam initiating breakdown. C cathode, A anode, RC return current conductor, ℓ guiding laser beam, e relativistic electrons.

ratus for the next microexplosion. In addition, a small nuclear fission reactor is needed for the startup operation to make the initial ignition. To reach larger yields than are possible with simple microexplosions, one can employ the concept of staging (Chapter 6.9), and the magnetic booster target staging concept explained in Chapter 8.17.

For missions within the solar system the very large exhaust velocities of $\gtrsim 10^8$ cm/sec, attainable with a thermonuclear microexplosion propulsion system, are undesirable, because the propulsion efficiency is highest if the final rocket velocity is of the same order as the exhaust velocity. If this condition is not met, most of the energy released goes into the jet and not into the rocket. For most missions within the solar system a final vehicle velocity of $\sim 10^2$ km/sec is sufficient, with a flight to Mars lasting about one week. This means that the specific impulse of thermonuclear microexplosions should be reduced ten-fold. If the jet power, which is the product of specific impulse and thrust, shall remain unchanged, a reduction in specific impulse has the benefit of increased thrust. The specific impulse can be decreased and the thrust increased simply by adding to the fireball some inert propellant, preferably hydrogen. The requirement to carry along some inert propellant

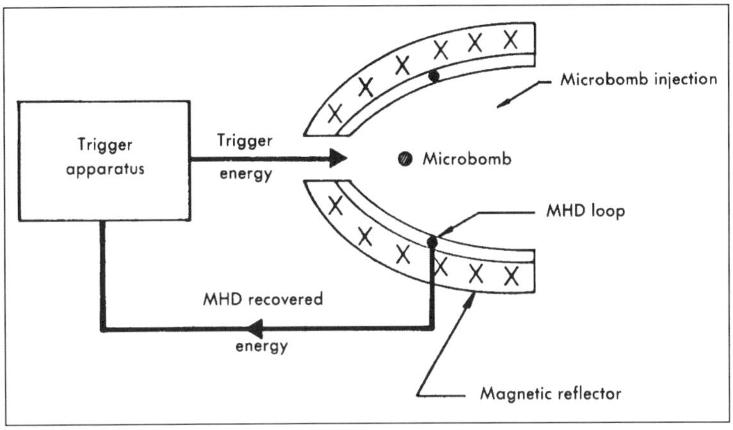

Figure 10.5: Schematic drawing of a nuclear microexplosion unit to be used for an efficient rocket propulsion system by which large payloads could be moved at great speed within the solar system.

has the additional benefit that it can be used to cool the space craft against residual waste heat coming from the rocket engine, thereby eliminating the need for a heavy radiator.

With thermonuclear microbomb rocket propulsion, payloads as large as millions of tons can be rapidly and economically moved within our solar system, and with transit times to Mars of about one week, to Jupiter of about one month and to Pluto less than a year. The implications this is going to have on astrophysics are at this time hard to assess but are most likely very great. For example, the establishment of a large research colony on Pluto, positioned at the edge of the solar system, would be of great scientific value.

In place of pure fusion microexplosions, or those making use of the autocatalytic fission-fusion process with natural uranium (or thorium) described in Chapter 6.2, one can also use the mini-nukes described in Chapter 7.14. There the energetic neutrons from the fission and fusion reactions lose their kinetic energy in the combustion products of the chemical high explosive, increasing both the specific impulse and thrust.

As for a reactor, one can there also use the configuration shown in Figure 8.26. A possible realization for rocket propulsion is shown in Figure 10.6, where as in Figure 10.3 the outer high current transmission line of Figure 8.26 is deformed into a coaxial circular strip line.

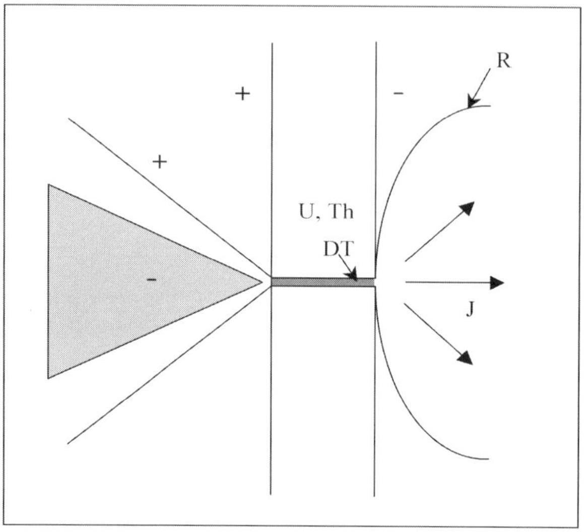

Figure 10.6: Circular stripline configuration for rocket propulsion, with plasma jet J and magnetic reflector R.

10.4 Interstellar Rocket Propulsion

With multistage thermonuclear microbomb rocket propulsion about 1/10 the velocity of light seems to be attainable. A study for a multistage unmanned interstellar probe reaching this velocity was made some time ago by the British Interplanetary Society under the name Project Daedalus. Its aim was a fly-by mission to Barnard's star. A propulsion unit is shown in Figure 10.7.

With $\sim 1/10$ the velocity of light, the trip would last several decades, a time which is quite reasonable, considering the enormity of such a mission.

If biological research can increase our life expectancy ~ 10 fold, a manned trip lasting a few decades would be comparable to a trip nowadays lasting a few years. A manned interstellar mission could therefore become a reality sooner than we might think.

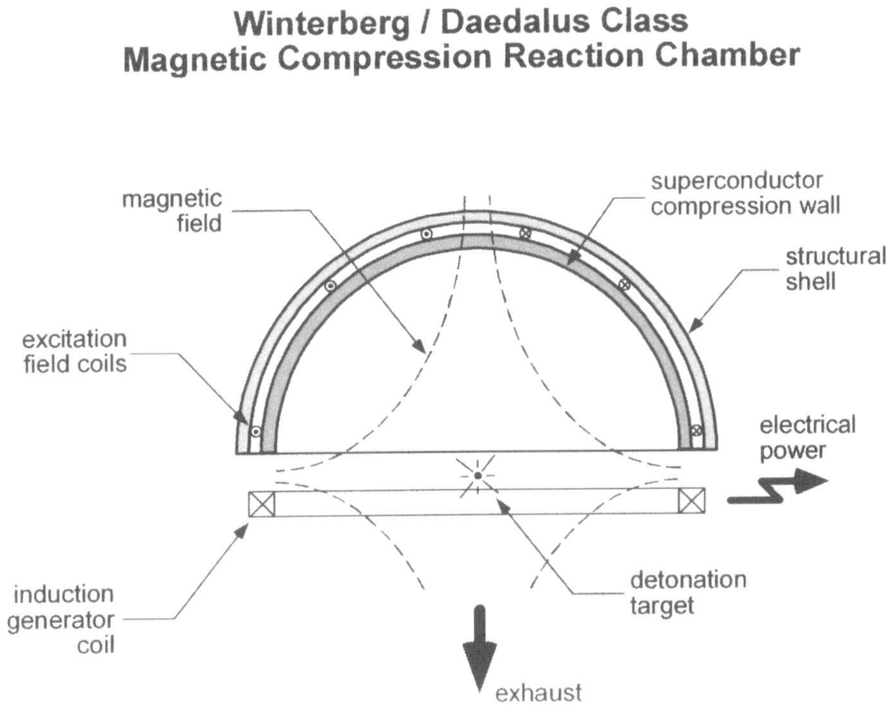

Figure 10.7: Courtesy of NASA Marshall Space Flight Center and University of Alabama in Huntsville.

With microbomb propulsion it should be possible to build interstellar space ships as large as lower Manhattan, flying with 1/10 the velocity of light to nearby solar systems. The basic configuration of such a star ship is shown in Figure 10.8. It consists of a many km high exponential tower. At the lower side of the tower are thousands of thermonuclear microbomb propulsion units. The small inner core of this star ship contains the payload with the crew. The outer part is mostly fuel. As the ship gains speed, the fuel in the layers labeled 1, 2, 3, ... is gradually used up, with the rocket engines at the end of these layers discarded. The exponential cross section of the star ship assures a maximum uniform distribution of the mechanical stress and allows it to drive continually at maximum acceleration, consistent with the maximum permissible structural stresses.

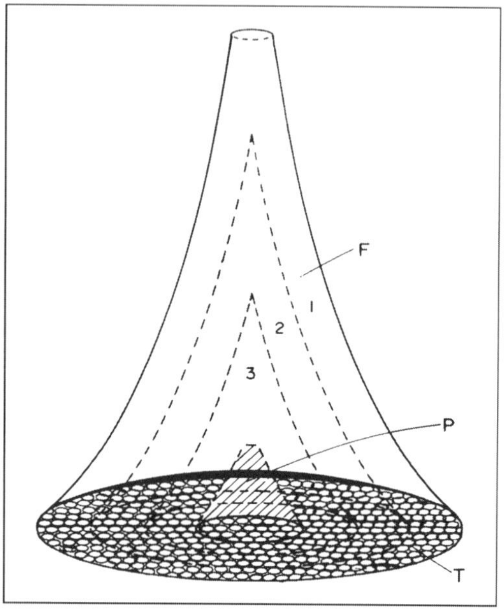

Figure 10.8: An "exponential tower" interstellar spaceship (schematic) using many propulsion units. The ship could have the mass of millions of tons and travel at one tenth the velocity of light. T is one of the many microexplosion propulsion units, and P is the payload. This configuration resembles the Hohmann "power tower" by space flight pioneer Walter Hohmann.

10.5 Thermonuclear Microexplosion-Driven Particle Accelerators

At present, the most important fundamental questions of physics demand that we study matter with ever larger particle accelerators. There are two areas where thermonuclear microexplosions can aid particle accelerators. First, with microexplosions, pulsed magnetic megagauss fields can be generated which could be used to bend beams of extremely energetic ions down to a much smaller radius than is possible with conventional accelerator techniques. Second, thermonuclear microexplosions could be used to drive collective type accelerators aimed at center of mass energies up to 10^5 GeV or more. And because of the large pulse power, thermonuclear microexplosion driven particle beams can be very intense.

Thermonuclear microexplosions can with ease generate pulsed magnetic fields up to $\sim 10^6$ G. One particular collective electron beam accelerator was shown in Fig. 8.6, there proposed for thermonuclear microexplosion ignition, but which can also be used to accelerate electrons to high energies. With a pulsed magnetic field of $\sim 10^6$ G, the electric field at the front of the radially inward compressed rotating electron beam would reach $\sim 3 \times 10^8$ V/cm, and a ~ 10 km long (as a consequence of magnetic insulation) accelerator would reach electron energies of 10^5 GeV.

Finally, there is the possibility of superintense beam accelerators. At the present, intense beam accelerators are Marx generators in conjunction with short pulse length transmission lines. Megajoule terawatt beams can be routinely produced this way. However, if the energy released in microexplosions is used to drive intense beam generators, pulsed beams with a power in excess of $\sim 10^{15}$ Watts may become possible, with the beam driven by microexplosions through direct magnetohydrodynamic energy conversion of the expanding fireball. Beams of this power and intensity would open up a new era of low energy nuclear physics with the prospect of large scale nuclear transmutations. The block diagram for such a superbeam accelerator is shown in Figure 10.9. It can be viewed as a beam amplifier, with the amplification done by microexplosions.

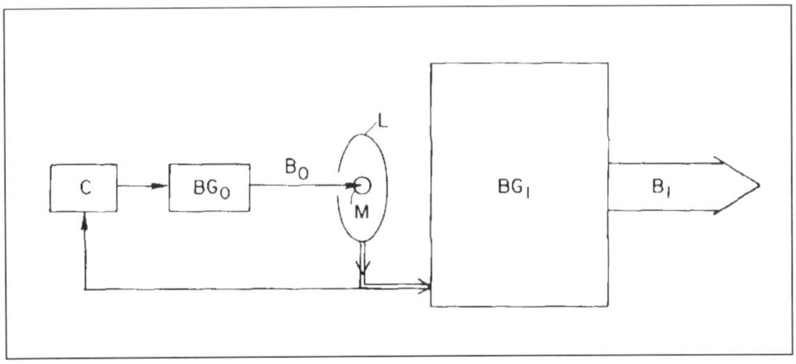

Figure 10.9: Flow diagram for a superbeam accelerator using a thermonuclear microexplosion reactor as an amplifier. C primary energy storage. BG_0 beam generator for primary beam. E_0 primary beam, M microexplosion target, L magnetohydrodynamic loop. BG_1 beam generator for superbeam B_1.

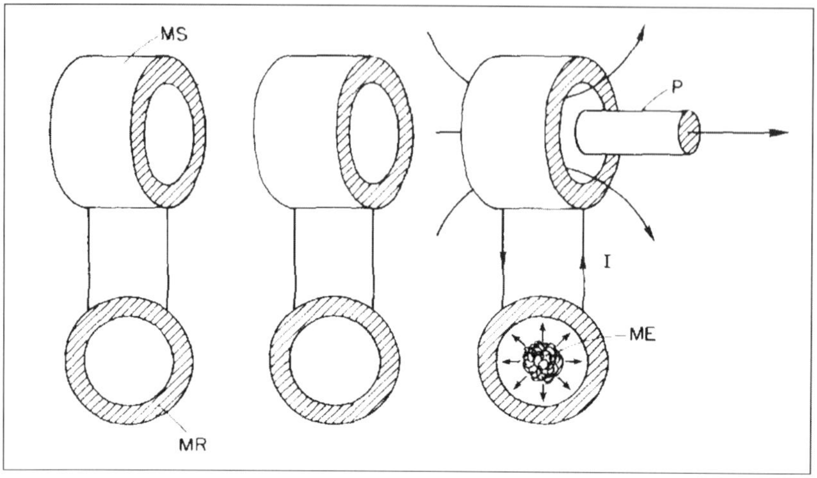

Figure 10.10: Electromagnetic gun driven by thermonuclear microexplosions to launch large masses into space. MS magnetic solenoid. MR microexplosion reactor cavity. ME microexplosion. I large current pulse. P projectile.

10.6 Thermonuclear Microexplosion Driven Space Launcher

The magnetic traveling wave macroparticle accelerator described in Chapter 8.11 was intended for impact fusion, with the goal of reaching velocities of 200 km/s for small projectiles. If driven by a chain of thermonuclear microexplosion reactors, an accelerator of this kind could launch large payloads into earth orbit, where the needed velocity is $\sim$ 10 km/s. The idea is explained in Figure 10.10. A sequence of microexplosion reactors induce large electric currents in a series of magnetic solenoids, forming the electromagnetic gun, propelling a cylindrical projectile to speeds up to 10 km/s. To puncture the atmosphere the projectile preferably has the form of a long slender cylinder.

10.7 Bibliography for Chapter 10

E. Teller in "Physics of High Energy Density", edited by P. Caldirola and H. Knoepfel, Academic Press, New York, 1971, p. 1ff.

F. Winterberg, Atomkernenergie-Kerntechnik **41**, 267 (1982).

F. Winterberg, "Vom griechischen Feuer zur Wasserstoffbombe", E. S. Mittler und Sohn, Herford und Bonn 1992, p. 110 ff. (in German).

F. Winterberg, Physics Today, September 1981, p. 9.

Fusion, August 1981, May-June 1986.

F. Winterberg, Physical Review **D 35**, 3500 (1987).

F. Winterberg, Raumfahrtforschung **15**, 208 (1971).

Chapter 11

Recent Developments

11.1 Chirped Laser Pulse Amplification

In laser amplifiers a small laser pulse is injected into the high gain medium, the medium with a large population inversion of "stored" photons possessing the same frequency, where the small laser pulse launches a photon avalanche (see Fig. 8.5). The avalanche can become so strong that it damages the high gain medium. This is in particular true if the high gain medium is made from glass. There the damage is caused by self-focusing, resulting from a nonlinear refractive index of the glass.

To overcome this problem one can employ what is called chirped laser pulse amplification. There the ultrashort laser pulse, intended to launch the avalanche in the high gain medium, is stretched (chirped) in time prior to its injection into the medium. It is the stretched out, lower intensity laser pulse which is injected into the high gain medium. Because of its reduction in intensity, the same is true for the avalanche in the high gain medium, reducing the danger for its damaging by the avalanche. The stretching out (chirping) of the short laser pulse is possible by the time-energy (resp. time-frequency) uncertainty relation: The shorter the duration of the pulse, the larger the spectral width of its frequency. Because of it, a short laser pulse is not exactly monochromatic. This permits to split it into a lower (towards the red) and higher (towards the blue) frequency part. This can be done by a (positive) dispersive element (the stretcher), consisting for example of a pair of diffraction gratings, or a long optical fiber with a frequency dependent refractive index, where the low frequency part of the split beam travels a longer time than the high frequency part before they reach the point where both components again merge and are injected into the high gain amplifier medium. Because this technique permits to stretch the original laser pulse thousandfold, the intensity of the avalanche in the high gain medium is reduced by the same factor. In emerging from the gain medium, the high- and low-frequency component are then recompressed by a (negative) dispersive element, the reverse as the one used for stretching the beam. With a thousandfold stretching and recompression, it is possible to amplify the initial laser pulse by a factor of the order 10^6, or about thousand times more than it would be otherwise possible. One can therefore generate a $\sim 10 - 100$ kJ, petawatt (10^{15} Watt) laser pulse, of the intensity and energy required for fast ignition. However, because of losses in the dispersive elements of the stretcher and compressor and the low efficiency of glass lasers in general, the overall laser efficiency is small.

A possible way to stretch and recompress a laser beam is shown in Fig. 11.1 and Fig. 11.2.

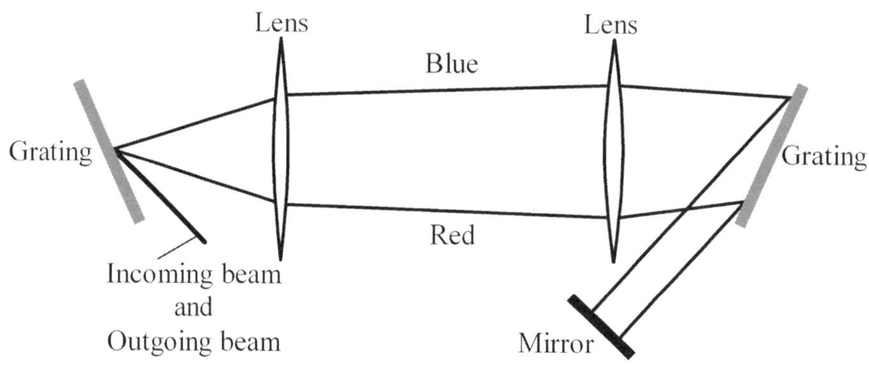

Figure 11.1: Diagramatic scheme to stretch a laser beam.

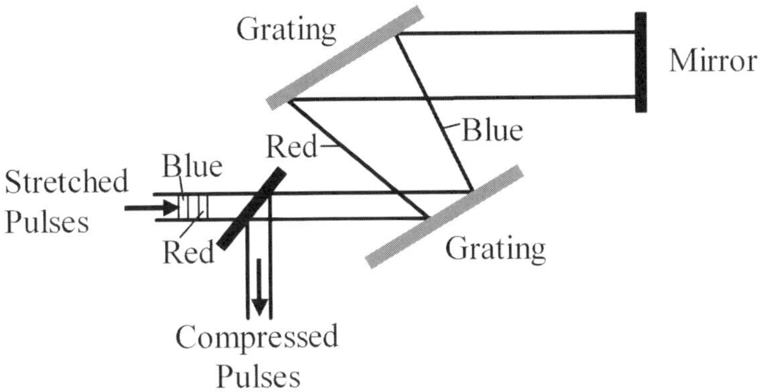

Figure 11.2: Diagramatic scheme to compress a laser beam.

11.2 Convergent Shock Wave Driven Megajoule — Petawatt Laser[1]

One can conceivably avoid the problem of laser beam damage for glass lasers, if the high gain medium is a dense optically transparent plasma. For thermonuclear applications short wave length lasers are there to be preferred. It seems that the most promising candidates are the noble gas ion lasers. The shortest reported laser wavelength is for Ne^{IV} with $\lambda = 2358$ Å, located in the ultraviolet. Since it will probably be difficult to produce four times ionized Neon, a better candidate is the Argon ion laser of singly ionized Argon. The shortest reported laser line for the Ar^+ ion laser is in the visible blue at $\lambda = 4879$ Å. Because the Ar^+ ion has fewer low lying energy levels than would normally exist in molecules, the blocking of the laser transition by population of the lower levels is less likely to occur.

One possibility to obtain a large population inversion is by heating liquid Argon to a temperature of $T = 90000$ °K, followed by its rapid expansion and cooling, whereby a high population inversion remains frozen in the Argon. Liquid Argon has a density of $\rho = 1.404$ g/cm^3. For $T = 9 \times 10^4$ °K, the pressure is $p = 2.55 \times 10^{11}$ dyn/cm^2 atm. The *ionization potential* of Argon

[1]Taken from a recently declassified paper

is $V_i = 15.7$ eV. With the Saha equation one then computes an ionization degree of $x \sim 50\%$. From this value it does not seem impossible to expect a population inversion of 10% for the upper laser level.

The population of the upper level can be enhanced by the so called entrapment of radiation which will occur at even lower densities than under consideration. The entrapment of radiation will make a photon be emitted and resonance-reabsorded many times inside the laser material. This entrapment of radiation will result in an increased effective life time for the upper laser level, of importance for a large population inversion.

It is proposed to heat liquid Argon by a convergent shockwave of a chemical explosive to a temperature of 9×10^4 °K. An expansion immediately following the heating, is likely to result in a lower temperature highly inverted Argon plasma.

The temperature behind a strong shock wave in a singly ionized plasma is given by

$$T = \frac{(\gamma - 1)}{(\gamma + 1)^2} \frac{A}{R} \mathrm{v}^2 \tag{11.1}$$

where A is the atomic weight, R the universal gas constant, γ the specific heat ratio, and v the shock velocity. For an Argon plasma $A = 40$. We furthermore put $\mathrm{v} = 8 \times 10^5$ cm/sec, the detonation velocity of hexogen. We thus have from (11.1) $T \sim 3 \times 10^4$ °K. This is lower by a factor 3 than the temperature of 9×10^4 °K, required to obtain a 50% ionization in Argon. Since at a temperature of 3×10^4 °K the degree of ionization would be only 2.5%, it is quite obvious that a higher temperature is required.

The temperature behind a strong shock wave rises as $r^{-0.4}$ for a cylindrical, and as $r^{-0.8}$ for a spherical convergent shock, where r is the radial distance of the shock wave from the center of convergence (chapter 5.3). Since the laser rod shall have cylindrical symmetry, a cylindrical convergent shock wave seems to be most suitable for our purpose. If the laser rod has a radius of $r_0 = 10$ cm, the high explosive has to be arranged in a cylindrical shell separated from the center of convergence by a radius $r_1 = 100$ cm in order to get a rise in temperatrue by about a factor 3. The situation is drawn schematically in Fig 11.3, with Fig. 11.4 illustrating the arrangement of the Argon ion plasma laser rod, a Q switched Argon ion laser to launch the avalanche in the Argon plasma rod with a high population inversion, and an optical lens to focus the laser beam on the thermonuclear material.

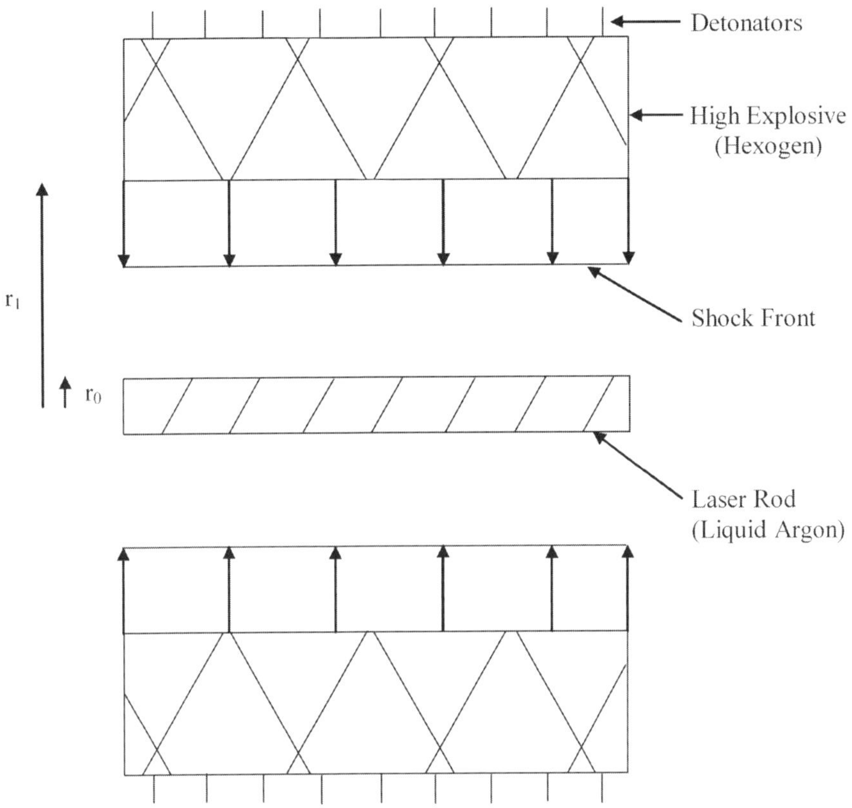

Figure 11.3: Cylindrical arrangement of high explosive and cylindrical laser rod for convergent shock wave heating.

The question of population inversion behind a strong shock wave requires a remark. It is clear that in complete thermodynamic equilibrium there is no population inversion. However, the situation is quite different in the material just behind a strong shock wave. Immediately behind the shock wave one has a uniform isotropic velocity distribution rather than a Maxwellian. If the population inversion is achieved by a resonant cross section of atomic particle collisions at a certain energy, and if the particles behind the shock front possess just this energy, by making a proper choice of r_0 and r_1, then a large population inversion is conceivable.

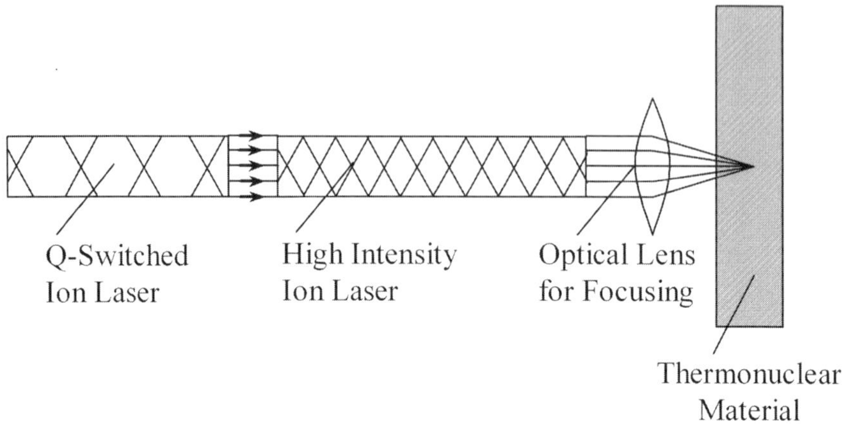

Figure 11.4: Arrangement of laser amplifier for the irradiation of the thermonuclear material.

Because the laser efficiency is at the most 10%, energy inputs in excess of 10^4 MJ are required, corresponding to an explosive charge in excess of one ton. Therefore such a laser is of limited interest for military applications, where it would have to compete with fission triggers in the kilogram range. Its primary significance seems to be in the Plowshare (peaceful use of nuclear explosives) field, where the size of the trigger is of minor significance but where emphasis is based on "cleanlines" of the nuclear charge. A second application may be for Orion-type nuclear pulse rocket propulsion, where the burnt high explosive can add to the overall momentum of the propulsion. Third, if one such laser can drive simultaneously a large number of thermonuclear micro-explosion chambers, it may in a one-shot throw-away operation, using a sequence of such throw-away lasers, perhaps be able to compete with the very expensive petawatt laser rep-rate-able ignition concept.

11.3 Impact Ignition

Because of the high cost for a 10-100 kJ petawatt laser, as required for the fast ignition of a DT target compressed to thousand times solid density by a 100 TW (10^{14} Watt) laser, it has been proposed by Murakami and Nagatomo to replace the petawatt laser with a small very fast moving projectile which upon impact on the highly compressed DT target would have the same effect. As it was shown in eq.(8.72) to reach the ignition temperature of the DT reaction requires an impact velocity of v $\sim 10^8$ cm/s. At this velocity and an impact energy equal of about 100 kJ $= 10^{12}$ erg, the projectile mass would have to be about 2×10^{-4} g. For a beryllium projectile this would imply a beryllium sphere with a radius equal to $r_0 = 3 \times 10^{-2}$ cm. According (8.78), the length of a electrostatic accelerator to reach a velocity of 10^8 cm/s, would be 24 km long. In the proposal by Murakami and Nagatomo a small flyer plate inside a cone stuck in the DT target is ablatively accelerated, by the same 100 TW laser pulse compressing the target. The precursor of this configuration is a configuration where a gold cone is stuck into the DT target, to facilitate the petawatt laser pulse to reach the center for the compressed target. In the modified impact ignition configuration the cone is enlarged, to permit that the flyer-plate can be accelerated to the final velocity over a larger distance. It is here where the advantage of all kinds of impact fusion schemes comes into play: A comparatively slow energy cumulation of the ignitor during its acceleration to a high velocity, that otherwise would not be possible. The two fast ignition configurations are shown side by side in Fig. 11.5.

Experiments done to verify this idea have so far only reached a velocity of 600 km/s, short of the required 1000 km/s. Because the flyer plate is accelerated inside a convergent conical duct, Taylor instability could become a serious problem to reach much higher velocities.

To avoid Taylor instability, one could use instead a convergent shock wave, which however, is mush less efficient. To increase the efficiency one may use several concentric flyer plates as in the multishell configuration described in chapter 5.6. This would increase the stability, with an efficiency in between the single flyer plate and a convergent shock wave.

11.4 Thermonuclear Microdetonation Macron Accelerator for Impact Ignition

Much higher macron-projectile velocities are possible if for the acceleration the thermonuclear energy released in an ignited magnetized plasma is used. For the realization of this idea one may ultimately use a thermonuclear detonation wave propagating down a high current pinch discharge channel. As

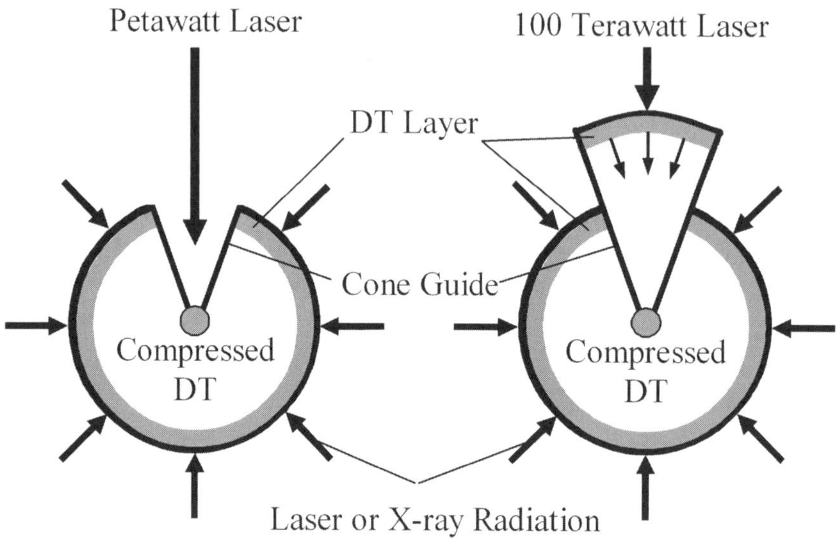

Figure 11.5: Fast ignition configuration with gold cone stuck in the DT target, with direct petawatt laser fast ignition and with impact ignition.

was shown in chapter 6.5, for the DT reation this requires a current of the order 10^7 A.

Following its ignition at one end of the pinch discharge channel, a detonation wave would propagate down the channel with a velocity which by order of magnitude is approximately 1/10 the velocity of light, the velocity of the DT fusion reaction α-particles.

In the plasma focus device kinetic energy is cumulated in the current sheet accelerated by the magnetic pressure acting on its rear. It is this cumulating process which permits the plasma focus to be driven by comparatively less expensive (lower voltage - longer discharge time) capacitors. But since one needs for a detonation wave propagating along a pinch discharge channel rather large currents, of the order 10^7 A, one may use here a Xram (Xram = Marx read backwards), where N magnetic coils are magnetized in series by a current I and are discharged in parallel, with the currents adding up to a current equal to NI. As in a Marx generator, an Xram permits a slow "charging" up, but in the Xram this must be done by opening switches, which is more difficult than the closing of switches as in a Marx generator. How a Xram drives a plasma focus gun is shown in Fig 11.6.

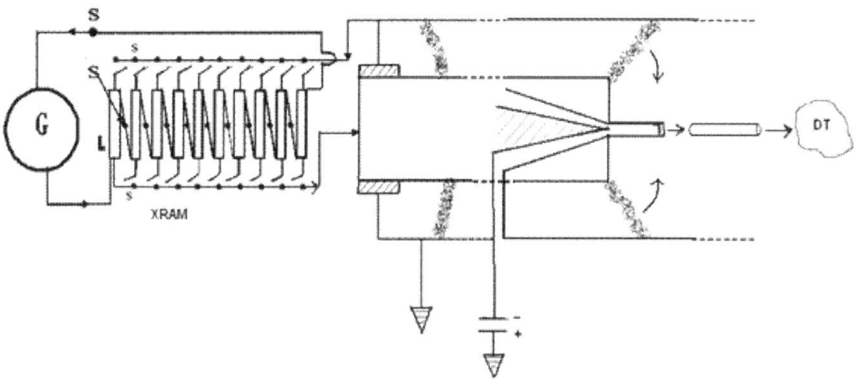

Figure 11.6: Plasma focus machine driven with an Xram, for the ignition of thermonuclear detonation wave along pinch confined liner, for the acceleration of a flyer plate projectile to $\sim 10^3$ km/s. G flywheel generator, L magnetic field coils, S switches, DT compressed DT target.

The opening of a switch is easier for a current I, as compared to a current NI. In addition, in a plasma focus machine, with its long co-axial "gun barrel" to accelerate the plasma sheet, a comparatively slow opening of the switches can be tolerated. Alternatively one may also use a bank of Marx generators, but this may be more expensive and less compact.

As shown in Fig. 11.6 and 11.7 to the right side end of the pinch column is attached a flyer plate, to be shot into a tube to direct the flyer plate onto the highly compressed DT target.

With the above given energy of 150 kJ, and a projectile velocity of 10^3 km/sec, the projectile mass would have to be 3×10^{-4} g. But since the energy delivered by the plasma focus fusion detonation can be easily made larger, it can deliver a buckshot of 10^3 km/s $\sim 10^{-4}$ g size macro particles, directed on to the highly compressed DT target. This greatly increases the chance that at least one of the particles hits and ignites the target.

The flyer plate acceleration can be described by the theory of explosive ballistics replacing a chemical detonation by a DT fusion detonation. If

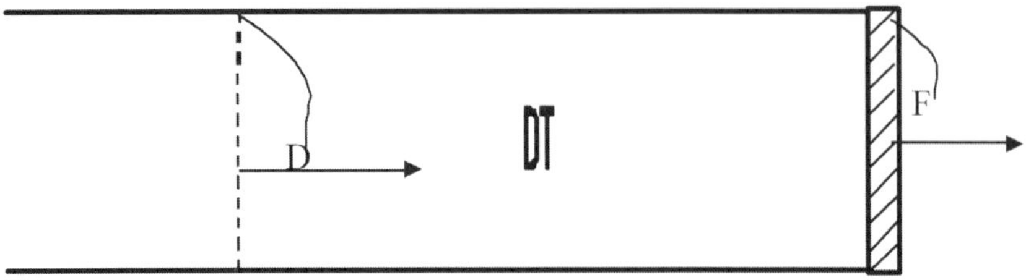

Figure 11.7: D thermonuclear detonation wave in DT in pinch magnetic field of the plasma focus, F flyer plate projectile.

W is the plasma velocity behind the supersonically moving thermonuclear detonation wave, a metallic flyer plate placed at the end of the detonating DT cylinder is accelerated to the velocity

$$V = W + \int_0^p \frac{dp}{a\rho} \tag{11.2}$$

where the 2nd term on the r.h.s. of (11.2) is the Riemann integral with a the velocity of sound in the burnt plasma behind the plate.

Experiments with chemical high explosives and metallic flyer plates have shown the $V \sim 2W$, about the same as for an incompressible plate of thickness d and density ρ under the action of a pressure p:

$$V = \frac{1}{\rho d} \int_0^p p \, dt \tag{11.3}$$

In shaped cylindrical high explosives the velocity of flyer plates placed at the end of the cylinder is about to be 1/2 as large due to the radial expansion of the detonation products behind the flyer plate. However, in the presence of a radially confining magnetic field the radial expansion is suppressed, but there the azimuthal field of the pinch discharge reduces the axial motion of the charged fusion products and with it the detonation velocity in the axial direction. Therefor, putting $V \sim W$ seems to be a good estimate. The DT fusion products (α-particles) have a velocity of 1.3×10^4 km/s, and one obtains (with $\gamma = 5/3$) that $V = W = D/(\gamma + 1) \sim 2.4 \times 10^3$ km/s.

11.5 Fast Ignition with Two Lasers

The proposed fast ignition of a highly compressed deuterium-tritium (DT) targets by petawatt lasers requires an energy of about 100 kJ. To lower the power of the laser, it is proposed to accomplish fast ignition with two lasers, one with lower power in the infrared, and a second one with high power in the visible to ultraviolet region. The infrared laser of lower power shall by its radiation pressure drive a large current in a less than solid density plasma placed inside a capillary, while the second high power-shorter wave length-laser shall ignited at one end of the capillary a magnetic field supported thermonuclear detonation wave in its blanket made from solid DT. The outer end of the capillary, together with its DT blanket, is stuck in the DT target.

This idea is described in Fig. 11.8, and has the following details and sequence of events:

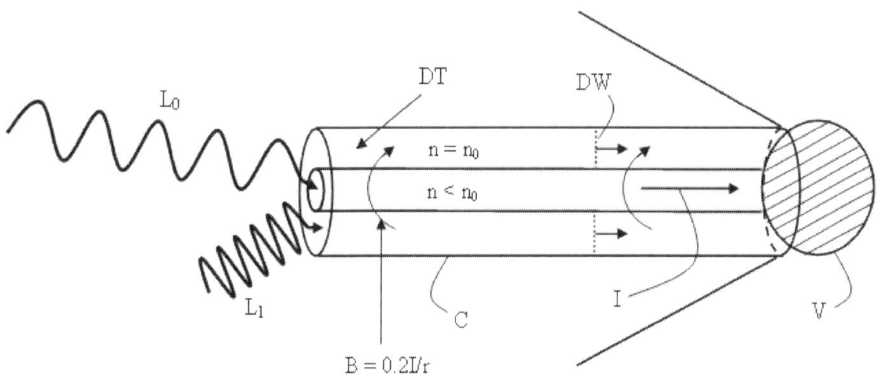

Figure 11.8: Detonation along capillary for fast ignition: L_0 longer , L_1 shorter-wave length laser pulse; $n = n_0 = 5\times10^{22}$ cm^{-3}, solid DT; C capillary, I current inside capillary, B magnetic field outside capillary; DW detonation wave front, V fast ignition volume for target.

1. A capillary C filled with a less than solid state density plasma ($n < n_0$) is surrounded by a solid state ($n = n_0 = 5 \times 10^{22}$ cm^{-3}) DT blanket B.

2. An infrared laser pulse L_0 is injected from the left into the left side end of the capillary C. The radiation pressure of the infrared laser pulse

accelerates the electrons inside the capillary to the right, generating a large electron current in that direction, with the return current flowing outside the DT containing blanket.

3. A powerful shorter wave length laser pulse L_1 heats up a section of the blanket at its one end to the ingnition temperature of the DT reaction, with the section larger than the Larmor radius of the DT charged fusion products (α-particles).

4. Provided the magnetic field set up by the electron current inside the capillary is large enough (of the order 10^7 A), the DT fusion reaction α-particles are entrapped inside the DT blanket, and for that reason can launch a thermonuclear detonation wave DW propagating to the right. If the other end of the capillary, together with its DT blanket, is stuck into the target, it can trigger in it a thermonuclear deflagration.

A coherent light beam with an average electric field $\left(\overline{E}^2\right)^{1/2}$ produces an electron drift motion of the plasma electrons with the drift velocity given by

$$\mathrm{v}_d = e^2\overline{E}^2/m^2c\omega^2 \tag{11.4}$$

where ω is the circular frequency of the laser light, e and m are the charge and mass of an electron, c the velocity of light. For (11.4) to be valid $\mathrm{v}_d << c$. For the laser radiation to be able to drive the electrons, the laser light frequency must be larger than the plasma frequency, i.e. $\omega >> \omega_p$, where $\omega_p = \left(4\pi n e^2/m\right)^{1/2}$ with n the electron number density.

Introducing the Poynting vector

$$S = (c/2\pi)\,\overline{E}^2 = P/A \tag{11.5}$$

where P is the power of the laser beam in ergs/s, and A the area of the capillary in cm^2, onto which the laser beam is projected. With (11.5) on can write for (11.4)

$$\mathrm{v}_d = (e/mc\omega)^2\,\frac{4\pi P}{A}. \tag{11.6}$$

If the plasma in the capillary is hydrogen plasma or singly ionized plasma, the total current induced by the laser beam is

$$I = nev_d A = c\left(\frac{\omega_p}{\omega}\right)^2 \frac{P}{I_A}, \omega > \omega_p \tag{11.7}$$

where $I_A = mc^3/e$ is the Alfvén current. In practical units it is equal to 17000 A. Assuming that $\omega = \omega_p$, one can write for (11.7)

$$P = I\frac{I_A}{c}. \tag{11.8}$$

For a current equal to $I = 10^7$ A, where the magnetic field is large enough to entrap the DT fusion reaction α-particles, one finds that $P \sim 10^{12}$ Watt.

For ω just slightly larger that ω_p, one can set $\omega \sim \omega_p$. In this case then, the current depends only on the laser power. However, the laser wave length λ can not be larger than $\sqrt{A}$, otherwise the laser beam can not propagate through the capillary. For the given example this means that $\lambda < 10^{-1}$ cm, permitting to use a high efficiency infrared laser.

For the current of $I = 10^7$ A, the Larmor radius r_L of the charged DT fusion reaction products (α-particles) is about 1/10 the radius of the capillary, sufficiently short to entrap the α-particles in the DT reaction burning zone, which is the condition for detonation along the capillary in the DT containing blanket. For the ignition of this magnetic supported detonation, a volume of the order $A\sqrt{A} \sim 10^{-3}$ cm^3 has to be heated up to a temperature of 10^8 °K, the DT ignition temperature, in less than the Lawson time $\tau_L = 10^{14}/n_0 = 2^{-9}$ sec. To heat a volume 10^{-3} cm^3 of solid DT to 10^8 °K requires about 10^2 kJ of laser energy, at a power of 5×10^3 Watt, well within the reach of exising laser technology.

The evolution of the novel fast ignition concept from the Kodama, Murakami, Nagatomo configuration is shown in Fig. 11.9. The initial breakthrough of the petawatt laser fast ignition concept made by Kodama, is illustrated in Fig. 11.9a. Inserting a metallic cone into the DT target, greatly increased the neutron yield. Because a much larger yield with this configuration seems possible only with a much larger petawatt laser, it was proposed by Murakami and Nagatomo to reach the same with about 10 times less laser power, utilizing the kinetic energy accumulation of a projectile accelerated to high velocities. Still better, seems to be an approach where energy is fed into the "ignitor" by thermonuclear reactions in a magnetized plasma at lower densities. Ultimately this would have to be done by a small magnetic field supported thermonuclear detonation.

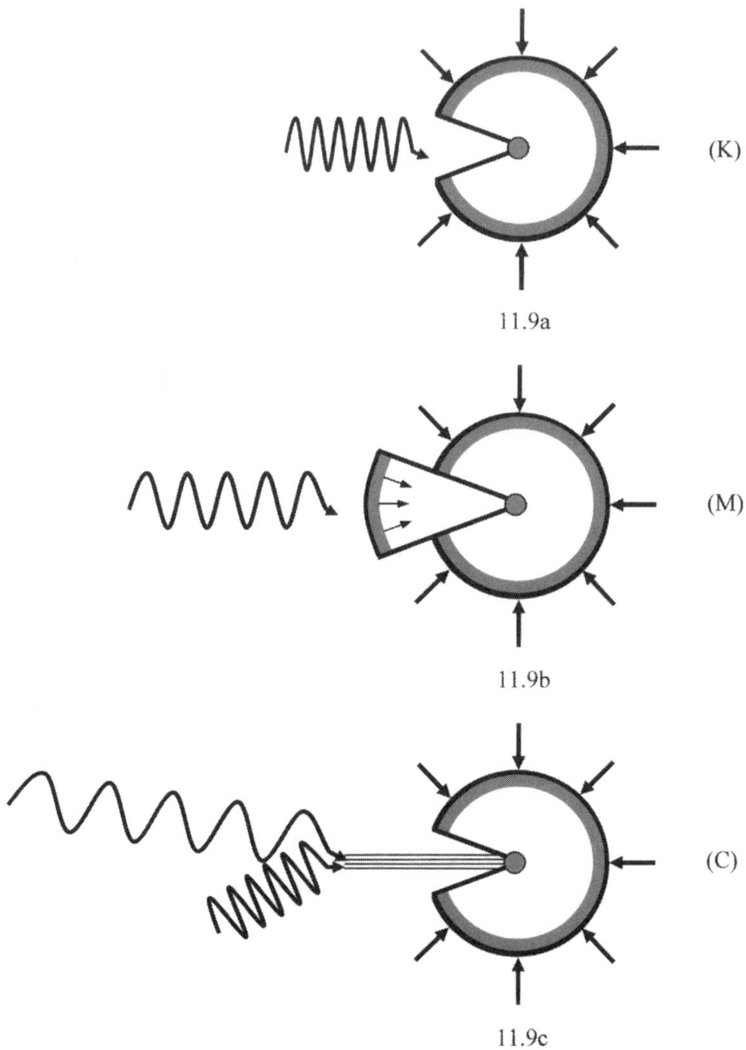

Figure 11.9: Evolution of the Kodama et al. (K), Murakami et al. (M) and Capillary (C) fast ignition configuration.

11.6 Conjectured Metastable Super-Explosives Formed under High Pressure for Thermonuclear Ignition

Under normal pressure the distance of separation between two atoms in condensed matter is of the order 10^{-8} cm, with the distance between molecules formed by the chemical binding of atoms of the same order of magnitude. As illustrated in Fig. 11.10, the electrons of the outer electron shells of two atoms undergoing a chemical binding form a "bridge" between the reacting atoms. The formation of the bridge is accompanied in a lowering of the electric potential well for the outer shell electrons of the two reacting atoms, with the electrons under the attractive force of both atomic nuclei. Because of the lowering of the potential well, the electrons undergo under the emission of eV-photons a transiton into lower energy molecular orbits.

Going to still higher pressures, a situation can arise as shown in Fig. 11.11, with the formation of electron bridges between inner shells, rep. shells inside shells. There the explosive power would be even larger. Now consider the situation where many closely spaced atoms are put under high pressure, making the distance of separation between the atoms so much smaller, that the electrons of the outer shells coalesce into one common shell surrounding both nuclei, with electrons from inner shells forming a bridge. Because there the change in the potential energy is much larger, the change in the electron energy levels is also much larger, and can be many keV. There then a very powerful explosive is formed releasing its energy in a burst of keV X-ray photons. This powerful explosive is likely to be very unstable, but it can be produced by the sudden application of a high pressure in just the moment when it is needed. Because an intense burst of X-rays is required for the ignition of a thermonuclear micro-explosion, the conjectured effect if it exists, has the potential to reduce the cost for the ignition of thermonuclear micro-explosions by orders of magnitude.

The energy of an electron in the ground state of a nucleus with the charge Ze is

$$E_1 = -13.6Z^2[\mathrm{eV}]. \tag{11.9}$$

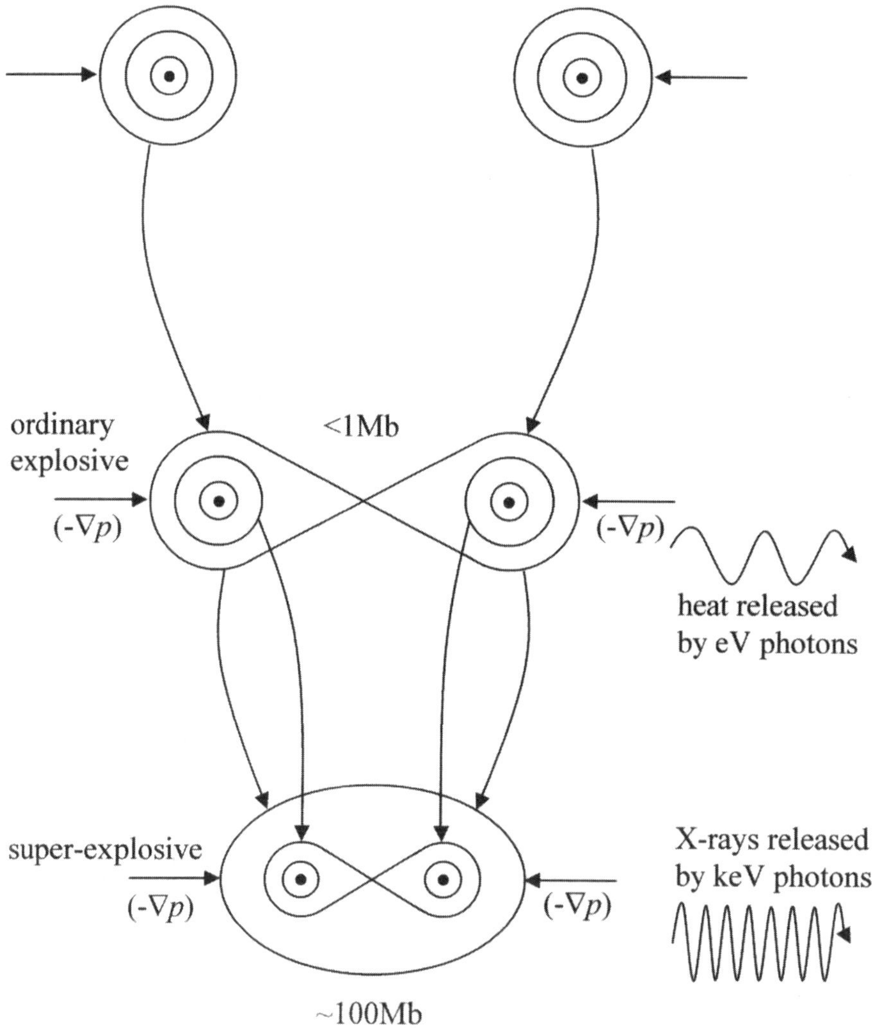

Figure 11.10: In an ordinary explosive the outer shell electrons of the reacting atoms form "eV" molecules accompanied by the release of heat through eV photons. In a super-explosive the outer shell electrons "melt" into a common outer shell with inner electron shells form "keV" molecules accompanied by the release of X-ray keV photons.

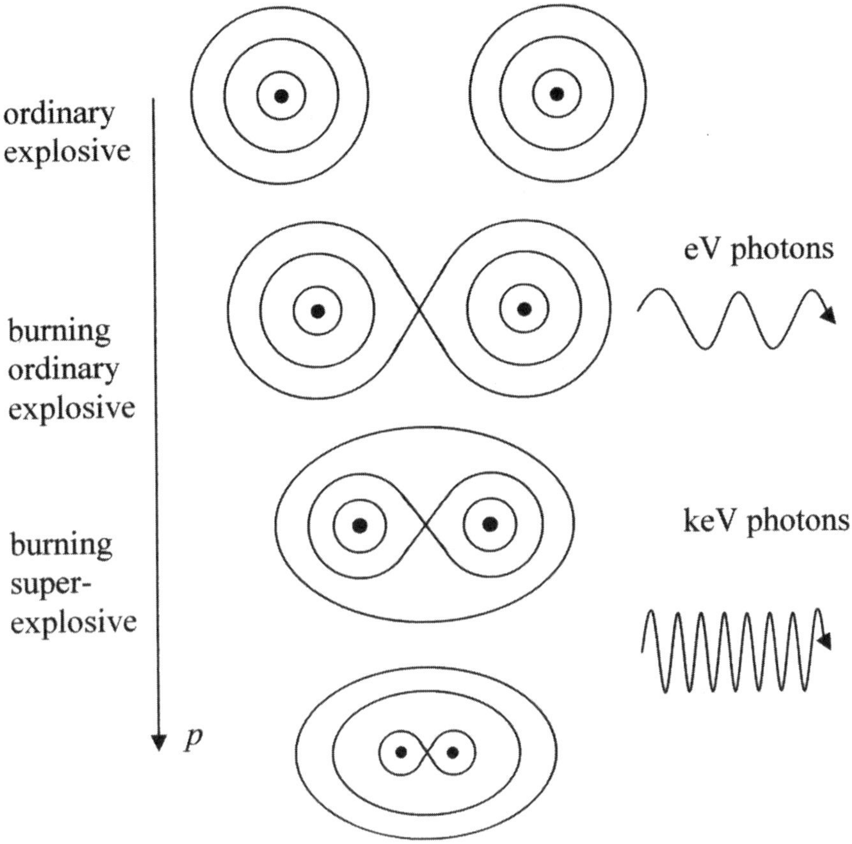

Figure 11.11: With increasing pressure electron-bridges are formed between shells inside shells melting into common shells.

With the inclusion of all the Z electrons surrounding the nucleus of charge Ze, the energy for all electrons is (see eq. (3.2)):

$$E_1{}^* = -13.6Z^{2.42}[\text{eV}], \tag{11.10}$$

with the outer electrons less strongly bound to the nucleus.

Now, assume that two nuclei are so strongly pushed together that they act like one nucleus with the charge $2Ze$, onto the $2Z$ electrons surrounding the $2Ze$ charge. In this case, the energy for the innermost electron is

$$E_2 = -13.6\,(2Z)^2\,[\mathrm{eV}], \qquad (11.11)$$

or if the outer electrons are taken into account,

$$E_2{}^* = -13.6\,(2Z)^{2.42}\,[\mathrm{eV}]. \qquad (11.12)$$

For the difference one obtains

$$\delta E = E_1{}^* - E_2{}^* = -13.62Z^{2.42}\left(2^{2.42} - 1\right) \sim 58.5Z^{2.42}[\mathrm{eV}]. \qquad (11.13)$$

Using the example $Z = 10$, which is a neon nucleus, one obtains $\delta E \sim 15$ keV. Of course, it would require a very high pressure to push two neon atoms that close together, but this example makes it plausible that smaller pressures exerted on heavier nuclei with many more electrons may result in a substantial lowering of the potential well for their electrons.

A pressure of $p \sim 100$ Mb $= 10^{14}$ dyn/cm^2 can be reached with existing technology in sufficiently large volumes, in at least three ways:

1. Bombardment of a solid target with an intense relativistic electron- or ion beam.

2. Hypervelocity impact.

3. Bombardment of a solid target with beams of by hypervelocity impact, followed by a convergent shock wave.

To 1: This possibility was considered by Kidder who computes a pressure of 50 Mb if an iron plate is bombarded with a 1 MJ (10 MeV - 10^6 A) relativistic electron beam, focused down to an area of 0.1 cm^2. Accordingly, a 2 MJ beam would produce 100 Mb. Instead of using an intense relativistic electron beam, one may use an intense ion beam, which can be produced by the same high voltage technique, replacing the electron beam diode by a magnetically insulated diode (chapter 8.5, Fig. 8.2). Using intense ion beams has the additional benefit that the stopping of the ions in a target is determined by a Bragg curve, generating the maximum pressure inside the target, not on its surface.

To 2: A projectile with the density $\rho \sim 20$ g/cm^3, accelerated to a velocity v= 30 km/s would, upon impact, produce a pressure of 100 Mb. The acceleration of the projectile to these velocities can be done by a magnetic traveling wave accelerator.

To 3: If, upon impact of either a particle beam or projectile, the pressure is less than 100 Mb, for example only of the order 10 Mb, but acting over a larger area, a tenfold increase in the pressure over a smaller area is possible by launching a convergent shock wave from the larger area on the surface of the target, onto a smaller area inside the target. The rise in pressure in a convergent spherical shock wave goes as $r^{-0.9}$ (see chapter 5.3), which means that 100 Mb could be reached by a ten-fold reduction in the radius of the convergent shock wave.

We assume an equation of state of the form $p/p_0 = (n/n_0)^\gamma$. For a pressure of 100 Mb $= 10^{14}$ dyn/cm^2, we may set $\gamma = 3$ and $p_0 = 10^{11}$ dyn/cm^2, p_0 being the Fermi pressure of a solid at the atomic number density n_0, with n being the atomic number density at the elevated pressure $p > p_0$. With $d = n^{-1/3}$, where d_0 is the lattice constant, one has

$$d/d_0 = (p_0/p)^{1/9} . \tag{11.14}$$

For $p = 10^{14}$ dyn/cm^2 one has $d/d_0 \sim 1/2$. Such a lowering of the inneratomic distance is sufficient for the formation of molecular states.

Calculations done by Muller, Rafelski, and Greiner show that for molecular states ${}_{35}$Br- ${}_{35}$Br, ${}_{53}$I- ${}_{79}$Au, and ${}_{92}$U- ${}_{92}$U, a two-fold lowering of the distance of separation leads to a lowering of the electron orbit energy eigenvalues by 0.35 keV, 1.4 keV and 10 keV respectively. At a pressure of 100 Mb= 10^{14} dyn/cm^2 where $d/d_0 \sim 1/2$, the result of these calculations can be summarized by (δE in keV)

$$\log \delta E \sim 1.3 \times 10^{-2} Z - 1.4 \tag{11.15}$$

replacing eq. (11.13), where Z is here the sum of the nuclear charge for both components of the molecule formed under the high pressure.

The effect the pressure has on the change in these quasi-molecular configurations is illustrated in Fig. 11.12 showing a $p-d$ (pressure-latice distance) diagram. This diagram illustrates how the molecular state is reached during the compression along the adiabat a at the distance $d = d_c$, where the pressure attains the critical value $p = p_c$. In passing over this pressure the electrons fall into the potential well of the two-center molecule, releasing their potential energy as a burst of X-rays. Following its decompression, the molecule disintegrates along the lower adiabat b.

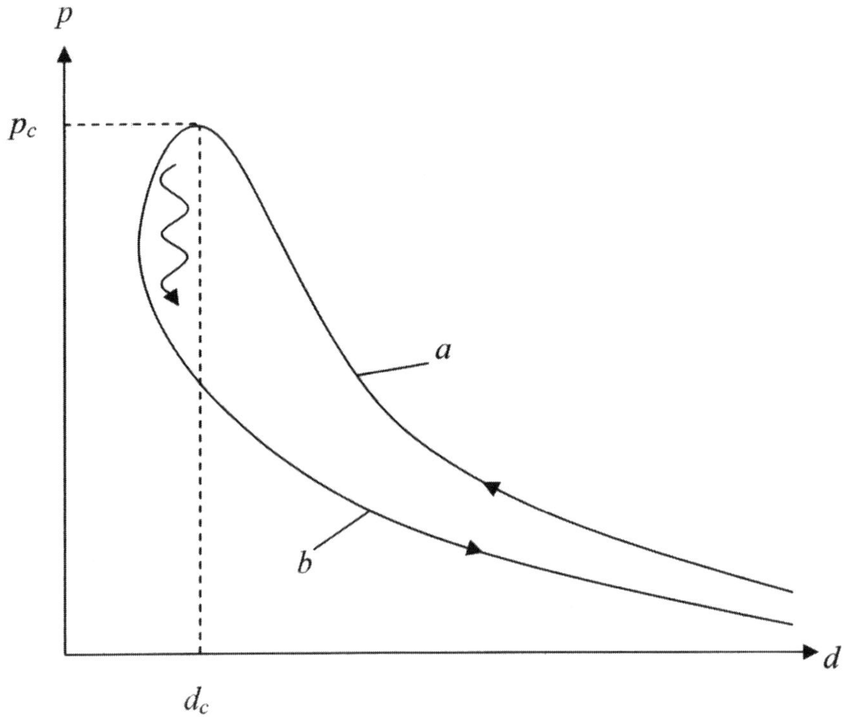

Figure 11.12: $p - d$, pressure — inner-atomic distance diagram for the upper atomic and lower molecular adiabat: a during the compression, and b during the decompression. $d = d_c$ is the critical distance for the formation of the molecular state.

The natural life time of an excited atomic (or molecular) state, emitting radiation of the frequency ν is given by

$$\tau_s \sim 3.95 \times 10^{22}/\nu^2[\text{sec}]. \tag{11.16}$$

For keV photons one finds that $\nu \sim 2.4 \times 10^{17}$ sec^{-1}, and thus $\tau_s \sim 6.8 \times 10^{-14}$ sec.

There are two conceivable ways the energy can be released:

1. In a shock front, if the rise time for the pressure τ_c is there shorter than τ_s. The rise time τ_c is about equal a mean free path divided by

the shock velocity v_s. In condensed matter the mean free path is by order of magnitude about equal the lattice constant d, hence the rise time

$$\tau_c \sim d/v_s. \tag{11.17}$$

Assuming that $v_s \sim 10^6$ cm/s, a typical value for the shock velocity in condensed matter under high pressure, and that $d \sim 10^{-8}$ cm, one finds that $\tau_c \sim 10^{-14}$ sec, and hence $\tau_c < \tau_s$. In reality the life time for an excited state is shorter than τ_s, and of the order of the collision time, which here is of the order of τ_c. This means that the X-ray pulse is released in the time (11.17). The time for the electrons to form their excited state in the molecular shell is much shorter and of the order $1/\omega_p \sim 10^{-16}$ sec, where ω_p is the solid state plasma frequency.

The release of the X-rays in the shock front is likely to accelerate the shock velocity, exceeding the velocity profile of the Guderley solution for convergent shock waves.

For the ignition of a thermonuclear reaction one may consider the following scenario illustrated in Fig. 11.13. A convergent shock wave launched at the radius $R = R_0$ into a spherical shell of outer and inner radius R_0, R_1, reaches near the radius $R = R_1$ a pressure of 100 Mb. After the inward moving convergent shock wave has reached the radius $R = R_1$, an outward moving rarefaction wave is launched from the same radius $R = R_1$, from which an intense burst of X-rays is emitted. One can then place a thermonuclear target inside the cavity of the radius $R = R_1$, with the target bombarded, imploded, and ignited by the X-ray pulse.

2. Instead of aiming at the release of the X-rays in the pressure spike of a shock wave, one may isentropically compress the super-explosive by a programmed pressure pulse, until the moment it reaches the critical pressure p_c, where the X-rays are explosively released. However, with the super-explosive made up from high Z-value atoms, the X-rays will be entrapped inside the super-explosive, with the result that the X-ray radiation energy is converted into black body radiation. To prevent this from happening, one may place a matrix small particles made up from the super-explosive in solid hydrogen. If the particles are small

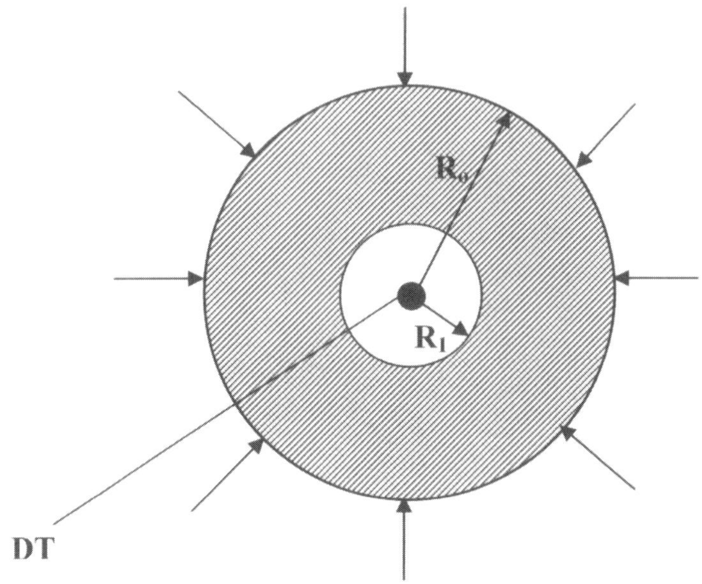

Figure 11.13: Inertial confinement fast ignition configuration.

enough to be transparent with regard to the X-rays released, the X-rays will heat up the surrounding hydrogen to high temperatures. There then, the energy released by the super-explosive will be transformed into thermal kinetic energy of the high temperature hydrogen, which in turn can be used for the implosion of a thermonuclear target.

If the change in pressure is large, whereby the pressure in the upper adiabat of Fig. 11.12 is large compared to the pressure in the lower adiabat, the X-ray energy flux is given by the photon diffusion equation

$$\phi = -\frac{\lambda c}{3} \nabla w. \tag{11.18}$$

Where w is the work done per unit volume to compress the material, with $w = p/(\gamma - 1)$. For $\gamma = 3$, one has $w = p/2$, whereby (11.18) becomes

$$\phi = -\frac{\lambda c}{6} \nabla p. \tag{11.19}$$

Assuming that the pressure e-folds over the same length as the photon mean free path, on has

$$\phi \sim (c/6)\, p. \tag{11.20}$$

For the example $p = 100$ Mb $= 10^{14}$ dyn/cm^2 one finds that $\phi \sim 5 \times 10^{23}$ erg/cm^{2}s $= 5 \times 10^{16}$ W/cm^2. Over an area of 0.1 cm^2 this is 5 petawatt, large enough for the ignition of a thermonuclear reaction. If the material to be compressed is made up of different atoms, the two components must form an alloy, and if this is not possible, should be a mixture of nano-particle powders.

11.7 Artificial Lightning as a Potent Inertial Confinement Fusion Driver

In an inertial confinement fusion a high gain is required. This poses a problem for laser fusion, where the photon flash from the thermonuclear micro-explosion is likely to destroy the optical laser ignition apparatus. Heavy ion beam ignition fares better, but it requires very large conventional particle accelerators, apart from the problem to stop the ion beam over a short distance, both in the direct and indirect (black body radiation induced) target implosion drive. Using a bank of Marx generators surrounding a DT target requires replaceable transmission lines, an unappealing feature, but the attainment of 30 MV would make possible to replace laser beams with light ion beams. This would eliminate the need for replaceable transmission lines, with the beams extracted from magnetically insulated diodes driven by Marx generators. The reason why high voltages are here important is that a beam gets stiffer, as the voltage driving the beam gets higher. A stiffer beam implies that the distance of the diode from the target can be made larger, of importance for a repetition rate operation. This illustrates the importance to aim for high voltages. Going to even higher voltages would still be better.

A voltage of 10^9 Volt would be ideal, because there a pulsed 10^7 Ampere GeV proton beam could be generated, with the current large enough to entrap in the target the fusion reaction alpha particles by the magnetic field of the beam. Gigajoule energies, which at these voltages can be released in less than 10^{-7} seconds, at a power of 10^{16} Watt, would make possible the ignition of a pure DD thermonuclear reaction, without any tritium. One gigajoule is on a logarithmic scale in the middle in between the one megajoule energy required for the ignition of a DT reaction, and the 10^3 gigajoule (fission bomb) energy used for the ignition of a DD reaction (Mike test). For an ignition energy of one gigajoule and a gain of $\sim 10^2$, an energy of 100 gigajoule $= 10^{18}$ erg would be released. Such a large yield would require to confine the explosion in a cavity with a radius of more than 10 meters. If the target is in the center of the cavity, the length of the beam reaching the target must be comparable to the diameter of the cavity. To direct the beam over such a distance onto a target, the beam must be stiff. This requires very high voltages.

In the DT reaction 80 percent of the energy goes into neutrons. In the DD reaction the proportion is much smaller. The DT reaction depends on tritium, which must be obtained by the breeding of lithium, a relatively rare element. Deuterium is abundant and everywhere available. All this suggests that the future of fusion is in the ignition and burn of deuterium and by implication in the attainment of very high voltages.

A typical lightning bolt has an energy of several hundred megajoule, discharging several coulombs with a current of 10 - 100 kA. Lightning occurs if the electric field between a cloud and the ground exceeds the breakdown of air at about 30 kV/cm. For a 300 meter long lightning this implies a potential difference of 10^9 Volt, and for a current of 100 kA a power of 10^{14} Watt. Most lightning discharges are from a negatively charged cloud to the ground, but at rare occasions from a positively charged cloud to the ground. In the rare cases the current may reach 300 kA discharging 300 Coulombs. At the potential difference of 10^9 Volt an energy of 300 gigajoule is released. This is equal to the energy released by 75 tons of TNT, with a power of 3×10^{14} Watt. By comparison, to ignite in liquid deuterium-tritium a thermonuclear micro-explosion, requires an energy of about 10 MJ with a power of 10^{15} Watt. This raises the question if one cannot make artificial lightning, comparable in energy and power of natural lightning, and use the thusly released energy to drive inertially confined thermonuclear microexplosions. One way it conceivably can be done is by charging to gigavolt potentials a magnetically insulated conductor levitated in ultrahigh vacuum (Winterberg, 1968, Winterberg, 2000). Here I will describe two schemes by which the same goal can be reached in a very different way.

The reaching out of high voltages is of importance in the quest for the ignition of thermonuclear reactions for two reasons:

1. With the energy E [erg] stored in a capacitor C [cm] charged to the voltage V [esu] equal to

$$E = (1/2)\, CV^2, \tag{11.21}$$

having an energy density

$$\epsilon \sim E/C^3 \sim V^2/C^2, \tag{11.22}$$

the energy is discharged in the time τ [sec]

$$\tau \sim C/c, \tag{11.23}$$

with a power P [erg/s]

$$P \sim E/\tau \sim cV^2. \tag{11.24}$$

This shows that for a given dimension of the capacitor measured in its length, and hence volume, the energy stored and power released goes in proportion to the square of the voltage.

2. If the energy stored in the capacitor is released into the energy of a charged particle beam, the current should be below the critical Alfvén limit

$$I = \beta\gamma I_A \tag{11.25}$$

where β =v/c, and $\gamma = (1 - \mathrm{v}^2/c^2)^{-1/2}$ is the Lorentz boost factor, with $I_A = mc^3/e$. For electrons I_A = 17 kA, but for protons it is 31 MA. Only for $I << \beta\gamma I_A$, can one view the beam as a beam of particles accompanied by the field of the particles, while for $I >> \beta\gamma I_A$ it is better viewed as an electromagnetic pulse carrying along with it some particles. For $I >> \beta\gamma I_A$, the beam can propagate in a space-charge and current-neutralizing plasma, but only if $I \leq \beta\gamma I_A$ can the beam be easily focused onto a small area, needed to reach a high power flux density. If a power of $\sim 10^{15}$ Watt shall be reached with a relativistic electron beam produced by a 10^7 Volt Marx generator, the beam current would have to be 10^8 Ampere with $\gamma \sim 20$ and $\gamma I_A \sim 3 \times 10^4$ Ampere, hence $I >> \beta\gamma I_A$. But if the potential is 10^9 Volt, a proton beam accelerated to this voltage and with a current of $I = 10^7$ A is below the Alfvén current limit for protons. And it would have the power of 10^{16} Watt, suficiently large to ignite the D-D thermonuclear reaction.

According to Paschen's law the electric breakdown field strength between two plane parallel conductors is only a function of the product pd, where p is the gas pressure and d the distance between the conductors. For dry air the breakdown field strength is 3×10^6 V/m, such that for a distance of 300 m the breakdown voltage would be 10^9 Volt. This is the voltage which under ideal conditions is reached in a lightning discharge. In reality the breakdown voltage is much smaller. The reason is that by a small initial inhomogeneity in the electric field, more negative charge is accumulated within the inhomogeneity, further increasing the inhomogeneity and eventually leading to a "leader", a small luminous discharge of electrons bridging part of the distance between the electrodes with a large potential difference. As a result a much larger electric field inhomogeneity is created at the head of the "leader", which upon repetition of the same process leads to a second "leader", followed by a third "leader", and son on, resulting in a breakdown between the electrodes by a "stepped leader", even though the electric field strength is less than the field strength for the breakdown by Paschen's law. What one has here is a growing electrostatic instability, triggered by a small initial electric field inhomogeneity.

A preferred point for the beginning of a stepped leader is the field inhomogeneity near the triple point where the conductor, the gas and the insulator, meet.

It is known, and used in power switches, that a gas jet under high pressure can blow out an electric arc discharge. Recognizing the breakdown below the Paschen limit as a growing electrostatic instability, it is conjectured that much higher voltages can be reached if the onset of this instability is suppressed by a gas flow, provided the stagnation pressure of the flow exceeds the electric pressure in between the electrodes with a large potential difference, thereby overwhelming the electric pressure of a developing electric field inhomogeneity. It is for this reason proposed to levitate a spherical conductor both by hydrodynamic and magnetic forces inside a Taylor flow (Taylor, 1922), a special drag-free spiral flow (see Fig. 1).

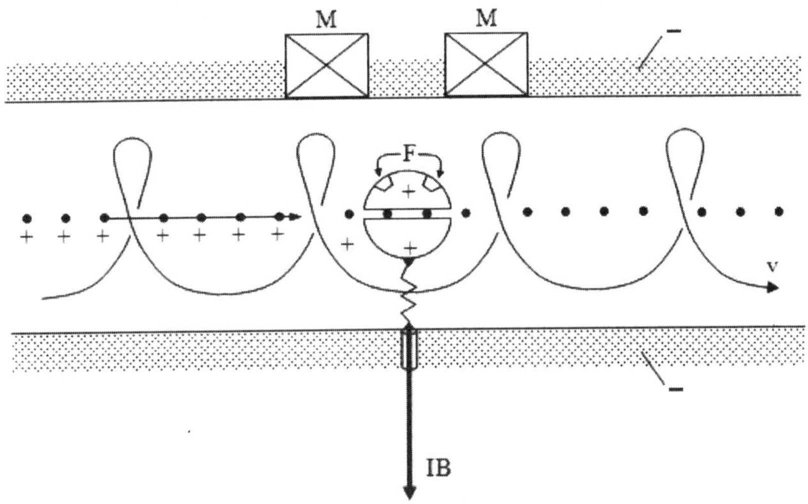

Figure 11.14: In drag-free Taylor flow magnetically levitated sphere to be charged up to ultrahigh potential by electrically charged pellets passing through the center of the sphere, **M** magnets, **F** ferromagnets, **IB** ion beam.

With the absence of drag forces on the sphere inside the horizontally flowing spiral Taylor flow, the sphere must only be levitated in the vertical direction by and externally applied magnetic field. By its levitation, the triple point as the source of a field inhomogeneity is eliminated.

If for air, at a pressure of 1 atmosphere, the breakdown voltage is 3×10^6 V/m, it would (according to Paschen's law) for a pressure of 300 atmospheres be equal to $\sim 10^9$ V/m, that is for meter-size distances of the order $\sim 10^9$ Volt.

For a meter-size sphere charged up to 10^9 Volt the electric field at its surface is $E \sim 10^7$ V/cm $\sim 3 \times 10^4$ [esu], with an electric pressure $E^2/8\pi \sim 4 \times 10^7$ dyn/cm^2 ~ 40 atmospheres. This field strength is below the field strength of $\sim 10^8$ V/cm where the disintegration of the conductor field ion emission sets in. At a pressure of 300 atmospheres, air (or some other gas) would have a density of the order $\rho \sim 1$ g/cm^3. For the stagnation pressure $p = (1/2)\,\rho \mathrm{v}^2$ of the Taylor flow moving with the velocity v [cm/s], to exceed the electric pressure, it is required that $(1/2)\,\rho \mathrm{v}^2 > E^2/8\pi$, from which one obtains v $\sim$ 100 meter/second. For a ten times smaller velocity one could reach 10^8 Volt, with the other parameters remaining the same. Instead of a gas under pressure one may also use a nonconducting fluid under normal pressure.

What remains is how to charge the sphere to such a high potential. There are two possibilities:

1. As in a Van de Graaff generator, by letting a charged nonconducting ribbon pass through the center of the sphere, releasing the charge in its center, or by replacing the ribbon with a stream of (positively) charged pellets.

2. By inductive charging with an externally applied rising magnetic field, releasing charges from the sphere to the fluid flowing through a hole passing through the center of the sphere.

The flow discovered by Taylor (Taylor, 1922) is the superposition of a uniform axial flow of constant velocity U with a constant swirl $W = (U/l)\,r$, where r/l is a measure for the intensity of the swirl. In cylindrical coordinates the stream function $\psi\,(r,z)$ of the Taylor flow satisfies the equation (Squire, 1956):

$$\frac{\partial^2 \psi}{\partial z^2} + \frac{\partial^2 \psi}{\partial r^2} - \frac{1}{r}\frac{\partial \psi}{\partial r} + \frac{4}{l^2}\psi = \frac{2U}{l^2} r^2 \tag{11.26}$$

with the velocity components in the z, r and ϕ directions given by

$$\frac{u}{U} = \frac{1}{r}\frac{\partial \psi}{\partial r} \tag{11.27}$$

$$\frac{v}{U} = \frac{1}{r}\frac{\partial \psi}{\partial z} \tag{11.28}$$

$$\frac{w}{U} = \frac{2}{lr}\psi \tag{11.29}$$

For a different problem the solution of (11.26), in terms of Bessel and Neumann functions, has been given (κ_1, κ_2 constants of integration) by Moore and Leibovich (Moore and Leibovich, 1971):

$$\psi = \frac{1}{2}Ur^2\left(1 + \kappa_1\frac{J_{3/2}(\xi)}{\xi^{3/2}} + \kappa_2\frac{N_{3/2}(\xi)}{\xi^{3/2}}\right) \tag{11.30}$$

where

$$\xi = \frac{2}{l}\sqrt{r^2 + z^2} \tag{11.31}$$

and

$$J_{3/2}(\xi) = \sqrt{\frac{2}{\pi\xi}}\left(\frac{1}{\xi}\text{sin}\xi - \text{cos}\xi\right), N_{3/2}(\xi) = \sqrt{\frac{2}{\pi\xi}}\left(\text{sin}\xi + \frac{1}{\xi}\text{cos}\xi\right) \tag{11.32}$$

The velocity components in the z, r and ϕ directions are

$$\frac{u}{U} = 1 + \kappa_1\left(\frac{J_{3/2}}{\xi^{3/2}} - 2\frac{r^2}{l^2}\frac{J_{5/2}}{\xi^{5/2}}\right) + \cdots \tag{11.33}$$

$$\frac{\nu}{U} = 2\kappa_1\frac{zr}{l^2}\frac{J_{5/2}}{\xi_{5/2}} + \cdots \tag{11.34}$$

$$\frac{w}{U} = \frac{r}{l}\left(1 + \kappa_1\frac{J_{3/2}}{\xi^{3/2}}\right) + \cdots \tag{11.35}$$

where the terms involving the Neumann functions and the second constant of integration κ_2 are represented by dots. The J-solution of (11.30), putting $\kappa_2 = 0$, has no singularity at $\xi = 0$, and is the solution of interest. With the appropriate boundary condition it is:

$$\psi = \frac{1}{2}Ur^2\left[1 - \left(\frac{2R/l}{\xi}\right)^{3/2}\frac{J_{3/2}(\xi)}{J_{3/2}(2R/l)}\right] \tag{11.36}$$

Now, if $J_{5/2}$ vanishes on the surface of a sphere placed in the Taylor flow, then all velocity components vanish on the surface of the sphere. And if $J_{5/2}$ is zero on the surface of the sphere, the circumferential shear vanishes as well. As a consequence, there is no boundary layer on the surface of the sphere and with it no drag. The pressure on the sphere is constant and the sphere stays at rest, except that it still is subject to a downward directed gravitational force, if U is directed horizontally. The downward force can be compensated by an externally applied magnetic field, making parts of the sphere from a ferromagnetic material.

We may add that in a beautiful experiment (Harvey, 1962) the Taylor solution had been verified. The result of this experiment is shown in Fig 11.15, where one can see how a spherical part of the fluid is trapped inside the axial flow with swirl.

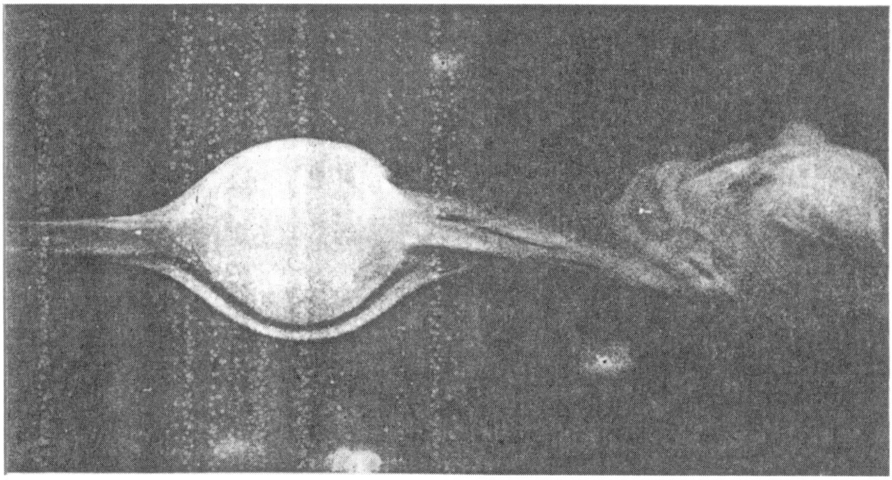

Figure 11.15: Experimental verification of the Taylor flow enclosing a non-moving spherical part unaffected by the flow.

One simple way the highly charged sphere can be discharged is over a spark gap, to be formed by moving the sphere towards the wall with the help of the magnetic field holding the sphere in the center of the Taylor flow, until the *pd* product becomes smaller than the value where breakdown sets in below the Paschen curve. If the sphere is positively charged, and if the discharge current is larger than the Alfvén current for electrons, this will favor a discharge into an intense ion beam below the Alfvén limit, suitable for thermonuclear ignition, with the self magnetic field of the beam even for the DD thermonuclear reaction.

To obtain an idea what kind of power is needed to drive the stepped leader stabilizing flow, we take as an example a flow velocity of v $\sim$ 30 m/s $= 3 \times 10^3$ cm/s and a density $\rho \sim 1$ g/cm^3. It has the stagnation pressure $p = (1/2)\,\rho \mathrm{v}^2 \sim 5 \times 10^6$ dyn/cm^2. If the cross section of the flow is one square meter $= 10^4$ cm^2, the power of the flow is $P = p\mathrm{v}\times 10^4$ erg/s $= 15$ MW. This power is quite large, but well within what is technically feasible.

Another problem is that the gas flow at high Reynolds numbers becomes trubulent. In a phenomenological description the turbulent flow can also be described by Taylor's solution, but the flow may there not be drag free. If the drag force on the sphere is not too large, it could be balanced by a magnetic force, very much as it is balanced against the gravitational force.

For gas flows of smaller density and smaller cross section, the idea could be tested by a comparatively modest effort.

Would it be not for breakdown, one could with a Marx generator reach in principle arbitrarily large voltages. In a Marx generator the buildup of the voltage is not fast enough to reach a gigavolt. It is the idea of the super Marx generator how this might be achieved.

To obtain a short discharge time with a single Marx generator, the Marx generator charges up a fast discharge capacitor, discharging its load in a short time. This suggests using a bank of such fast discharge capacitors as the elements of a Marx generator, each one of them charged up by one Marx generator to a high voltage. One may call such a two stage Marx generator a super Marx generator (see Fig.11.16). If N fast capacitors are charged up by N Marx generators in parallel to the voltage V, the closing of the spark gap switches in the super Marx generator adds up their voltages to the voltages NV. In the super Marx generator, the Marx generators also serve as the resistors in the original Marx circuit. One may also disconnect the Marx generator from the super Marx after the charge-up is completed.

By connecting the high voltage terminal of the super Marx generator to a Blumlein transmission line, a very high voltage pulse with a fast rise time can be generated. At these high voltages ion beams are favored over electron beams, because electron beams are there above the Alfvén limit. To assure that all the ions have the same charge to mass ratio, the gas or liquid must be hydrogen or deuterium, otherwise the beam will spread out axially, losing its maximum power.

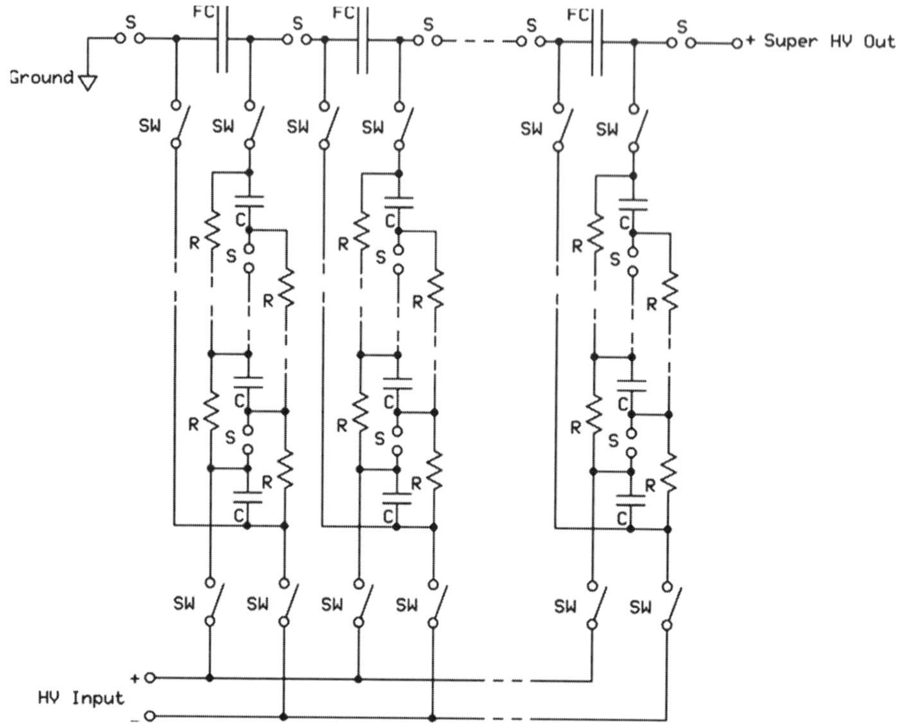

Figure 11.16: In a super Marx generator, N Marx generators charge up N fast capacitors FC to the voltage V, which switched into series add up their voltages to the voltage NV.

Instead of making the breakdown in hydrogen gas, one may let the breakdown happen along a thin liquid hydrogen jet, establishing a bridge between the high voltage terminal of the Blumlein transmission line and the thermonuclear target.

Unlike the traditional spherical implosion type targets for the DT reaction, for the DD reaction cylindrical targets are preferred. Only there is a micro-detonation in pure deuterium possible. This is so because the large electric current of the intense hydrogen beam generates a large azimuthal magnetic field which entraps the charged fusion products within the cylindrical target, the condition for burn and detonation. In a z-pinch, a magnetic field supported detonation along a pinch discharge channel is possible, for pinch currents of the order 10^7 Ampere. The same is true if the large pinch current of a $\sim 10^7$ Ampere is replaced by an ion beam with the same current. For a detonation to occur in deuterium, the burn of the tritium and He^3 fusion products of the DD reaction is important.

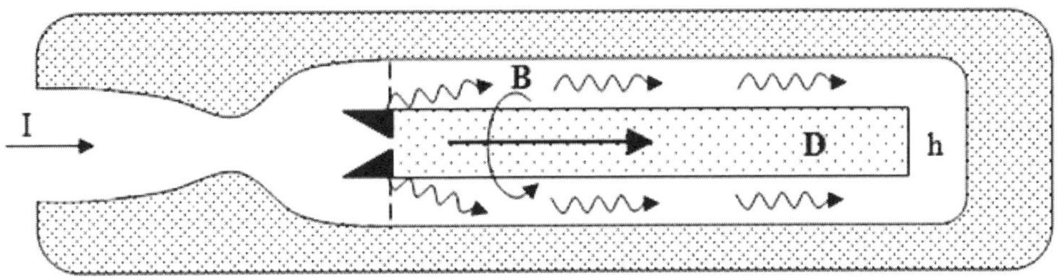

Figure 11.17: Pure deuterium fusion micro-detonation ignited with an intense proton beam. **B** solid deuterium rod, **h** hohlraum, **I** proton beam, **B** magnetic field.

In a possible configuration shown in Fig.11.17, the liquid (or solid) D has the shape of a cylinder, placed inside a cylindrical hohlraum **h**. A GeV proton beam **I** coming from the left, in entering the hohlraum dissipates part of its energy into a burst of X-rays, compressing the deuterium cylinder, and part of it igniting a detonation wave propagating down the cylinder.

If the rod has the length z and the density ρ, the ignition condition for deuterium requires that $\rho z > 10$ g/cm^2, at a temperature $T \sim 10^9$ K. Normally, the ρz condition is given by a ρr condition for the radius r of a deuterium sphere. Here however, the radial entrapment of the charged DD fusion reaction products, ensured by the magnetic field of the proton beam current in excess of 10^7 Ampere, replaces the $\rho r > 10$ g/cm^2 condition, to a $\rho z > 10$ g/cm^2 condition, which is much easier to achieve. The yield of the deuterium explosion then only depends on the total mass of the deuterium. (For the DT reaction one must have $\rho r \geq 1$ g/cm^2 and $T > 10^8$ K.)

With both the beam and the target (initially) at a low temperature, the stopping length is determined by the electrostatic two stream instability. In the presence of the strong azimuthal magnetic field, it is enhanced by the formation of a collisionless shock. The stopping range of the protons by the two stream instability is here given by

$$\lambda = \frac{1.4c}{\epsilon^{1/3}\omega_i} \tag{11.37}$$

where c is here the velocity of light, the ω_i the proton ion plasma frequency. Furthermore $\epsilon = n_b/n$, where n_b is the proton density in the proton beam, and n the deuterium target number density. If the cross section of the beam is 0.1 cm^2, one obtains for a 10^7 Ampere beam that $n_b = 2 \times 10^{16}$ cm^{-3}. For a 100 fold compressed deuterium rod one has $n = 5 \times 10^{24}$ cm^{-3} with $\omega_i = 2 \times 10^{15}$ s^{-1}. One there finds that $\epsilon = 4 \times 10^{-9}$ and $\lambda \sim 1.2 \times 10^{-2}$ cm. At a deuterium number density $n = 5 \times 10^{24}$ cm^{-3}, the deuterium density is $\rho = 17$ g/cm^{-3}. To have $\rho z \geq 10$ g/cm^{-3}, thus requires that $z \geq 0.6$ cm. With $\lambda < z$, the condition for the ignition of a thermonuclear detonation wave is satisfied.

The ignition energy is given by

$$E_{ing} \sim 3nkT\pi r^2 z \tag{11.38}$$

For 100 fold compressed deuterium, one has $\pi r^2 = 10^{-2}$ cm^2, when initially it was $\pi r^2 = 10^{-1}$ cm^2. With $\pi r^2 = 10^{-2}$ cm^2, $z = 0.6$ cm, $KT \sim 10^{-7}$ erg ($T \sim 10^9$), one finds that $E_{ign} \sim 10^{16} = 1$ gigajoule. This energy is provided by the 10^7 Ampere - 10^9 Volt proton beam lasting 10^{-7} seconds. The time is short enough to assure the cold compression of deuterium to high densities. For a 10^3 fold compression, found feasible in laser fusion experiments, the ignition energy is reduced to 100 MJ.

11.8 Bibliography for Chapter 11

D. Strickland and G. Mourou, Opt. commun. **56**, 219 (1985).

F. Winterberg: Can a Laser Beam Ignite a Hydrogen Bomb, S-RD-1 (Secret Restricted Nuclear Weapons Data) NP-18252, Washington DC: US Atomic Energy Commission, classified 1970, declassified 2007.

N. G. Basov, S. Yu Guskov and L. P. Feoktistov, J. Sov. Laser Res. **13**, 396 (1992).

M. Tabak et al. Phys. Plas. **1**, 1626 (1994).

R. Kodama et al., Nature **412**, 798 (2001).

R. Kodama et al., Nature **418**, 933 (2002).

R. Kodama et al., Nuclear Fusion **44**, S276-S283 (2004).

M. Murakami and H. Nagatomo, Nucl. Instr. and Methods Phys. Res. A **544**, 67 (2005).

M. Murakami et al., Plasma Phys. Control. Fusion **47**, B815 (2005).

F. Winterberg, Plasma Phys. Control. Fusion **50**, 035002 (2008).

J. H. Eberle and A. Sleeper, Phys. Rev. **176**, 1570 (1968).

O. S. Lieu and E. H. Shin, Appl. Phys. Lett. **20**, 511 (1972).

F. Winterberg, Z. Naturforsch. **63a**, 1 (2008).

B. Müller, J. Rafelski and W. Greiner, Phys. Lett. **47B**, 1 (1973).

Xiao and Thadhani, J. Appl. Phys. **96**, 2000 (2004).

R. E. Kidder, in "Physics of High Energy Density", Academic Press, New York, p. 348 (1971).

J. K. Harvey, J. Fluid Mech. **14**, 585 (1962).

F. K. Moore and S. Leibovich in "Research on Uranium Plasmas and their Technological Applications" K. H. Thom and R. T. Schneider eds., Technical Information Office, NASA Sp-236, 1971 p. 95-103.

H. B. Squire, Surveys in Mechanics, G. K. Batchelor and R. M. Davies eds., Cambridge University Press, 1956.

G. I. Taylor, Proc. Roy. Soc. A, **102**, 180 (1922).

F. Winterberg, Phys. Rev. **174**, 212 (1968).

F. Winterberg, Phys. Plasmas **7**, 2654 (2000).

Chapter 12

The Future

12.1 What Kind of Burn

While it is difficult to predict the future, some general observations can be made. The achievement of DT ignition and burn is assured in the near future, but beyond that no firm prediction can be made, because this will depend on economic and environmental considerations, quite apart from unforeseen political factors.

There are a few ways the future may go:

1. DT burn coupled to U238 or Th232 burn, by surrounding a DT sphere with a shell of U238 or Th232, with compression and fast ignition of the DT done as shown in Fig. 12.1.

2. Ignition of a small DT trigger, with the trigger igniting a much larger amount of D. This can be done by something resembling the Teller-Ulam configuration, to compress and ignite the DT and D spheres, with both of them placed inside a "hohlraum". A configuration of this kind is shown in Fig. 12.2.

3. Pure DD burn with the attainment of gigavolt potentials.

Replacing conventional nuclear reactors with fusion-fission micro-explosion reactors would eliminate the need for the enrichment of uranium or the production of plutonium to run conventional nuclear reactors. Unlike hybrid fusion-fission reactors, where a burning fusion plasma is surrounded by a subcritical fission reactor, there is no danger of a possible meltdown.

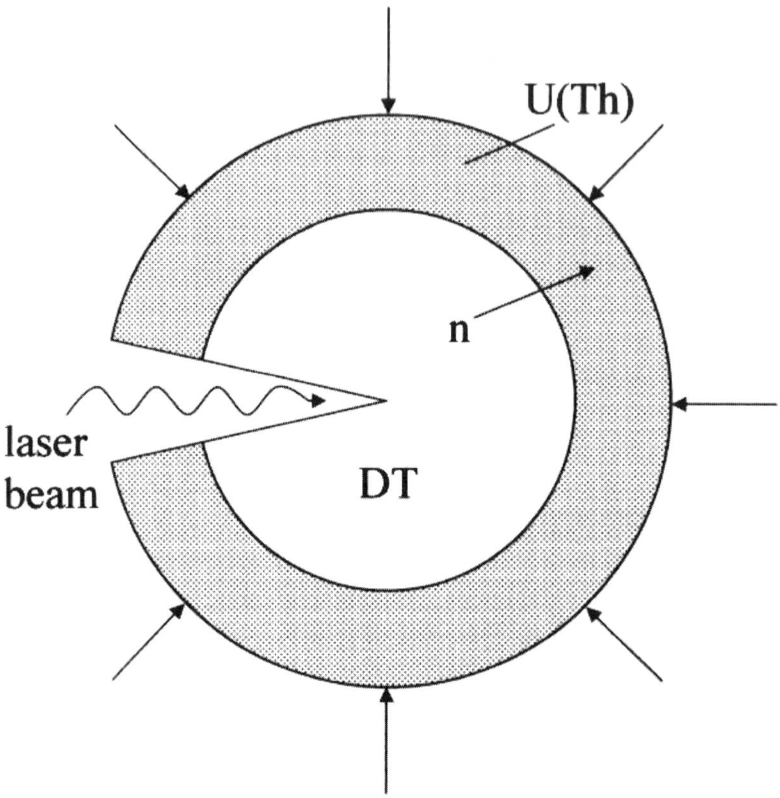

Figure 12.1: Coupled DT fast ignition → n → U(Th) fission burn.

A DT triggered deuterium burn would greatly reduce the need for lithium to breed tritium. This could lead to a future of an almost pure deuterium fusion burn, with the neutrons released by the DD reaction sufficient to breed enough tritium for the DT triggers. Pure DD burn would completely eliminate the dependence on lithium.

12.2 Driver Development

Considering the huge size of the laser driver for the National Ignition Facility (NIF), one major future thrust will be the search for a more compact driver.

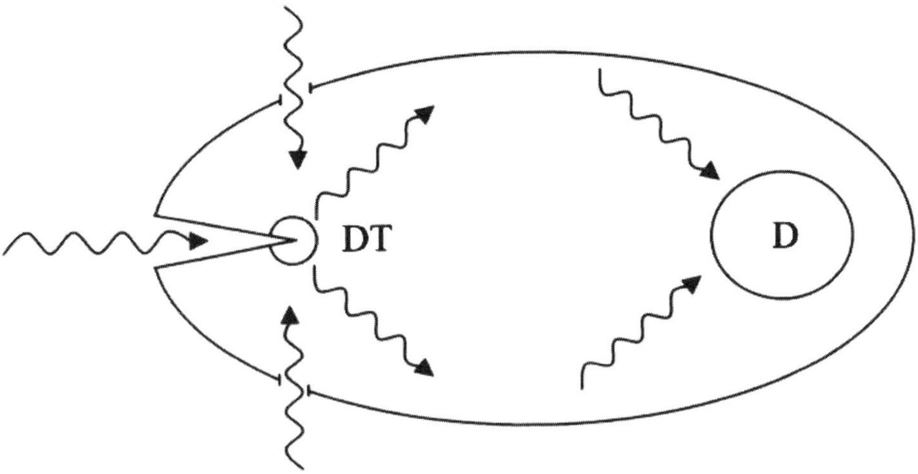

Figure 12.2: Micro-Teller-Ulam configuration with DT replacing the fission bomb trigger, with the fast ignition of the DT by laser- or particle beams.

And because of the in general poor laser efficiency, also the search for a more efficient driver. Both of these requirements are of special importance for micro-explosion nuclear-pulse rocket propulsion. For earth-bound power plants, the driver must at least be efficient. Efficient but large drivers are heavy ion-and macro-particle accelerators. Both of them are many miles long. This makes them unsuitable for micro-explosion rocket propulsion, but both solve the "stand-off problem", to keep the micro-explosion at a safe distance away from the wall of the nuclear combustion chamber. With laser drivers they only seem to share their good stand-off property, because for laser drivers there is the problem that the photon flash of a micro-explosion can destroy the optical system of the laser. This problem does not arise with particle accelerators.

Macroparticle accelerator drivers are ideally suited for magnetized target fusion, where the needed impact velocity of ~ 50 km/s would make possible cm-size targets. This is a velocity 4 times smaller than the velocity of 200 km/s required for non-magnetic field assisted impact fusion. It means that for magnetic field assisted impact fusion one needs only a 16 times smaller in length macroparticle accelerator.

12.3 GeV Intense Relativistic Ion Beam Drivers

One exceptionally interesting driver concept is the intense GeV ion beam driver, where the beam is extracted from a magnetically insulated GeV capacitor, or from one of the devices described in Chapter 11.7. For the first case, the capacitor can be realized by a levitated superconducting torus, enclosed in a large magnetic field generated by large toroidal currents flowing through the superconducting torus. To prevent vacuum breakdown, the torus would have to be levitated in ultrahigh vacuum. Since such a vacuum exists in space, this may lead to an ideal driver for nuclear micro-explosion rocket propulsion. Deforming the torus into a large elongated configuration as shown in Fig. 12.3, the thusly deformed torus can become the outer surface of the entire space craft, serving as a huge capacitor to draw the energy for an ion beam to trigger a thermonuclear micro-explosion. As in the Orion-type nuclear pulse propulsion concept, the micro-explosion would take place at a prescribed distance behind the space craft. The large magnetic mirror at the end of the magnetically insulated superconducting space craft would softly repel the expanding plasma fire ball of the micro-explosion and replace the pusher plate of the Orion nuclear pulse propulsion concept without the need for a shock absorber. In addition to providing thrust, the expanding plasma ball can (as shown in Fig. 12.3) induce a large current in a ring electrode to drive a magnetic field coil at the front of the space craft. Thermionic emitters placed at the inner side of this coil would during the rise of the magnetic field draw electrons from the space craft and discharge them into the vacuum of space, positively charging the space craft up to a GeV electrostatic potential, providing the driver energy for subsequent micro-explosions.

A laser beam directed from the space craft onto the micro-explosion target, in vaporizing the hydrogen in the small rocket chamber which is part of the target, emits through the Laval nozzle a jet towards the space craft, and establishes a conducting path for the ion beam to ignite the target. Because the intense GeV ion beam current can be well above the critical current to magnetically entrap the DT (or DD) fusion alpha particles, this reduces the requirement for target compression.

With the generation of intense GeV proton beams as described in Chapter 11.7, thermonuclear burn in an earth bound DD fusion reaction power plant could also be realized.

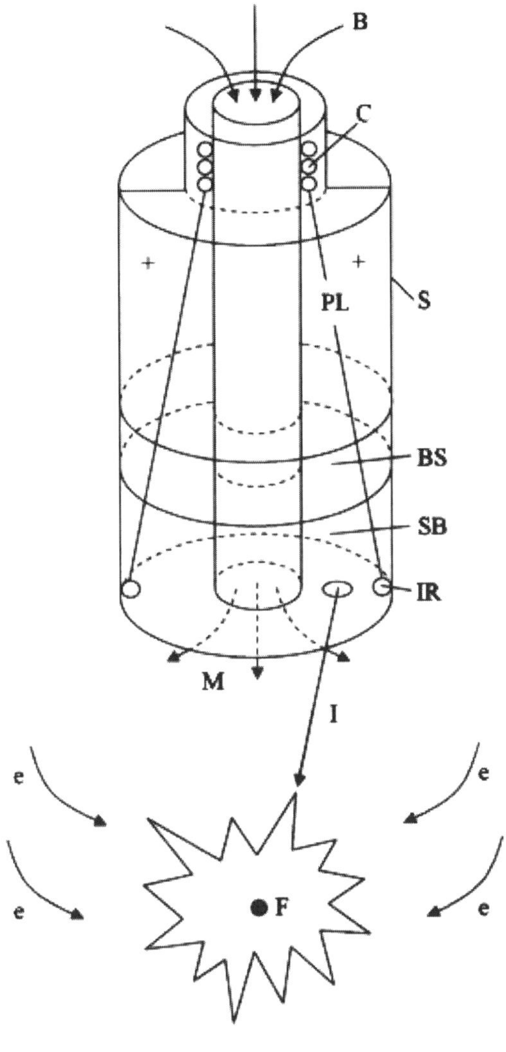

Figure 12.3: Advanced Deuterium Fusion rocket Propulsion for Manned Deep Space Mission. A superconducting "Atomic" Space Ship, positively charged to GeV potential, and with azimuthal currents and magnetic mirror **M** by magnetic field **B**. **F** fusion minibomb in position to be ignited by intense ion beam **I**, **SB** storage space for the bombs, **BS** bioshield for the paylod **PL**, **C** coils pulsed by current drawn from induction ring **IR**.

Appendix A

Comparison of the Recently Proposed Super Marx Generator Approach to Thermonuclear Ignition with the DT Laser Fusion-Fission Hybrid Concept by the Lawrence Livermore National Laboratory

A.1 Introduction

Since 1954 I have been actively involved in inertial confinement fusion research at a time this research was still classified in the US. I had independently discovered the basic principles and presented them in 1956 at a meeting of the Max Planck Institute in Goettingen organized by von Weizsaecker. To reach high temperatures, these principles are the Guderley convergent shock wave and the Rayleigh imploding shell solutions. The abstracts of the meeting still exist and are in the library of the University of Stuttgart.

The meeting was overwhelmed by the optimism to ignite a deuterium plasma, with the Max Planck Institute proposing a stellarator-like magnetic confinement configuration. At that time the ignition of a thermonuclear reaction with Guderley's convergent shock wave solution seemed only possible with the deuterium-tritium (DT) reaction, and for that reason was not considered worth for funding. However, following the publication of a paper by Trubnikov and Kudryavtsev [1], presented at the 1958 2nd United Nations Conference for the Peaceful Use for Atomic Energy, showing the importance of the electron synchrotron losses from a magnetized plasma, the hopes for a viable deuterium fusion plasma magnetic confinement configuration had been given up in favor of deuterium-tritium magnetic confinement configurations. But because a burning deuterium-tritium plasma is primarily the source of energetic neutrons, ideally suited for the fast fission of natural uranium or thorium, it was obvious to combine fusion with fission, with fusion producing neutrons and fission producing heat. Such an approach however, would not eliminate the generation of fission products as in a pure fission reactor, still posing a similar environmental nuclear waste disposal problem.

The proposed Super Marx generator pure deuterium micro-detonation ignition concept can be compared to the Lawrence Livermore National Ignition Facility (NIF) Laser DT fusion-fission hybrid concept (LIFE) [2]. In a Super Marx generator a large number of ordinary Marx generators charge up a much larger second stage ultra-high voltage Marx generator, from which for the ignition of a pure deuterium micro-explosion an intense GeV ion beam can be extracted. A typical example of the LIFE concept is a fusion gain of 30, and a fission gain of 10, making up for a total gain of 300, with about 10 times more energy released into fission as compared to fusion. This means a substantial release of fission products, as in fusion-less pure fission reactors. In the Super Marx approach for the ignition of a pure deuterium micro-detonation a gain of the same magnitude can in theory be reached [3]. If feasible, the Super Marx generator deuterium ignition approach would make lasers obsolete as a means for the ignition of thermonuclear micro-explosions.

A laser DT inertial confinement fusion reactor configuration requires a high gain, typically of the order 10^3, to make up for the poor laser efficiency. But the intense photon flash from the high gain fusion micro-explosion, entering the optical laser system with the velocity of light, can destroy the entire laser ignition apparatus, unless the laser is separated by a safe distance from the micro-explosion, which has its own technical problems.

The Super Marx generator concept for the ignition of a pure deuterium fusion reaction, if feasible, would not only bypass the fusion-fission hybrid concept, but would make the entire laser fusion approach obsolete and with it most of the rest, like the ordinary Marx generator electric pulse power approach, where the X-rays emitted from an array of exploding wires compresses and ignites a DT pellet.

A.2 Solution in between Two Extremes

Up until now nuclear fusion by inertial confinement has only been achieved by using a fission explosive as a means (driver) for ignition. This is true not only for large thermonuclear explosive devices, like the 1952 pure deuterium Mike Test (carried out in the South Pacific with the Teller-Ulam configuration), but also for small deuterium-tritium (DT pellet) micro-explosions, (experimentally verified with a fission explosive at the Nevada Test Site by the Centurion-Halite experiment). From this experience we know that the ignition is easy with sufficiently large driver energies, but which are difficult to duplicate with lasers or electric pulse power. The problem therefore is not the configuration of the thermonuclear explosive, but the driver, be it for the ignition of pure deuterium (D) as in the Mike Test, or for the ignition of deuterium-tritium (DT), as in the Centurion-Halite experiment, because for sufficiently large driver energies the target configuration is of secondary importance.

I claim that substantially larger driver energies can be reached by a "Super Marx generator". It can be viewed as a two-stage Marx generator, where a large number of ordinary Marx generators assumes the role of a first stage. If the goal is the much more difficult ignition of a pure deuterium micro-explosion, the Super Marx generator must in addition to deliver a much larger amount of energy (compared to the energy of the most powerful lasers), also generate a magnetic field in the thermonuclear target that is strong enough to entrap the charged DD fusion products within the target. Only then is the condition for propagating thermonuclear burn fulfilled. For this to happen, a 100 MJ-1GeV-10^7 Ampere proton beam is needed. This is the sought after solution, in between the two extremes, shown in Fig. A.1. It hopefully can be realized with the Super Marx [3].

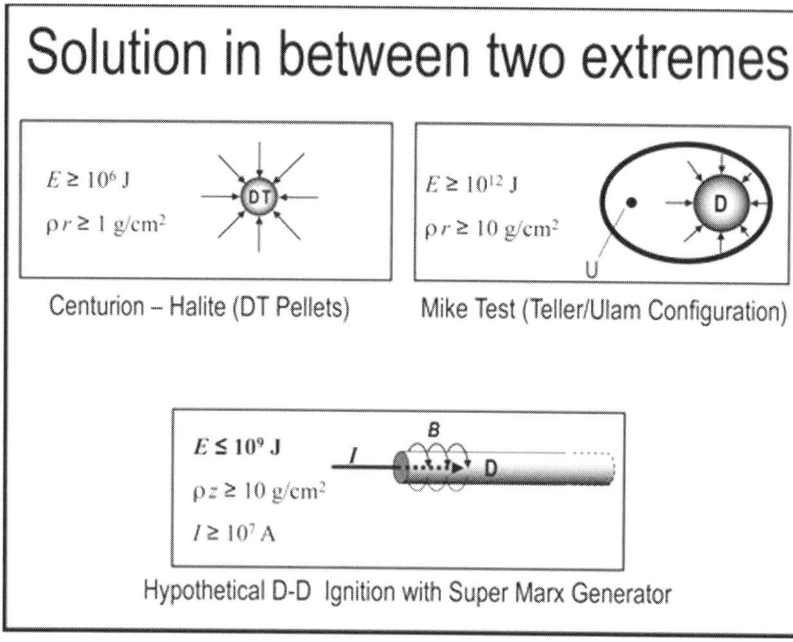

Figure A.1: Ignition of a deuterium target by a GeV-10 MA proton beam.

A.3 From the Marx to the Super Marx

Fig. A.2 shows the circuit of an ordinary Marx generator, and Fig. A.3 that of a Super Marx. An artist's view of a mile-long Super Marx generator, charged up by about 100 ordinary Marx generators, and connected to the chamber in which the thermonuclear target is placed, is shown in Figs. (A.4-A.6).

As shown in Fig. A.7, the Super Marx is made up of a chain of co-axial capacitors with a dielectric which has to withstand the potential difference of 10^7 Volt between the inner and outer conductor.

Following their charging up the Super Marx generator, the Marx generators are disconnected from the Super Marx. If the capacitors of the Super Marx can hold their charge long enough, this can be done by mechanical switches.

To erect the Super Marx, its capacitors C are switched into series by circular spark gap switches S. The capacitors of the Super Marx are mag-

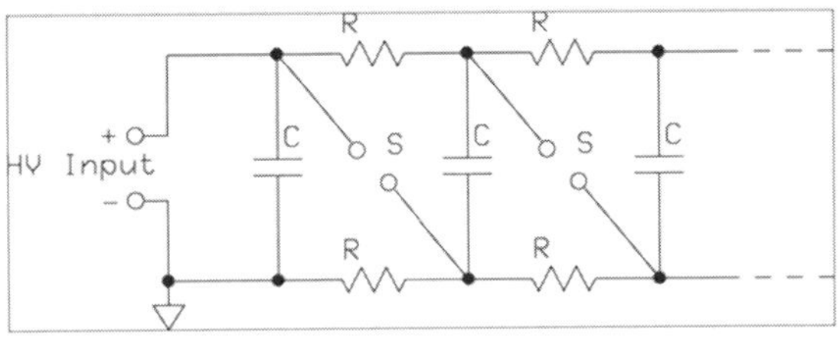

Figure A.2: In an ordinary Marx generator n capacitors C charged up to the voltage v, and are over spark gaps switched into series, adding up their voltages to the voltage $V = nv$.

netically levitated inside an evacuated tunnel, and magnetically insulated against the wall of the tunnel by an axial magnetic field B, generated by super conducting external magnetic field coils M, and in addition by an azimuthal field through the axially flowing currents during the erection of the Super Marx. The magnetic insulation criterion requires that $B > E$, where B is measures in Gauss and E in electrostatic cgs units. If $B = 3 \times 10^4$ Gauss for example, magnetic insulation is possible up to $E = 3 \times 10^4$ esu $\cong 10^7$ V/cm, at the limit of electron field emission. To withstand a voltage of 10^9 Volt between the outer positively charged surface of the capacitors in series and the tunnel wall, then requires a distance somewhat more than one meter.

The capacitance of one co-axial capacitor with the inner and outer radius, R_0 and R_1 of length l and filled with a dielectric of dielectric constant ϵ is

$$C = \epsilon \frac{l}{2\ln(R_1/R_0)} \text{cm} \tag{A.1}$$

Assuming a breakdown strength of the dielectric larger than 3×10^4 V/cm, and a potential difference of 10^7 Volt between the inner and outer conductor, the smallest distance of separation d between both conductors has to be $d = R_1 - R_0 \cong 2 \times 10^2$ cm. If for example $l = 1.6 \times 10^3$ cm, $R_1 = l/2 = 8 \times 10^2$ cm, and $\epsilon \cong 10$, one finds that $C \cong 2 \times 10^4$ cm. For these numbers the energy e stored in the capacitor ($V = 10^7$ Volt $\cong 3 \times 10^4$ esu) is

$$e = (1/2)CV^2 \cong 10^{13} \text{erg} \tag{A.2}$$

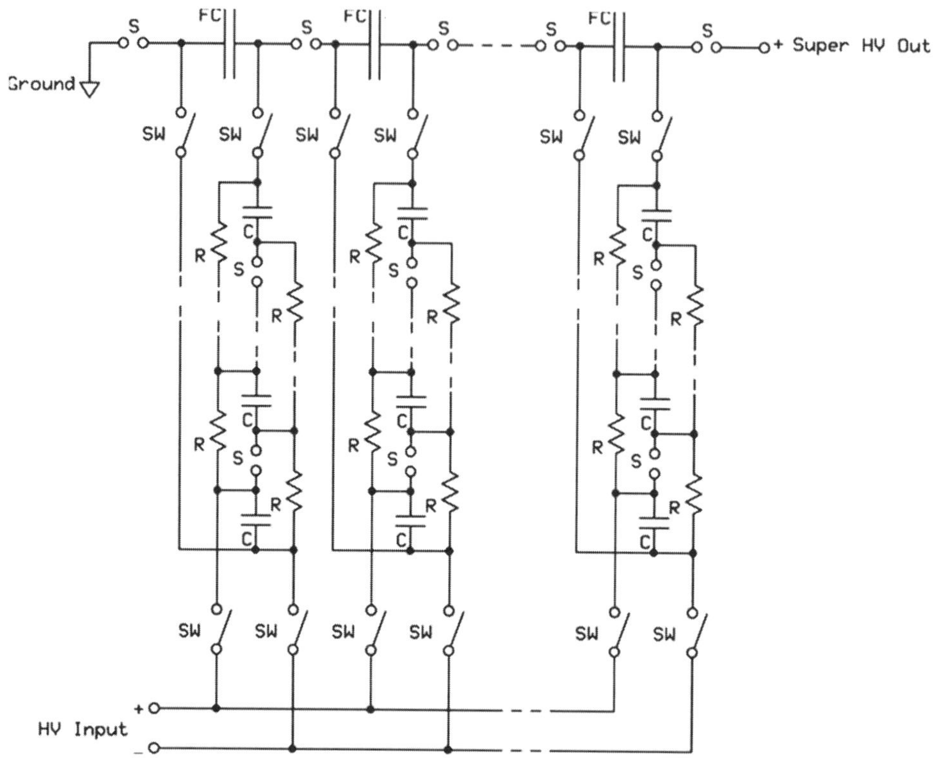

Figure A.3: In a Super Marx generator, N Marx generators charge up N fast capacitors FC to the voltage V, which switched into series add up their voltages to the voltage NV.

which for the 100 capacitors of the Super Marx add up to $e \sim 10^{15}$ erg. About 10 times more energy can be stored, either if the radius of the capacitor is about 3 times larger, or with a larger dielectric constant, or with a combination of both. This means that for about 100 capacitors an energy 10^{16} erg = 1 GJ can be stored in the mile-long Super Marx.

Another idea proposed by Fuelling [4], is to place the ordinary Marx generators of the 1st stage inside the coaxial capacitors of the Super Marx. The advantage of this configuration is that it does not require to disconnect the Marx generators from the capacitors of the Super Marx prior to its firing. Because the charging and discharging of the Super Marx can there be done

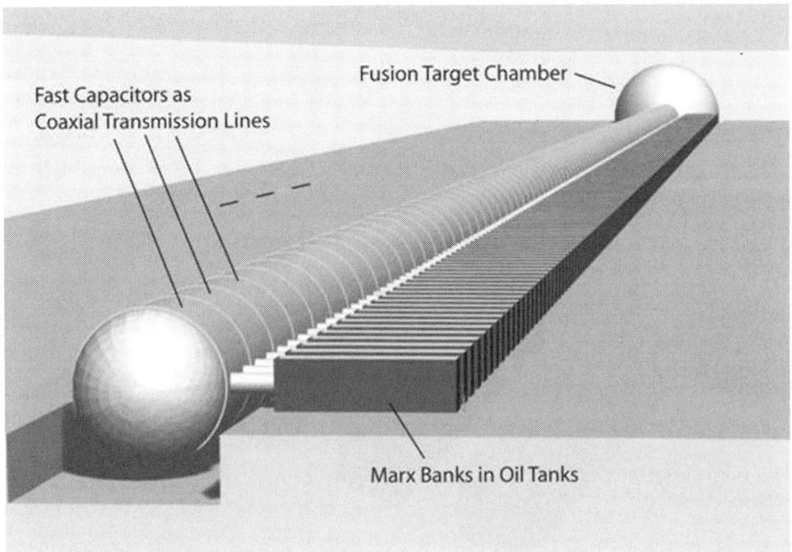

Figure A.4: Artistic perception of a 1.5 km long Super Marx generator, composed of 100 x 15 m long high voltage capacitors each designed as a magnetically insulated coaxial transmission line. The coaxial capacitors/transmission lines are placed inside a large vacuum vessel. Each capacitor/transmission line is charged by two conventional Marx generators up symmetrically to 10 MV (± 5 MV). After charge-up is completed, the Marx generators are electrically decoupled from the capacitors/transmission lines. The individual capacitors/transmission lines are subsequently connected in series via spark gap switches (i.e. the "Super Marx" generator), producing a potential of 1 GV.

very fast, one can use compact water capacitors where $\epsilon \cong 80$. And instead of magnetic insulation of the capacitors of the Super Marx against the outer wall one can perhaps use transformer oil for the insulation. Giving each inner segment of the Super Marx enough buoyancy, for example by adding air chambers, these segments can be suspended in the transformer oil. There the outer radius of the co-axial capacitors is much larger. This permits to store in the Super Marx gigajoule energies.

If the danger of electric breakdown in the transformer oil should still persist, one might try to prevent the breakdown oil, by a rapid spiraling flow

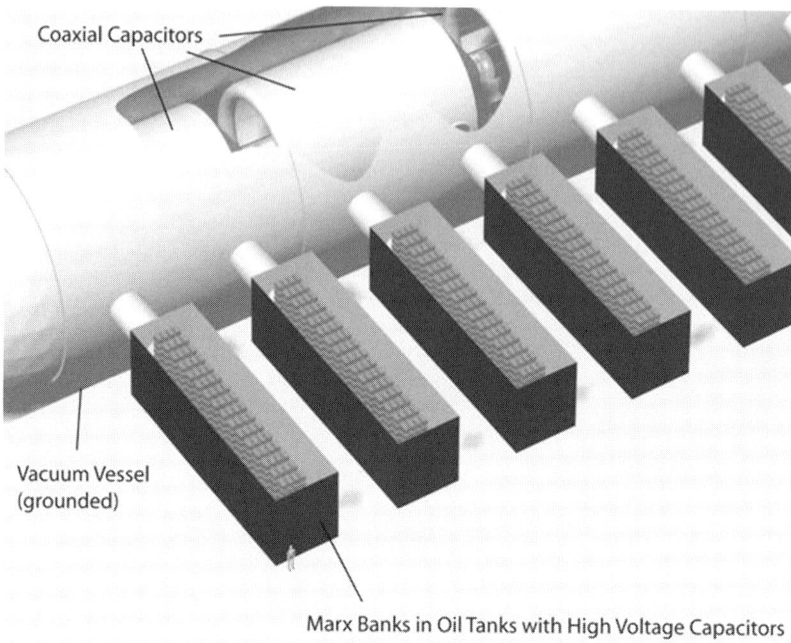

Figure A.5: Detail view of a section of the Super Marx generator. Two conventional Marx banks charge up one coaxial capacitor/transmission line element to 10 MV.

of the transformer oil, disrupting the formation of stepped leaders leading to the breakdown [5].

A.4 Connecting the Super Marx to the Load

As shown in Fig. A.8, the last capacitor of the Super Marx is guided to the load by a cylindrical Blumlein transmission line. Because it is uncertain if a Blumlein transmission line can withstand a voltage of 10^9 Volt long enough, one may consider a different configuration shown in Fig. A.9a, where a superconducting ring with a large toroidal current is chosen as the last capacitor. There, because of the large azimuthal magnetic field set up by the toroidal current magnetically insulates the 10^9 Volt charged up capacitor against breakdown to the wall. This idea was previously proposed by the

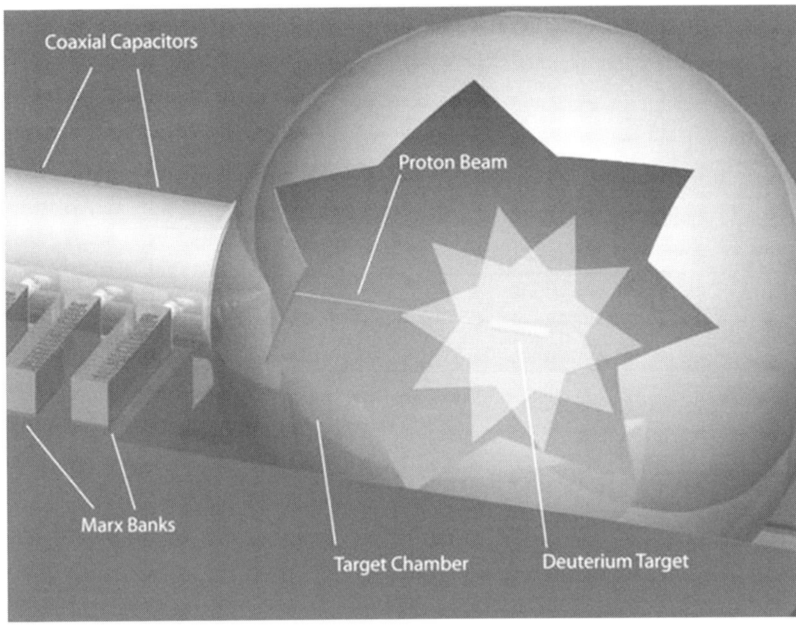

Figure A.6: Injection of GeV 10 MA proton beam, drawn from Super Marx generator made up of magnetically insulated coaxial capacitors into chamber with cylindrical deuterium target.

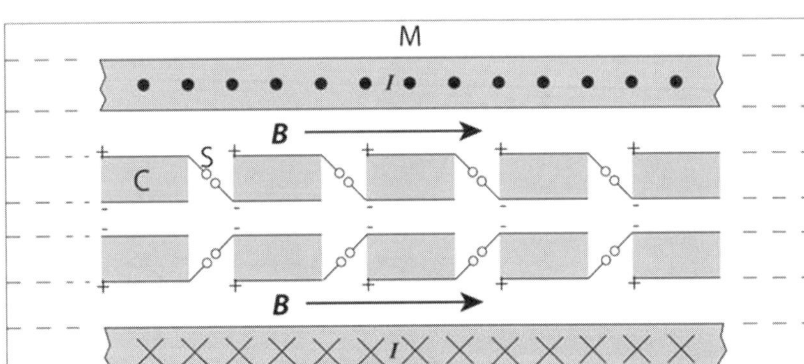

Figure A.7: Showing a few elements of the Super Marx generator.

author in a concept to charge up the forces to gigavolt potentials by a beam of charged particles [6, 7].

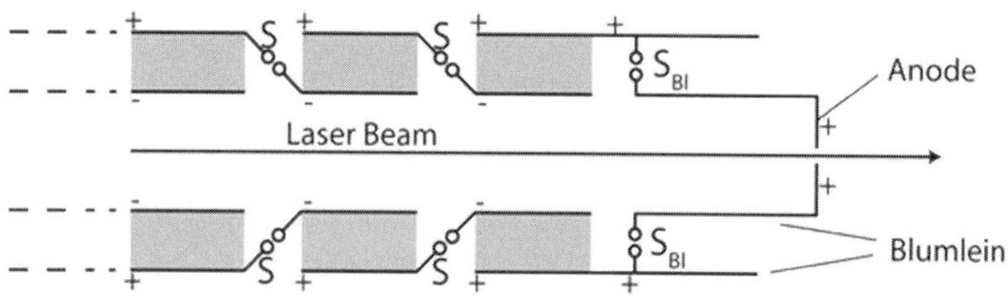

Figure A.8: Connection of the last capacitor of the Super Marx to the Blumlein transmission line.

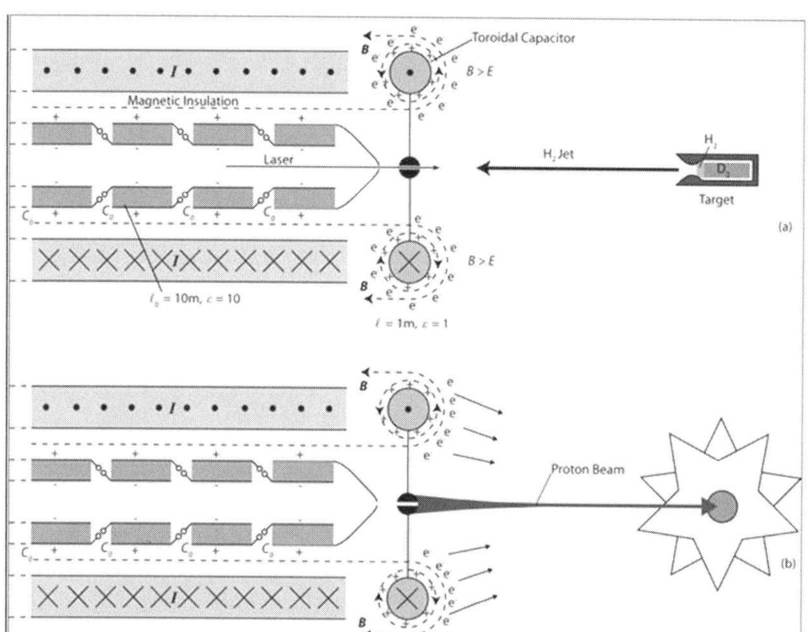

Figure A.9: The superconducting toroidal capacitor (a) and its discharge onto the target (b).

The load is the deuterium target shown in Fig. A.10. It consists of a solid deuterium rod covered with a thin ablator placed inside a cylindrical hohlraum. To the left of the hohlraum and the deuterium rod is a mini-rocket chamber filled with solid hydrogen.

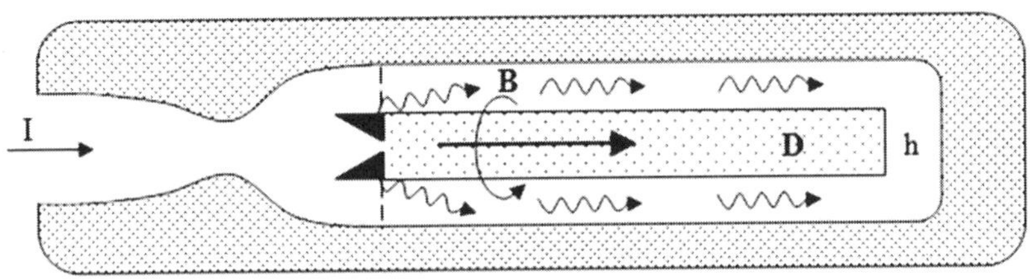

Figure A.10: Possible deuterium micro-detonation target: I ion beam, D deuterium cylinder, B magnetic field, h cylindrical hohlraum.

The method of discharging the energy from the Blumlein transmission line to the target then goes as follows (see Fig. A.11): 1. A short laser pulse is projected into the mini-rocket chamber, with the laser beam passing through a small hole in the center of the Blumlein transmission line. 2. By heating the hydrogen in the mini rocket chamber to a high temperature, a supersonic hydrogen jet is emitted through the Laval nozzle towards to the end of the Blumlein transmission line, forming a bridge in between the Blumlein transmission line and the target. 3. A second laser pulse then traces out an ionization trail inside the hydrogen jet, facilitating an electric discharge between the Blumlein and the target, with the space charge neutralizing plasma pinching the proton beam down to a small diameter. The large magnetic field of the 10^7 Ampere discharge current favors the creation of a proton beam in the hydrogen rich jet, not only because of the Alfvén limit for electrons, but also because of the large radiation friction of GeV electrons going in proportion to $\gamma^2 = (1 - \mathrm{v}^2/c^2)^{-1}$ [8]. Since for GeV protons $\gamma \sim 1$, but for electrons $\gamma \geq 10^3$, the friction force on the electrons is larger by many orders of magnitude. For a GeV - 10^7 A proton beam, focused down to an area of 0.1 cm^2, one computes a proton number density in the beam equal to $n_b \cong 2 \times 10^{16}$ cm^{-3}. To have a non-relativistic electron return current in the plasma of the hydrogen jet, the number density of the jet must be $n_\epsilon \geq n_b$. This condition should be satisfied, since the jet comes from the dense laser heated solid hydrogen in the mini rocket chamber attached to the target. If for the discharge to the target instead of the Blumlein transmission line the magnetically insulated superconducting torus, shown in Fig. A.9a, is used, the charging and discharging of the torus is done from a spherical electrode in the center of the torus as shown in Fig A.9b.

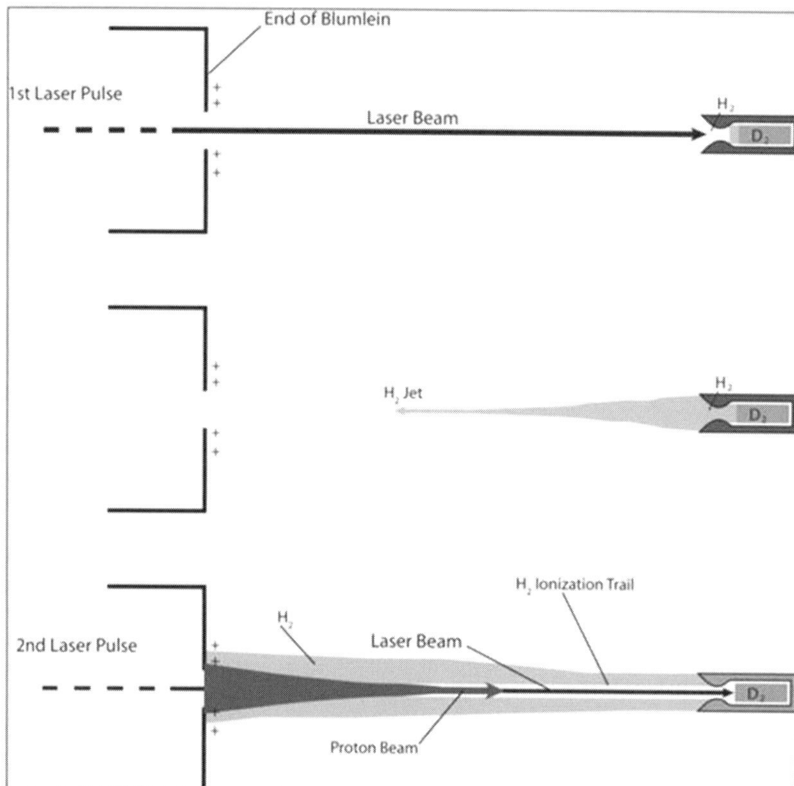

Figure A.11: Sequence of events to bombard the target by the proton beam from the Blumlein transmission line.

A.5 Thermonuclear Ignition and Burn

For the deuterium-tritium thermonuclear reaction the condition for propagating burn in a sphere of radius r and density ρ, heated to a temperature of 10^8 K, is given by $\rho r \geq 1$ g/cm^2. This requires an energy of about 1 MJ.

For the deuterium reaction this condition is $\rho r \geq 10$ g/cm^2, with an ignition temperature about 10 times larger. That a thermonuclear detonation in deuterium is possible at all is due to the secondary combustion of the T and He^3 DD fusion reaction products [9]. The energy required there would be about 10^4 times larger or about 10^4 MJ, for all practical purposes out of reach for non-fission ignition. However, if the ignition and burn is along a deuterium cylinder, where the charged fusion products are entrapped within the cylinder, the condition $\rho r \geq 10$ g/cm^2 can be replaced by

$$\rho z \geq 10 \mathrm{g/cm}^2 \tag{A.3}$$

where z is the length of the cylinder. The entrapment is possible if a large current is flowing through the cylinder, generating a large azimuthal magnetic field. The condition to entrap the charged fusion products by the azimuthal magnetic field is there given by the inequality

$$r_f < r_c \tag{A.4}$$

In (A.4) r_f is the charged fusion product Larmor radius, and r_c the radius of the deuterium cylinder, where

$$r_f = \frac{M\mathrm{v}c}{ZeB} \tag{A.5}$$

with B the magnetic field strength, M and Z the mass and charge, and v the velocity of the charged fusion products. By order of magnitude one has v$\sim c/10$. A current I [A] flowing through the deuterium cylinder produces a magnetic field which at the surface of the cylinder of radius r_c is

$$B = \frac{0.2I}{r_c} \tag{A.6}$$

Inserting (A.5) and (A.6) into (A.4) one has

$$I > \frac{5M\mathrm{v}c}{Ze} \tag{A.7}$$

This inequality is well satisfied if $I \geq 10^7$ A.

If the GeV - 10^7 Ampere proton beam passes through background hydrogen plasma with a particle number density n, it induces in the plasma a return current carried by its electrons, where the electrons move in the same direction as the protons. But because the current of the proton beam and the return current of the plasma electrons are in opposite directions, they repel each other. Since the stagnation pressure of the GeV - 10^7 Ampere proton beam is much larger than the stagnation pressure of the electron return current, the return current electrons will be repelled from the proton beam, towards the surface of the proton beam.

The stagnation pressure of a GeV proton beam is (M_H proton mass)

$$p_i \cong \rho_i c^2 = n_b M_H c^2 \tag{A.8}$$

For $n_b \cong 2 \times 10^{16}$ cm^{-3} one obtains $p_i \cong 3 \times 10^{13}$ dyn/cm^3. For the electron return current one has (m electron mass)

$$p_e = n_e m \mathrm{v}^2 \tag{A.9}$$

With the return current condition $n_e e\mathrm{v}_e = n_i e\mathrm{v}_i$, where for GeV protons $\mathrm{v}_i \cong c$, one has

$$\mathrm{v}_e/c = n_i/n_e \tag{A.10}$$

Taking the value $n_e = 5 \times 10^{22}$ cm^{-3}, valid for uncompressed solid deuterium, one obtains $\mathrm{v}_e \cong 10^4$ cm/s and hence $p_e \cong 5 \times 10^3$ dyn/cm^2. This is negligible agains p_i, even if n_e is 10^3 time larger, as in highly compressed deuterium. The assumption, that the magnetic field of the proton beam is sufficiently strong to entrap the charged fusion products within the deuterium cylinder, is therefore well justified.

If the charged fusion products are entrapped within the deuterium cylinder, and if the condition $\rho z > 10$ gm/cm^2 is satisfied, and finally, if the beam energy is large enough that a length $z > (10/\rho)$ cm is heated to a temperature of 10^9 K, a thermonuclear detonation wave can propagate down the cylinder. This then leads to large fusion gains.

The stopping length of single GeV protons in dense deuterium is much too large to fulfill inequality (A.3). But this is different for an intense beam of protons, where the stopping length is determined by the electrostatic proton-deuteron two-stream instability [10]. In the presence of a strong azimuthal magnetic field the stopping length is enhanced by the formation of a collisionless shock [11]. For the two-stream instability alone, the stopping length is given by

$$\lambda \cong \frac{1.4c}{\epsilon^{1/3}\omega_i} \tag{A.11}$$

where c is the velocity of light, and ω_i the proton ion plasma frequency, furthermore $\epsilon = n_b/n$ with n the deuterium target number density. For a 100-fold compressed deuterium rod one has $n = 5 \times 10^{24}$ cm^{-3} with $\omega_i = 2 \times 10^{15}$ s^{-1}. One there finds that $\epsilon = 4 \times 10^{-9}$ and $\lambda \cong 1.2 \times 10^{-2}$ cm. This short length, together with the formation of the collision-less magnetohydrodynamic shock, ensures the dissipation of the beam energy into a small volume at the end of the deuterium rod. For a deuterium number density $n = 5 \times 10^{24}$ cm^{-3} on has $\rho = 17$ g/cm^3, to have $\rho z > 10$ g/cm^2, then requires $z \geq 0.6$ cm. With $\lambda < z$, the condition for the ignition of a thermonuclear detonation wave is satisfied.

The ignition energy is given by

$$E_{ing} \sim 3nkT\pi r^2 z \tag{A.12}$$

where $T \sim 10^9$ K. For 100-fold compressed deuterium, one has $\pi r^2 = 10^{-3}$ cm^2, when initially it was $\pi r^2 = 10^{-1}$ cm^2. With $\pi r^2 = 10^{-3}$ cm^2, $z = 0.6$ cm cm one finds that $E_{ing} \leq 10^{16}$ erg or ≤ 1 GJ. This energy is provided by the 10^7 Ampere-GeV proton beam lasting 10^{-7}s. The time is short enough to ensure the cold compression of deuterium to high densities. For a 10^3-fold compression, found feasible in laser fusion experiments, the ignition energy is ten times less. In hitting the target a fraction of the proton beam energy is dissipated into X-rays by entering and bombarding the high Z material cone, focusing the proton beam onto the deuterium cylinder. The X-rays released fill the hohlraum surrounding the deuterium cylinder, compressing it to high densities, while the bulk of the proton beam energy heats and ignites the deuterium cylinder at its end, launching in it a detonation wave. Both the energy and the magnetic field are supplied by the proton beam from the Super Marx generator, more than what all other inertial confinement fusion drivers are capable. At 10^9 Volt, a 10^7 Ampere proton beam current is below the Alfven limit for proton.

A.6 Conversion of the Explosively Released Energy

As shown in Fig. A.6, the deuterium micro-explosion takes place inside an evacuated cavity. If this cavity has the radius R and if it is filled with a magnetic field of strength B, it contains the magnetic energy

$$e_M = \frac{4\pi}{3} R^3 \frac{B^2}{8\pi} = \frac{1}{6} R^3 B^2 \tag{A.13}$$

Let us assume that $R = 15$ meter $= 1.5 \times 10^3$ cm, and that $B = 2 \times 10^4$ G (which can be reached with ordinary electromagnets), one finds that $e_M =$ 23 GJ.

The rapidly expanding fully ionized fire ball of the deuterium micro-explosion pushes the magnetic field towards the cavity wall, and if the wall is covered with induction coils, the released fusion energy is converted into an electromagnetic pulse lasting for the time $\tau \cong R/a$ where $a \cong 10^8$ cm/s is the expansion velocity of the fireball. For $R = 1.5 \times 10^3$ cm, one finds $\tau \cong 1.5 \times 10^{-5}$ s. This time is long enough for the pulse to be stretched out and converted into useful electromagnetic energy.

For an ignition an energy of 100 MJ and a yield of 23 GJ, the fusion gain would be $G = 230$, about the same as for the LIFE concept. However, since even in pure deuterium burn neutrons are released through the secondary combustion of the tritium D-D fusion reaction products, a much higher overall gain is possible with an additional fission burn, as in the LIFE concept [2].

A.7 Other Possibilities

The energy of up to a gigajoule, delivered in $\sim 10^{-7}$ seconds at a power of 10^{16} Watt, opens up other interesting possibilities.

1. If instead of protons heavy ions are accelerated with such a machine at gigavolt potentials, these ions will upon impact be stripped off of all their electrons, in case of uranium all of its 92 electrons. Accordingly this would result in a 92 fold increase of the beam current to $\sim 10^9$ Ampere. With such an ultrahigh current, a very different fusion target, shown in Fig. A.12, seems possible, where a solid deuterium rod is

placed inside a hollow metallic cylinder. The inner part of the beam I_1 will directly pass through the deuterium inside the cylinder, while the outer part of the beam I_0 will be stopped in the cylindrical shell, there depositing its energy and imploding the shell onto the deuterium cylinder, at the same time compressing the azimuthal magnetic beam field inside the cylinder. If ignited at the position where the beam hits and ignites the cylinder, will lead to a deuterium detonation wave propagating down the cylinder.

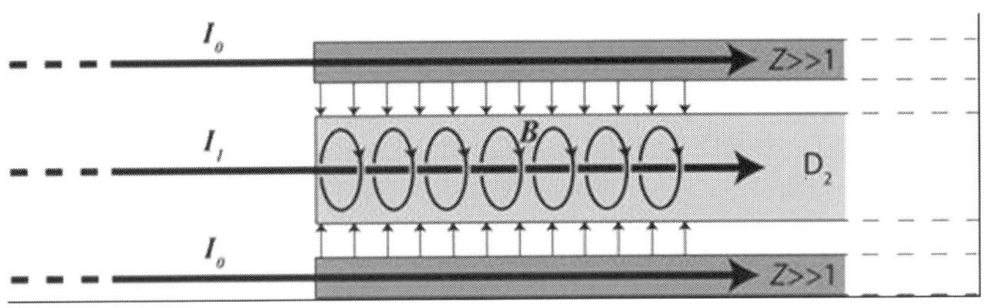

Figure A.12: Bombarding a cylindrical, deuterium containing target, with an intense heavy ion beam.

2. At a beam current of $\sim 10^9$ Ampere, will lead to a large, inward directed magnetic pressure. At a beam radius of 0.1 cm, the magnetic field will be of the order of 2×10^9 Gauss, with a magnetic pressure of 10^{17} dyn/cm^2 $\sim 10^{11}$ atmospheres. At these high pressures the critical mass of fissile material (U235, Pu 239, and U233) can be reduced to $\sim 10^{-2}$ g [12-14]. This would make possible micro-fission explosion reactors not having the meltdown problem of conventional fission reactors.

3. In general, the attainable very high pressures would have many interesting applications. One example is the release of fusion energy from exotic nuclear reactions, like the pB^{11} neutron-free fusion reaction, conceivably possible under very high pressures.

A.8 Discussion

The fusion/fission LIFE concept proposed by the Livermore National Laboratory is an outgrowth of the DT laser ignition project pursued at the

National Ignition Facility. Ignition is there expected in the near future. With its large fission component, it is difficult to see how the LIFE can compete with conventional fission reactors. Like them it still has the fission product nuclear waste problem. For this reason it hardly can without fission solve the national energy crisis, for what it has been billed by California Governor Schwarzenegger [15].

The proposed Super-Marx concept is by comparison a much more ambitious project, because it recognizes that the fundamental problem of inertial confinement fusion is the driver energy, not the target. And that only with order of magnitude larger driver energies can real success be expected. This in particular is true, if the goal is to burn deuterium. Unlike deuterium which is everywhere abundantly available, the burn of deuterium-tritium depends on the availability of lithium, a comparatively rare element.

It is also worthwhile to compare the Super Marx generator approach for thermonuclear micro-explosion ignition with the work by Basko and his group [16, 17]. They hope to achieve the same with heavy ion particle accelerators. And they too propose cylindrical targets, with magnetic fields (both axial and azimuthal), to entrap the charged fusion products. They can make a good case for the ignition of DT microexplosions, much better than what is possible with lasers, but because it is difficult to reach petawatt megajoule energies with a space charge limited particle accelerator, they think the ignition of the deuterium reaction is possible only with the help of a DT tablet, very much as in my 1982 concept [9]. In the absence of such a DT ignitor, their calculations predict beam energies larger than 100MJ for compression and ignition, with comparable gains and yields as in the Super Marx generator approach. They share the conclusion, that the ignition of thermonuclear micro-explosions, most likely requires much larger energies than those hoped for, and more in line what I had thought to be attainable with a levitated superconducting capacitor back in 1968 [6], with the prospect of gigajoule energies released in about 10^{-7} seconds. Another drawback of their proposal is that it requires to produce a large magnetic field in the target. While in the proposed Super Marx generator approach a magnetic field of the required magnitude is generated by the ion beam, it must in their approach be set up by an auxiliary discharge over a disposable transmission line.

A.9 Bibliography for the Appendix

[1] B. A. Trubnikov and V. S. Kudryavtsev, 2nd United Nation Conference on the Peaceful Use of Atomic Energy, Paper P/2213.

[2] LIFE: Clean Energy from Nuclear Waste, Lawrence Livermore National Laboratory (LLNL),
http://lasers.llnl.gov/missions/energy_for_the_future/life/

[3] F. Winterberg, J. Fusion Energy **28**, 290 (2009).

[4] S. Fuelling, private communication.

[5] F. Winterberg, J. Fusion Energy, (2008), to be published, online at:
http://www.springerlink.com/content/6262561738342487.

[6] F. Winterberg, Phys. Rev. **174**, 212 (1968).

[7] F. Winterberg, Phys. of Plasmas **7**, 2654 (2000).

[8] L. O. Landau & E. M. Lifshitz, The Classical Theory of Fields, Pergamon Press, New York 1971, p.194.

[9] F. Winterberg, J. Fusion Energy **2**, 377 (1982).

[10] O. Buneman, Phys. Rev. **115**, 503 (1959).

[11] L. Davis, R. Lüst, A. Schlüter, Z. Naturforsch. **13a**, 916 (1958).

[12] F. Winterberg, Nature **241**, 449 (1973).

[13] G. A. Askaŕyan, V. A. Namiot, M. S. Rabinovich, JETP Lett. **17**, 424 (1973).

[14] F. Winterberg, Z. Naturforsch. **28a**, 900, 1973.

[15] A. Schwarzenegger, Press Release of 11/10/2008 by Office of the Governor of California.

[16] M. M. Basko, M. D. Churazov, and A. G. Aksenov, Laser and Particle Beams (2002), **20**, 411-414.

[17] A. G. Aksenov, M. D. Churazov, GSI Annual Report 2002 - 'High Energy Density Physics with Intense Ion and Laser Beams,' edited by Dieter H. H. Hoffmann (2002), page 66.

About the Author

Friedwardt Winterberg, born in 1929 in Germany, received his PhD under Werner Heisenberg in 1955. On the invitation of the US Government he came to the United States in 1959, and is since 1963 Professor of Physics at the University of Nevada. He is an elected member of the International Academy of Astronautics in Paris, France, and an Honorary Member of the German Aero-Space Society Lilienthal-Oberth. His pioneering work in nuclear fusion by inertial confinement earned him the 1979 Hermann Oberth Gold Medal of the Wernher von Braun International Space Flight Foundation, and in 1981 a citation by the Nevada Legislature.

Index